Life Science
70.00

Genetics of Cellular, Individual,
Family, and Population Variability

Genetics of Cellular, Individual, Family, and Population Variability

Edited by
CHARLES F. SING
AND
CRAIG L. HANIS

New York Oxford
OXFORD UNIVERSITY PRESS
1993

Oxford University Press

Oxford New York Toronto
Delhi Bombay Calcutta Madras Karachi
Kuala Lumpur Singapore Hong Kong Tokyo
Nairobi Dar es Salaam Cape Town
Melbourne Auckland Madrid

and associated companies in
Berlin Ibadan

Published by Oxford University Press, Inc.,
200 Madison Avenue, New York, New York 10016

Library of Congress Cataloging-in-Publication Data
Genetics of cellular, individual, family, and population variability/
edited by Charles F. Sing and Craig L. Hanis.
p. cm.
Includes bibliographical references and index.
ISBN 0-19-506625-1
1. Medical genetics. 2. Population genetics.
I. Sing, Charles F. II. Hanis, Craig L.
RB 155.G3898 1993 616'.042—dc20 92-20412

9 8 7 6 5 4 3 2 1

Printed in the United States of America
on acid-free paper

Preface

This volume has been assembled as a tribute to William Jackson Schull. Jack has been mentor, colleague, and friend of each of the authors. His contributions to the fields of human genetics, epidemiology, radiation biology, neurobiology, and public health have been enormous. Most important, his insights have contributed immeasurably to the maturation of the broad range of scientific ideas that is represented by the chapters making up this volume. Although the range of topics is broad, a unifying theme links these contributions. Each author addresses the role of genetic variation at either the cellular, individual, family, or population level.

Throughout their careers, the contributors to this volume have dedicated themselves to solving one (in some cases several) of the important questions central to human genetics. The volume begins with a discussion of the sources and rates of mutation which ultimately give rise to the vast amount of extant genetic variation. This is followed by presentations of current understanding of how genetic variation is maintained within and among populations. The volume ends with discussions of the implications of such variation for understanding the evolution of our species. In between are discussions of the insights that may be obtained through the careful study of genetic variation in cells, individuals, families, and populations.

Each author summarizes the evolution of knowledge in a representative area of research. There are no complete explanations or final answers to be found in these presentations. This was not the intention. Rather, this volume demonstrates the connectedness of a variety of approaches and how insights in one area can contribute to understanding in another. For example, our knowledge about the evolution of DNA sequences influences how we study and interpret the associations between particular mutations and phenotypic variation. This volume will also serve as a reminder of the many important questions that deal with crossing levels from cells to populations that remain unsolved and deserve our most careful study. As is expected from such a group of authors, many insights into the research challenges that we might anticipate in the years ahead are woven into the fabric of every chapter.

One conclusion will be patent from reading this volume. The authors emphasize synthesis of ideas in contrast to description. The intense molecular investigation of the human genome that is in progress will provide raw material for substantial insights into each of the central problems of human genetics. This development, however, has been coupled with challenges to integrate both mathematically and biologically these new data into the mainstream of research on humans. This volume presents a strong and convincing message that the importance of quantitative sciences will increase in the next decade as a consequence of our expanded knowledge of the micro world of DNA and cellular biochemistry. Finally, the chapters included here serve to chronicle progress that has been made toward understanding our species through the application of genetic perspectives.

We gratefully acknowledge the editorial assistance of Betty Hsiao and Debbie Theodore. Special thanks goes to Debbie for her devotion to organizing all details of this effort from start to end. She helped to maintain a manageable order out of a project that was often at the edge of chaos.

Ann Arbor–Houston
June 1992

C. F. S.
C. L. H.

Contents

IV POPULATION VARIABILITY

Contributors

Susan H. Blanton
Health Sciences Center
Department of Pediatrics
University of Virginia
Charlottesville, Virginia

Eric Boerwinkle
Center for Demographic and Population Genetics
Graduate School of Biomedical Sciences
University of Texas Health Science Center at Houston
Houston, Texas

Ranajit Chakraborty
Center for Demographic and Population Genetics
Graduate School of Biomedical Sciences
University of Texas Health Science Center at Houston
Houston, Texas

Aravinda Chakravarti
Department of Human Genetics
Graduate School of Public Health
University of Pittsburgh
Pittsburgh, Pennsylvania

Lawrence Chan
Department of Cell Biology
Baylor College of Medicine
Houston, Texas

Stephen P. Daiger
Medical Genetics Center
Graduate School of Biomedical Sciences
University of Texas Health Science Center at Houston
Houston, Texas

Robert E. Ferrell
Department of Human Genetics
Graduate School of Public Health
University of Pittsburgh
Pittsburgh, Pennsylvania

D. Michael Hallman
Center for Demographic and Population Genetics
Graduate School of Biomedical Sciences
University of Texas Health Science Center at Houston
Houston, Texas

Craig L. Hanis
Center for Demographic and Population Genetics
Graduate School of Biomedical Sciences
University of Texas Health Science Center at Houston
Houston, Texas

Alfred G. Knudson, Jr.
Fox Chase Cancer Center
Institute for Cancer Research
Philadelphia, Pennsylvania

Bruce R. Levin
Department of Biology
Emory University
Atlanta, Georgia

Wen-Hsiung Li
Center for Demographic and Population Genetics
Graduate School of Biomedical Sciences
University of Texas Health Science Center at Houston
Houston, Texas

Shyan-Woei Liu
Department of Cell Biology
Baylor College of Medicine
Houston, Texas

Gregory Livshits
Institute of Molecular Evolutionary Genetics
The Pennsylvania State University
University Park, Pennsylvania

Jean W. MacCluer
Department of Genetics
Southwest Foundation for Biomedical Research
San Antonio, Texas

James V. Neel
Department of Human Genetics
University of Michigan Medical School
Ann Arbor, Michigan

Masatoshi Nei
Institute of Molecular Evolutionary Genetics
The Pennsylvania State University
University Park, Pennsylvania

Tatsuya Ota
Institute of Molecular Evolutionary Genetics
The Pennsylvania State University
University Park, Pennsylvania

Masanori Otake
Department of Statistics
Radiation Effects Research Foundation
Hiroshima, Japan

Sharon L. Reilly
Department of Human Genetics
University of Michigan Medical School
Ann Arbor, Michigan

Charles F. Sing
Department of Human Genetics
University of Michigan Medical School
Ann Arbor, Michigan

Peter E. Smouse
Center for Theoretical and Applied Genetics
Cook College
Rutgers University
New Brunswick, New Jersey

James N. Spuhler
Genetics Group
University of California
Los Alamos National Laboratory
Los Alamos, New Mexico

Emöke J. E. Szathmary
Office of the Dean
Faculty of Social Sciences
The University of Western Ontario
London, Ontario, Canada

Kenneth M. Weiss
Department of Anthropology
The Pennsylvania State University
University Park, Pennsylvania

Weijun Xiong
Department of Cell Biology
Baylor College of Medicine
Houston, Texas

I

CELLULAR VARIABILITY

1

Medieval *mappaemundi* and the Conceptual Map of Genetics: Changing Views of Cancer Biology and Other Thoughts

KENNETH M. WEISS

MAPPAEMUNDI: MEDIEVAL DIAGRAMS OF TRUTH

Science is always driven by ideas stimulated by an excess of inexplicable facts. One hallmark of an area on the frontier of knowledge is that many different explanations seem to fit the facts about as well. We try to resolve this dilemma by integrating developments, arising from different parts of our field, to form a synthetic new understanding. In this chapter I will discuss this phenomenon in the context of modern developments in genetics, and as a case in point, in regard to the genetics of cancer. To do so, I will use the metaphor of a map.

We usually view maps as factual guides to location and relationships, but to medieval Christians, a world map, or *mappamundi*,[1] was a symbolic orientation to ideas as much as it was a geographical work. The mapmaker, often a cleric, was "explaining the world as well as drawing it. This was important, for the world of the Middle Ages was mysterious, its lands unexplored, its seas unfathomable, its forest haunted by unpredictable occupants and its nights sinister with malevolent darkness" (Moir, 1979).[2]

One school of medieval mapmakers conceived of the earth in a T–O configuration, as shown in Figure 1.1, a woodcut after the 7th century church scholar (and expert on paradise) St Isidore of Seville: the circle of the earth,

1. Actually, *mappa mundi* is a solecism; the word *mappa* refers to a napkin or tablecloth; more proper terms would be *carta* or *tabula geographica*; however, some of the early *mappaemundi* were used as alter cloths, and the term *mappamundi* was well established in the middle ages. *Mappamundi genetica* completes the felony by being made up, so to speak, of whole cloth.

2. For what I say about the medieval maps, I draw on Beavan and Philott (1873), Bagrow (1964), Boorstin (1985), Campbell (1981), Noble (1981), Wilford (1981), and Harley and Woodward (1987).

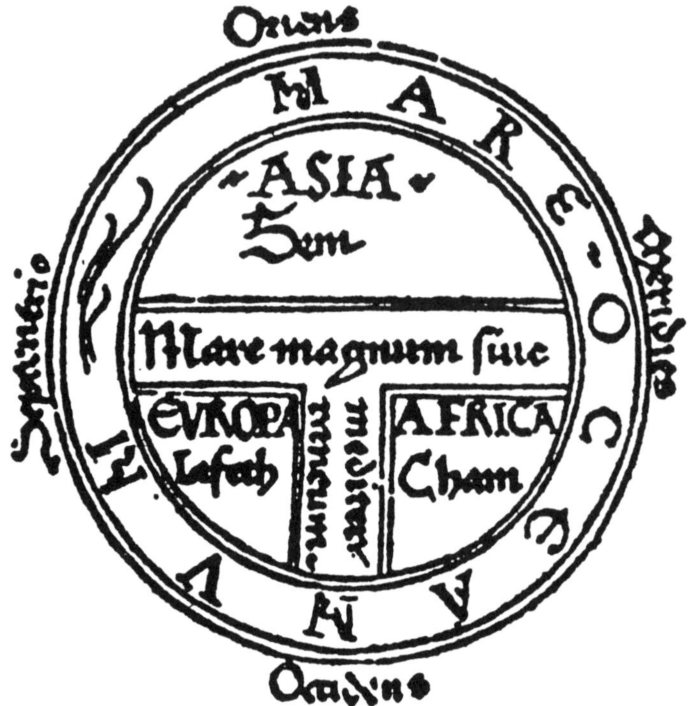

Figure 1.1. T-O style map from woodcut by St Isidore of Seville showing medieval European view of the organization of the earth.

the *orbis terrarum*, is surrounded by a circumfluent "ocean-river" delimiting the known world, with the major land areas divided by the T-shaped pattern of the rivers Don and Nile, and the Mediterranean Sea.

On 13th century European maps (see Figures 1.2–1.4), some true geographic features were found, but the most important features were related to Biblical meaning, such as the Garden of Eden, Rivers of Paradise, and Noah's Ark. At the center of these maps lies the holy city of Jerusalem.

We can think of the conceptual world of genetics in a similar way. Because Dr Schull has traveled much of the genetic as well as geographic map, indeed helping to pave many roads of the former, and because he has also been interested in the medieval Church, I thought that in this volume to honor his work, it would be appropriate to use the medieval *mappamundi* as a device for discussing our conceptual map of genetic variation, our *mappamundi genetica*, as it has developed over the past 20 years or so.

To do this, I will associate cardinal points on the conceptual map with cardinal ideas in three realms of genetic variation: among individuals, among cells, and among genes (see Figure 1.5). I also want to note by example how, at any time, there are many ideas of how the conceptual map should be interpreted, only some of which prove to be correct. Finally, I will ask to what Jerusalem these ideas may have led us.

Figure 1.2. Psalter *Mappamundi*, now in British Museum.

I will illustrate these points in regard to a long-standing problem in cancer biology and genetics, namely to explain the age patterns of onset of cancer in terms of a *genetic model.* Despite decades of work in this area it is still difficult to decide among competing models. However, recent progress in the cardinal areas in genetics may make it possible to refine our understanding of this interesting and important problem.

Christ was symbolically associated with East, where lay the Garden of

Figure 1.3. *Mappamundi* from Hereford Cathedral, England, reproduced from commercially available reprint sold by the Cathedral.

Eden and perfection. For this reason, medieval maps were "oriented" with the East, or Orient, at the top, representing Christ sitting upon the circle of the earth (Isaiah 40:22).[3] If East represents the complete wisdom of God, this is an area to which science can lay no claims. Though all truth may flow from the rivers of Paradise, we mortals cannot go upstream, and I dare

3. Of course, Isaiah is in the Old Testament.

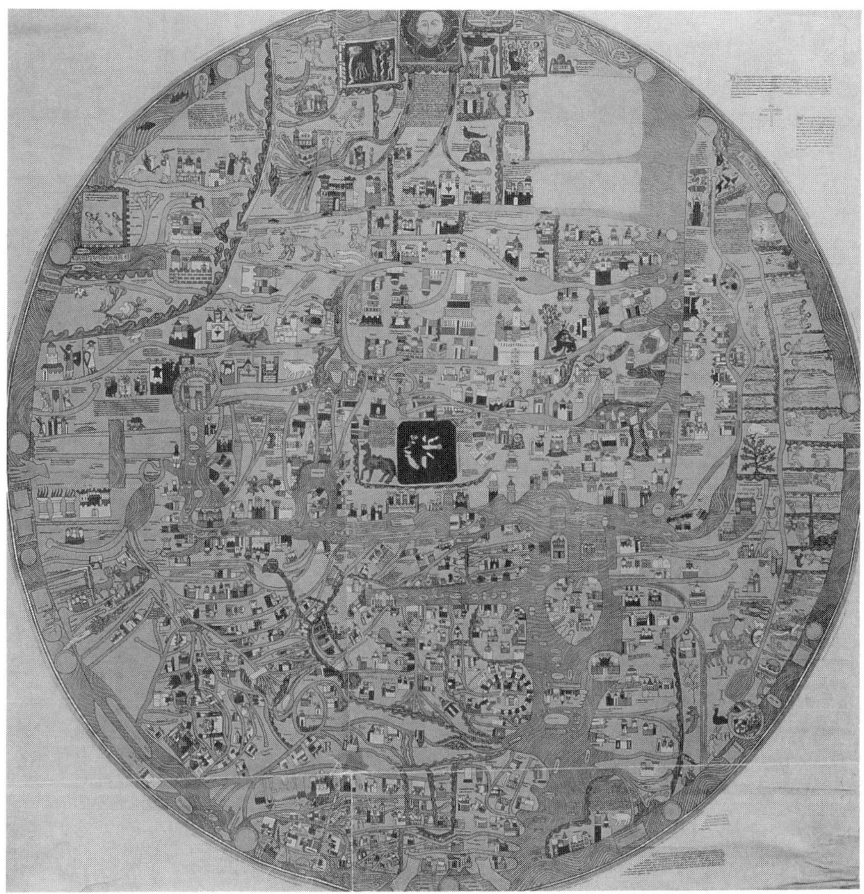

Figure 1.4. *Mappamundi* from Ebstorf, Germany (original destroyed in World War II). Reproduced from Bagrow, 1964.

not attempt to identify any idea in genetics with the Eastern point on our conceptual map. We must begin elsewhere.

SOUTH: VARIATION AMONG INDIVIDUALS

The South is associated with the left, or "sinister," hand of Christ on the Ebstorf map (Figure 1.4); there, *mappaemundi* placed a bestiary of mutant peoples reported by travelers, or imagined, to live in this remote reach of the world.[4] On the *mappamundi genetica* I use South to represent allelic variation among the individuals in a population. Technology has recently

4. Actually, the word *north* is more typically associated with *left* geographically, being to the left of east, but this is not the orientation of the *mappaemundi*.

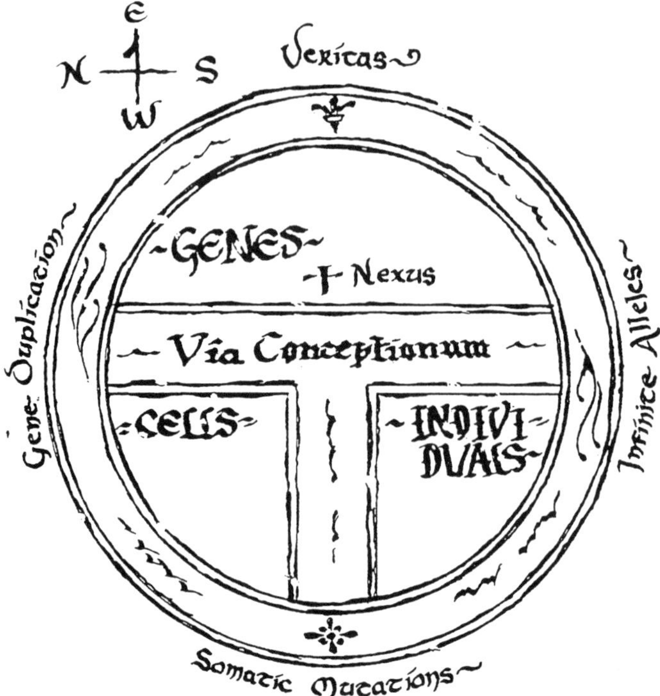

Figure 1.5. *Mappamundi Genetica*, ca. 1989, in the author's own not-very-monastical hand.

allowed us to travel much closer to the genome, and there we too have discovered a diversity of mutants that had not previously been thought to exist.

Prior to the discovery of the sequence-coding nature of DNA, mutation was typically viewed as a "pressure," or rate, of recurring transitions among a limited number of possible allelic variants of a gene (e.g., from, say, A to a and back: $A \rightleftarrows a$). The dynamics of allele frequencies produced by such mutation pressures and other evolutionary forces comprised the heart of population genetics theory during much of this century.

However, the stream of new data on genetic variation that began to accumulate with the use of electrophoresis in the 1950s revealed much more variation (higher heterozygosity) than had been expected on the basis of the existing theory (e.g., see Nei, 1987). We have learned that over even modest stretches of DNA so many different nucleotide changes are possible that almost every person is a unique heterozygote and each new mutation can be considered for practical purposes to be unique.

The idea that new alleles arising at a locus are nearly always distinct has profoundly changed our concept of variation. In a landmark paper, Kimura and Crow (1964) explored the evolutionary consequences of such an idea, now known as the "infinite alleles" or "unique mutation" model, which has taken a central place in the development of modern population genetics (Gale, 1990; Hartl and Clark, 1989; Li and Graur, 1991; Nei, 1987).

Infinite alleles and genetic epidemiology

This level of heterozygosity forces genetic epidemiologists to face the befuddling possibility that genetic disease may be due to different genotypes from family to family, and/or to genotypes at different loci, a pattern that may be intractible to work out with epidemiological methods (e.g., Sing and Moll, 1989). Hardly a week goes by without the announcement of some new example of disease heterogeneity, including even classic single-locus diseases like phenylketonuria (PKU) (e.g., Daiger et al., 1989; Rey et al., 1988), or cystic fibrosis (CF) (e.g., Roberts, 1990). This level of complexity casts doubt on our ability to do genetic counseling or screening (e.g., Roberts, 1990) or to infer "causation" for quantitative disease risk factor (Sing and Moll, 1989).

It may be possible to take advantage of, rather than to be paralyzed by, this new level of variation. New mutations arise on a chromosomal background that reflects a specific history of prior mutations, and these haplotype backgrounds can often be traced to or identified with the geographic or ethnic origins of the new mutation. For example, the haplotype backgrounds of European PKU mutants show that most cases arise on chromosomes that are descendants of only a few original pathogenic European mutations (e.g., Daiger et al., 1989; Rey et al., 1988). Similar patterns are found for basically all diseases that have been adequately studied, including the abnormal hemoglobin genes (Weatherall, 1991) and others (Weiss, 1993). Knowing what mutations are specific to which ethnic groups may help in genetic counseling.

Clustering mutant haplotype sequences in this geographically and historically sensible way is a form of *cladistic* analysis. The same kind of analysis can be applied to genes that affect a *quantitative* trait. The haplotype sequences at such a locus can be arranged into a phylogenetic tree, and a nested analysis of variance can identify which cluster of sequences share by common ancestry mutations of statistically significant effect. Such analysis has been done on the genetics of lipid and enzyme levels (e.g., Templeton et al., 1987, 1988). Essentially, evolution provides structure we can use to simplify the dimensions of existing genetic variation.

Clonal disease

When a cladistic analysis identifies a mutation as being causal within a given population, cases arising due to that mutation (including any of the background haplotypes descended from the original mutant chromosome) are *clonal diseases* in the population, just as cancers are cellularly clonal diseases within individuals.

We can sometimes infer the nature of causation of a disease if we can assume that cases in a given population are clonal and the disease is in fact genetic. In such populations, affected individuals are like affected sibs (or, for recessive disorders, like inbred individuals). Among sets of such affected individuals, the regions flanking the pathogenic allele will share sequence homozygosity, to an extent dependent on their common ancestry that those individuals share common ancestry; that is, to the degree that recombination has not broken up the sequences found on the original mutant chromosome (e.g., T. Cox et al., personal communication, 1990). The original haplotype sequence may be reconstructable, and the pattern of shared seqence among those affected may be useful in fine mapping of the causal locus itself.

Natural experiments may sometimes be found that help reduce the heterogeneity problem, if clonal cases of a disease can be ascertained. I think diabetes mellitus (type II) is largely clonal is Amerindians and populations admixed with them (Weiss et al., 1992). One or more predisposing mutations occurred early in the settlement of the New World and quickly rose to high frequency, so that many or most currently affected Amerindians share clonal copies of these mutations. Given the history of about 1,000 generations of evolution in the New World, these chromosomes may share a region largely identically by descent (i.b.d.) surrounding the gene(s) responsible for this susceptibility, that has not seriously been interrupted by subsequent recombination. Such a region could be on the order of 1–5 megabytes of sequence in length (Connor and Weiss, unpublished results).

The genetic architecture of phenotypes

We can begin to generalize about the complexity of genetic control of quantitative traits, that is, their genetic causal *spectrum* of effects, or *genetic architecture*. Most of the existing data have come from studies of isogenic lines in agricultural and experimental genetics (Lande, 1981; Paterson et al., 1988, 1991; Weiss, 1993; Wright, 1968; recent data summarized in Connor and Weiss, in prep b). Typically, the alleles responsible for variation in a quantitative trait have an approximately exponential distribution of absolute additive effects relative to the mean. That is, alleles of large effect are quite rare in the population, and most alleles have very small effect. Similarly, although many loci may segregate alleles that affect the trait, only a few have any alleles with large effect.

The human data are very limited, but generally tell a similar story. The best data are for the cholesterol (see Figure 1.6), triglyceride, and lipoprotein levels (e.g., Humphries et al., 1987; Sing and Boerwinkle, 1987; Sing and Moll, 1989), where a substantial fraction of the genetic variation is at loci whose allelic effects are too small to identify individuals (classic *polygenes*). However, a number of other loci have alleles with specifically detectable effects on phenotypic variance, although almost all of these effects are small (and the alleles rare).

Presumably, variability in such traits has been maintained by a balance among drift, stabilizing selection, and mutation. Genes with major effect will, if they affect fitness, generally be kept to low, or rise to high, frequency depending on whether they are deleterious or beneficial. For example, the

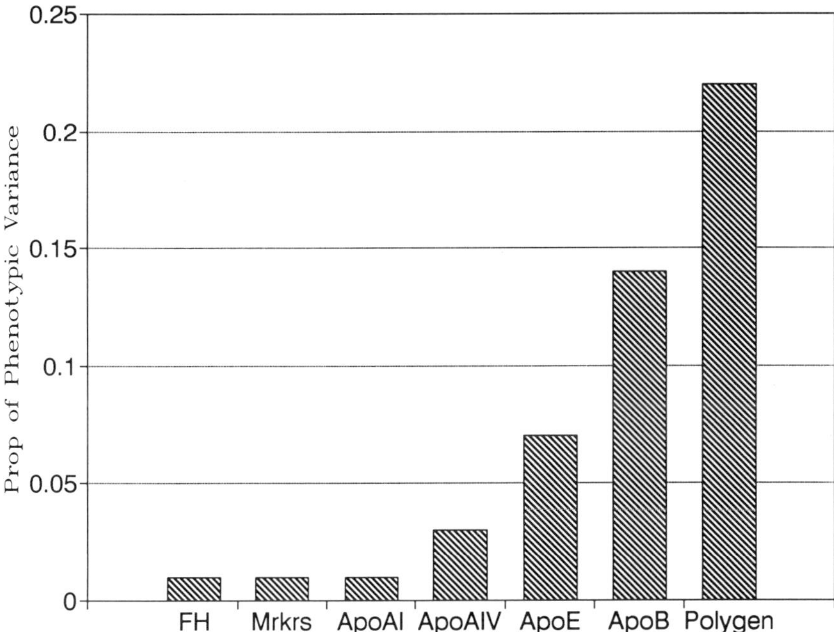

Figure 1.6. Genetic determination of serum cholesterol levels, showing distribution of frequency of alleles having statistically detectable effects. Redrawn from Sing and Boerwinkle, 1987.

mutations responsible for Familial Hypercholesterolemia (FH) inactivate the LDL Receptor (LDLR) protein, inhibiting the removal of cholesterol from the bloodstream. Many different LDLR mutations are known; some inactivate the LDLR locus, and have dramatic effects on cholesterol levels, while others modify the LDLR protein and have less lipid effect, especially in heterozygotes. The LDLR inactivation mutations are a set of heterogeneous, but individually very rare alleles, that collectively account for only a small (though epidemiologically serious) fraction of overall cholesterol variation.

Conditions for the evolution of a stable equilibrium phenotype distribution for a quantitative trait have been worked out, beginning at least with Wright (1931), and in recent years largely by several authors (e.g., Barton and Turelli, 1989; Lande, 1987; Turelli, 1987). Generally it is assumed that extreme phenotypes are selected against, with more common "normal" phenotypes having essentially "full" Darwinian fitness. For genes responsible for chronic diseases like cancer, including many cholesterol-raising mutations, the fitness effects of the mutation are very very slight (Weeks et al., in prep; Weiss, 1993). When translated into hazard effects, the nearly neutral nature of most mutations can be seen.

Presumably the same is true for traits not related to disease per se, such as those that control various enzyme levels, height, obesity, and so on. What this means is that to a great extent the distribution of allele frequencies and allelic effect will be that produced by mutation and genetic drift, and can be approximated by the neutral theory of evolutionary genetics (e.g., Gale, 1990; Hartl and Clark, 1989; Nei, 1987).

The infinite alleles model has widespread implications, not the least of these being that differences in alleles and their frequences between populations are driven by somewhat different population genetic processes than were at the core of recurrent-mutation models. This has transformed evolutionary population genetics, but these changes are not the subject of this chapter. I will return to the infinite alleles model later. But first our journey across the genetic map takes us westward.

WEST: VARIATION AMONG CELLS

In the far West, the late 13th century *mappamundi* from Hereford (Figure 1.3) depicts the pillars or Hercules, which the classic mythical figure created at the Straits of Gibraltar to mark the limits of the known world, on his way to the island of Erytheia to perform one of his twelve labors. Daylight also fades in the West, reducing our ability to resolve pattern and order.

In our schematic map, this is where I place the population of cells in an organism. My analogy to the setting sun and an ocean of uncertainty are the effects of somatic mutation, which break down the order of the genome during of course of life. Like other voyages into the unknown, we generally associate somatic mutation with disorder, degeneration, and disease.

Somatic "mutation" is not new to biology. In a sense, heritable biological alteration during the lifetime was the essence of Lamarck's ideas on 'evolution' by the inheritance of acquired characteristics, ideas even Darwin toyed with. This changed once Weissmann showed that the germ line and the soma (body tissues) were separated, and the genetic constancy of the individual became a central dogma in biology. However, we now know of a variety of mechanisms that occur somatically to change the genome, most of which have been implicated in disease.

The idea of somatic mutation was "in the air" in the late 1960s. Two major classes of disease, autoimmunity and cancer, were considered prime examples. Both tumors and autoimmune diseases increase with age, and involve attacks against normal tissue, or failure to attack abnormal "self" tissue by the immune system. If this were due to the occurrence of somatic mutations, then so might be the phenomenon of aging itself. Neither somatic gene rearrangement in the production of antibody specificity, nor the role of the HLA system in self–nonself recognition had yet been discovered in the early 1970s, but the clonal nature of immunocyte specificity was known, as was the general nature of self–nonself immune discrimination.

A changing *hazard function* (age-specific risk of death) is the *sine qua non* of senescence. Based on what was then known, Nobel immunologist Macfarlane Burnet (e.g., 1965, 1976) suggested that somatic mutations might lead to the growth of "forbidden clones" of aberrant cells which the system of "immune surveillance," itself aging and subject to somatic mutation, could not detect and remove. Because anti-self antibodies could attack different tissues in diverse ways, autoimmunity was a candidate for over-arching theories of aging. These ideas were extensively elaborated by a British radiobiologist gadfly of the time, P.R.J. Burch, who showed that the suggested processes (plus others of his invention) could generate the kinds of hazard function associated with senescent diseases, especially but not restricted to cancers (Burch, 1965; 1969).

Cancers were ideal candidates for such diseases. Mutagens are carcinogens, occur stochastically, and more than one event seemed needed to produce cancer. Cancers also turned out to be clones, so that the causal process could be related to the way mutations could *transform* a single somatic cell. This all suggested that the age pattern of cancer could be modeled by multistage waiting-time models; that is, the *probability* a cancer arises at a given age, t, the hazard function, depends on how many somatic transforming events are required to transform a cell, the somatic mutation rate, and perhaps other factors like the effects of growth (cell-division) promotors. If one could specify a genetic model for this process, one could test whether the observed schedule of age-specific cancer incidence or mortality rates was consistent with that model.

Models of carcinogenesis

Two basically different approaches were taken to this problem. One is to assume a model, and *estimate* the values of its key parameters from the data. For example, one common model supposes that a cancer requires that b independent mutational events have occurred in a single stem cell lineage to produce a cellular genotype that leads to a neoplastic clone (tumor). Here oversimplifying, if we assume that these events occur uniformly with risk per unit time μ, then the resulting hazard function can be approximated by a Weibull probability function in which $h(t) \approx b\mu^b t^{b-1}$. This has the form of a straight line if plotted on a log-t log-h scale. For a rigorous discussion of these and related methods see Moolgavkar (1991) or Whittemore and Keller (1978).

A systematic survey of international cancer hazard functions in 1969 showed that such a model fit closely to a wide variety of worldwide site-specific cancer data (Figure 1.7), and from this it was estimated that there were five to seven steps on the unholy staircase to cancer (Cook et al., 1969). This has long and widely been generally taken as a satisfactory statistical explanation for cancer.

The second, and more biological, approach to cancer modeling was also developed at about the same time. This method is to *stipulate* rather than estimate the number of transforming events, and to use the model to estimate aspects of those events (e.g., reviewed recently by Moolgavkar, 1991). In this context, in 1971 A.G. Knudson proposed his somatic mutation model to

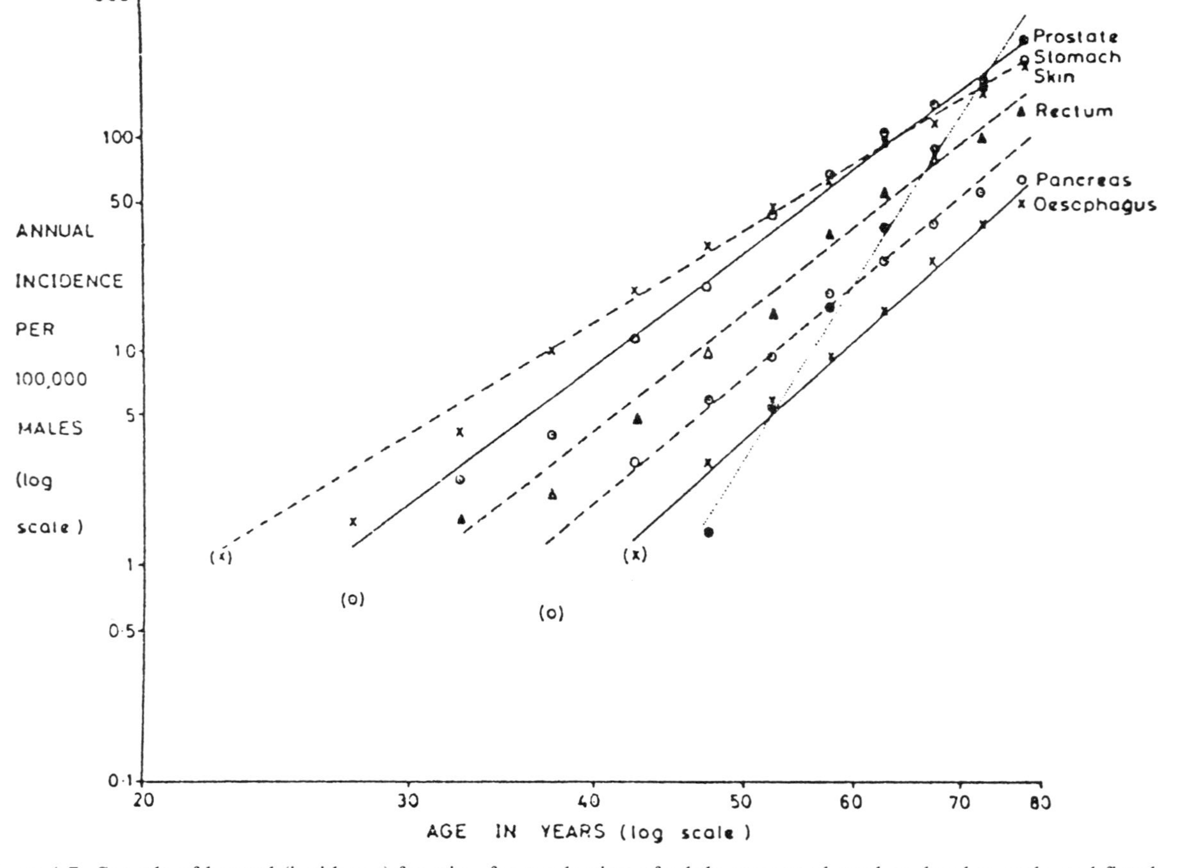

Figure 1.7. Sample of hazard (incidence) function for a selection of adult cancers plotted on log-log scale and fitted to Weibull function. Reproduced from Cook et al., 1969.

explain the epidemiology of *retinoblastoma* (RB) (Knudson, 1971). If multistage mutational somatic processes are responsible for cancer, an individual who inherited one of these mutations would have to await fewer somatic events before a tumor arose than individuals in whom all these events must occur somatically. This should be reflected as differences in the age-pattern between heritable and sporadic forms of the same tumor.

The data suggested to Knudson that RB was a two-stage tumor. The bilateral, multifocal nature of tumors in heritable RB, and their earlier age of onset, fit a one-stage process, while the unilateral, unifocal, somewhat later-occurring sporadic cases suggested a two-stage waiting process. Knudson showed that the age patterns of these two forms of RB were consistent with this hypothesis.

Knudson chose his test case perceptively: a flood of studies has shown in great detail that every essential element of Knudson's idea was correct. The RB gene has been mapped to chromosome 13, and the second transforming event also involves this gene. A tumor arises when a retinal cell contains no functional copies of the gene; that is, at the cellular level the RB gene acts in a recessive way. The discovery of the RB locus led to the finding of a whole class of cellularly recessive cancer genes, one of which is RB, called "anti-oncogenes" or tumor-suppressing genes, that appear to be generally involved in carcinogenesis (e.g., Friend et al., 1988; Sager, 1989; Scrable et al., 1990).

Many processes can generate loss of both copies of a suppressor gene, but one in particular has been important in recent cancer research. Through various forms of aberrant mitosis, a somatic cell that is genetically heterozygous at a suppressor (or any other) locus can produce daughter cells that are homozygous for one or the other of its alleles at that locus. *Mitotic recombination* is one such process. This led to a simple, and subsequently widely applied test to detect tumor suppressor genes in action. The test is to identify sequences in the inherited genotype that are heterozygous in or near a particular locus, as well as random heterozygous markers, and to test tumor tissue from the same individual for *loss of heterozygosity* or LOH, mutations. Such a test can be done from tumor tissue and blood or other adjacent normal tissue, using standard restriction fragment length polymorphism (RFLP) or polymerase chain reaction (PCR) methods.

LOH mutations have been found in a number of cancers, one of the first being RB. Although LOH is a normal, if rare, mitotic phenomenon, its occurrence at critical loci seems to be implicated in carcinogenesis. When tumor tissue is compared to adjacent normal tissue from the same individual, a whole array of LOH mutations is discovered. The mutations are spread across the genome, and extensive studies of colorectal tumors have found somatic LOH and other mutations to have occurred on essentially every chromosome (Figure 1.8) (Vogelstein et al., 1989; also Astrin and Costanzi, 1989; Baker et al., 1989). The findings are similar, though with fewer current data, for other tumors including melanoma (Fountain et al., 1990), lung (Weston, et al., 1989), kidney (Morita et al., 1991), nervous system (Bigner

A

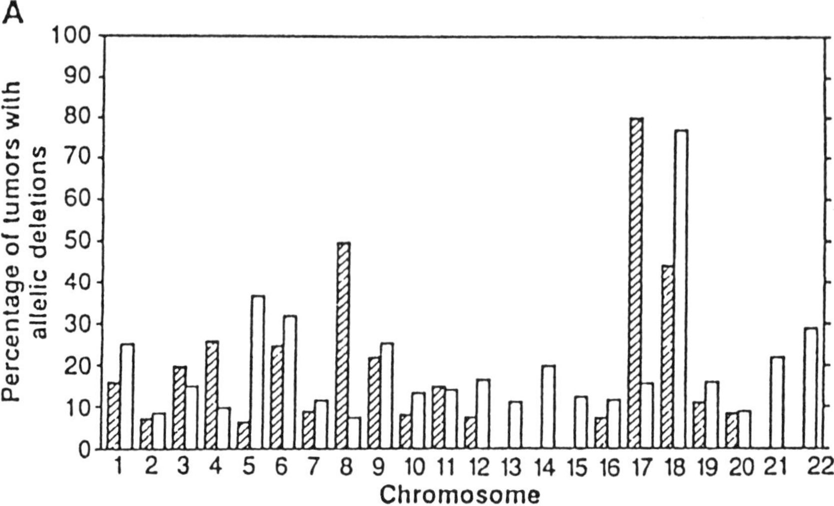

B

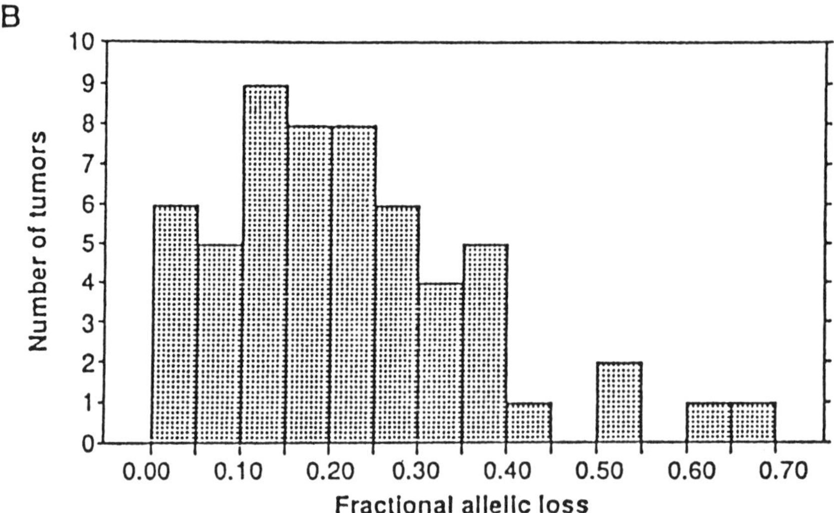

Figure 1.8. Distribution among colon cancers of chromosomal regions in which detectable loss of heterozygosity has occurred. (A) frequency of allelic deletions in individual chromosomes among tumors; (B) frequency of allelic deletions in individual tumomrs. The Fractional Allelic Loss (horizontal axis) is the percent informative markers suffering allelic deletion. From Vogelstein et al., 1989.

et al., 1988), breast (Devilee et al., 1989), prostate (Carter, et al., 1990)—in short, for most tissues. This has made the finding of a scattering of LOH mutations general, routine, and—one hallmark of understanding—predictable.

The studies done to date indicate that several specific suppressor loci, some typical to the tumor, others in common with other tumors, seem to have suffered LOH mutations in a high fraction of cases. For colorectal cancer these include two genes associated with Familial Adenomottors Polyposis (FAP) on chromosome 5 (loci denoted MCC and APC) (Kinzler et al., 1991; Nishisho et al.; 1991), the "p53" locus on chromosome 17, and one or more loci on chromosome 18. The p53 locus, like the RB locus as well, is implicated in many tumors, and a whole panoply of different mutations have been identified (e.g., Hollstein et al., 1991). And p53 may positively or negatively regulate steps in the cell cycle (Levine et al., 1991).

This LOH test does not screen for genes of dominant effect, but another diverse set of dominantly acting carcinogenic mutations exist. At these loci, a mutation can be effective in a cell even if just one of the homologues is affected. Such loci are called *oncogenes*. Few if any truly dominant heritable oncogene mutations have been found, presumably because they are incompatible with successful embryogenesis. But they can be found to have occurred somatically, just as LOH mutations do.

The set of loci that have somatically mutated in a given tumor is known as the *allelotype* of that specific case of the tumor. The work done to date is clear in suggesting that the allelotypes vary among individuals affected with cancer of the same organ, even if some loci are often included in the mutated set. Indeed, a substantial fraction of all loci tested (including many random markers) has experienced a change. This raises an important warning that should clearly be stated at this early time in such investigations. We do not yet have adequate data on the age-specific distribution of allelotypic differences between pairs of randomly chosen *normal* cells in an individual, and thus we do not yet know directly what fraction of the cancer allelotypes are causal and what fraction simply spontaneous somatic mutational "noise." It is also possible some sites have mutated after tumor initiation, during tumor progression and evolution, and hence are not causal per se. However, the colorectal allelotypes were correlated with tumor type, but *not* with patient age, suggesting that they are causal.

Given the millions of stem cells in a given tissue, their rapid division rates, and the length of a human life, there is room for an enormous amount of genetic diversity to accumulate via somatic mutations. The number of cell divisions in the colorectal epithelium in a normal lifetime can be on the order of 10^{10}, comparable to the number of individuals in the history of a species! Thus, the allelotypes found in cancer are not surprising, and we might even consider whether the theory for the evolution of a genetic basis for phenotypic diversity can be applied to the generation of cancer within a lifetime.

I will return to this topic below. But now let us move away from the chaos of intra-population and intra-individual variation, to a more orderly topic.

NORTH: VARIATION AMONG GENES

The Ebstorf map shows the right hand of Christ in the North. The right is the antithesis of the left, representative of "right," goodness, and order. On the conceptual map of genetics, if mutation represents disorder, I place order here, as represented by order in the genome.

Neither Darwin nor his followers could provide a convincing mechanism to generate the variation on which natural selection could work, especially to create major new biological functions. For this, neither Goldschmidt's "hopeful monsters" caused by occasional major mutation, nor simple sequence evolution do the trick in a very convincing way (though recent progress in developmental genetics shows that large, even large orderly, changes can be brought about by relatively simple mutants, e.g., in homeotic genes). Natural selection is generally conservative, screening out mutations that too greatly alter the function of existing genes in a complex organism. Fine tuning is its role.

By 1970, accumulating evidence from a variety of sources began to suggest the needed mechanism. It had long been known from anatomy that whole organisms have "metameric" (modular) structures, characterized by repeated variations on a theme; examples are somites and their vertebral segments, teeth, and—important for this chapter—the histological processes by which mature tissues develop from less differentiated stem cells. Sequence data had also begun to show similar patterns among gene products. For example, related genes, such as those for tRNA and rRNA were repeated in the genome. Genes with similar functions, such as actin and myosin, or the globin genes, were found to have similar amino acid sequences and structures.

Around 1970–1971 Susumu Ohno generalized these facts to suggest that higher organisms evolved by the duplication and subsequent sequence divergence of genes (Ohno et al., 1968; Ohno, 1969, 1970). Investigators have since found evidence of gene duplication ranging from the repeated use of exons to the thousands of copies of *Alu* sequence, and that the map of our chromosomes is populated with linkage groups containing members of multi-gene families. These genes clearly have common evolutionary ancestry. Part of the variation in genome size may be due to the very different number of copies of repeat sequences (which themselves propagate by duplication-like mechanisms), plus a matrix of non-coding, presumably non-functional DNA whose length evolution is not understood (Lewin, 1990). Gene families have now been shown to involve essentially every major physiological system that has been studied in any detail. The genome itself is thus one large phylogenetic tree of genes. This pattern is found so pervasively and fundamentally in our biology that in my view metameric "logic," which can be described by the rubric *duplication with variation* should be considered a fundamental principle of evolution (Weiss, 1990b).

These comments are directly applicable to almost any trait. We know they apply to the globin genes, the genes for collagen, lipid transport, the immune system, color vision, and the like. The same is true with regard to cancer. The same specific genes as well as types of genes are implicated in the variety of cancers that have been studied. This suggests that differentiation in different tissues may be controlled by the same, or homologous, regulatory genes. Multigene families associated with the fundamental processes regulating cell division and differentiation are associated with cancer; examples are the *ras* and *myc* genes, repressor genes like RB, "p53" and others, the growth hormone family, the protein kinase family, the thyroid receptor gene family, and others (Baker et al., 1989; Hanks, 1988; Klein, 1988; Lane and Benchimol, 1990; Nigro et al., 1989; Sager, 1989; Scrable et al., 1990; Vogelstein et al., 1988; Weston et al., 1989).

Bishop has depicted the physiology of cell growth and differentiation schematically as a "circuitry" diagram (Figure 1.9, Bishop, 1987). The lines represent enzymatic pathways controlled by members of a variety of gene families like those just mentioned. Missteps anywhere in the diagram have been found in different forms of cancer. Yet, the diagram itself is *generic*, and applies to different tissue types (Bishop, 1987; Weinstein, 1988; Sager, 1989). The total number of genes (and consequently, mutations) potentially involved may be large; each gene identified in Bishop's diagram, for example, will have one or more regulating genes, and so on.

In this evolutionary sense cancers may be *homologous* diseases between tissues. Perhaps such homology is what we should expect: once mechanisms evolved for generating terminally differentiated tissues, that is, to create a renewable tissue "bed" out of less fully committed generating "stem" cells, such mechanisms would have been re-used where needed, as in the evolution of the different specialized tissues in the body. We know that normal tissue differentiation involves members of gene families, much of the knowledge

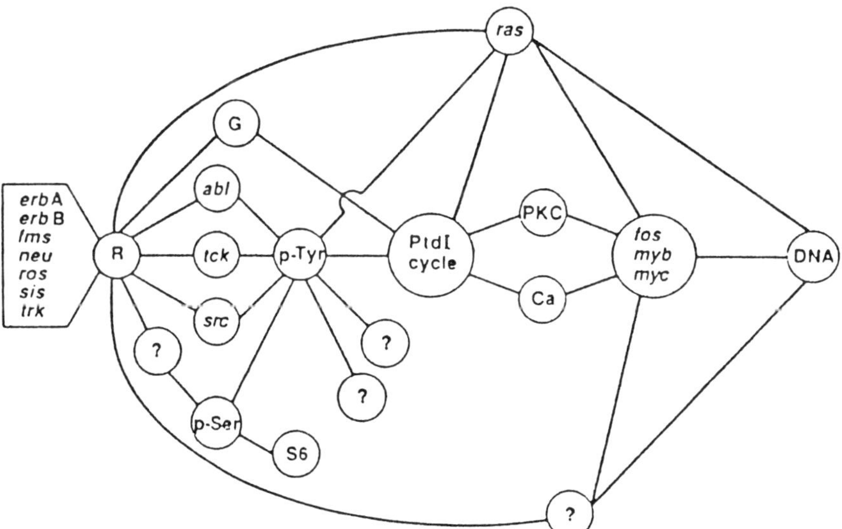

Figure 1.9. Schematic "circuitry" diagram for the regulation of cellular proliferation. The functions of various proto-oncogenes are shown in terms of their relation to each other. This diagram is partially hypothetical (not all relationships have been worked out in detail) and incomplete. G, regulatory protein; R, receptor; p-Ser, p-Tyr phospho- serine and tyrosine; S6 a ribosomal protein in response to mitogenic signals, PKC protein Kinase C; PtdI, phosphatidylinositol. From Bishop, 1987.

having come in fact from cancer research. In that sense, a stomach and lung are, I believe, homologous organs.

Our improved understanding of gene duplication, somatic mutation, and infinite alleles—the cardinal ideas on our conceptual genetic map—help us to explain the genetic basis of normal and abnormal control of most complex phenotypes we might be interested in, cancer being but one interesting and medically important example. However, when problems like this are not fully understood, as in the case with cancer, it is common that many different theories can explain the same data, even by appealing to the same general principles. This is the state of the use of such principles to explain the age patterns of cancer, a topic to which I would like to turn briefly.

PRESTER JOHN AND PIED PIPERS

It's easy to bid one rack one's brain—
I'm sure my poor head aches again,
I've scratched it so, and all in vain.
Oh for a trap, a trap, a trap!
—Browning, 1845

In 13th century Christendom, fears and zealtory led to expeditions into poorly known parts of the world in the quest for righteousness, which for us can symbolize the quest for major theory in science. Among the Biblical descendants of Cain was Gog, who ruled a people, the Magog, somewhere in Asia. The Bible prophesies that the Magog horde will descend upon Christendom with the forces of the Apocalypse on the Day of Judgment (Ezek. 38:1–9). In the third century BC, Alexander the Great was said to have built a great stone wall to help conain the Magog empire, shown on *mappaemundi* in remote parts of the far Northeast.

But the fearful medieval world was unconvinced, and was quite receptive when reassuring rumors appeared in the 13th century of a powerful Christian king, Prester John. He was the ruler of a devout empire somewhere in Asia, who promised to bring his forces to the aid of Europe on Judgment Day (this reflected the European struggles with Islam at the time). A supposed letter to this effect in Prester John's own hand to the Pope was widely circulated in Europe. Efforts to locate Prester John's kingdom were reflected in many medieval maps, in what has been called the greatest hoax in the history of geography (Cortesao, 1969). After each unsuccessful attempt to find him, his kingdom was relocated on subsequent maps. No such hero existed, but the search for Prester John stimulated decades of European voyages of discovery and trade in Asia.

Another relevant legend is that of the Pied Piper who, in Robert Browning's poem, led the children of the town of Hamelin into the mountain because the mayor had cheated him of his payment for ridding the town of rats (*Browning*, 1845). This child's story derives from medieval legend growing ultimately out of the Children's Crusade of AD 1212, in which thousands of children were led to their deaths in pursuit of a righteous cause. Much was learned about the map of the Asia Minor from the crusades (though they are not explicitly depicted on the *mappaemundi*).

In the poem as quoted above, when the Mayor of Hamelin pleads for an answer to the rat problem, the Pied Piper appears in response. In search of answers, scientists, like mayors, are sometimes enticed down paths leading nowhere, or at least "elsewhere" (Eric Boerwinkle, personal communication, 1991). The history of science shows how difficult it is at any time to distinguish which among competing ideas will prove successful. An example from cancer biology illustrates this.

Multiple roads to the same observation

I referred earlier to PRJ Burch, who became captivated by the idea that somatic mutations could explain diverse diseases and aging, influenced by Burnet and others regarding the role of the immune system in normal as well as abnormal regulation. The potential universal appeal of somatic mutation led Burch to apply general stochastic hazard models to almost any age-related trait. His fanciful speculations to rationalize this view led him far into the mountain, even given the imperfect knowledge of his time—he suggested a marker-recognition system ubiquitous enough to explain essentially all aspects of tissue size, cell number, and maintainance. But his speculations are interesting in that Burch did correctly sense the importance of intercellular communication by circulating agents, including the immune system, of ligand-receptor recognition as a trigger for cell-growth regulation, of tissue-specific cell-surface antigens and of gene transcription regulated by sequence-specific protein-DNA complexes. He toyed with the idea of somatic loss of heterozygosity, even with evolution by gene-duplication. All these are major themes in the modern molecular biology of cancer.

Burch's papers were read by his contemporaries in the area, he was Knudson's source for the equations used to test his somatic mutation model for RB, and he is still cited in the same way (Erlandsson et al., 1988). Many highly qualified investigators continue to use similar mathematical logic to models that could provide unitary explanations for aging (e.g., Finch, 1990) and in particular, for cancer.

As mentioned above, the Weibull model is another that fits cancer data very well (Cook et al., 1969). If a population is assumed to consist of a set of individuals who are heterogeneous in their susceptibility (for genetic or environmental reasons), such that their level of heterogeneity modifies the rate at which their six-hit process occurs, a mixture Weibull model can statistically explain secular trends in the shape of the hazard function for some cancers, such as lung cancer, that have been affected by changing environments (e.g., Manton et al., 1986; Manton and Stallard, 1984, 1988).

The fit of a two-stage model was mentioned earlier in the context of the discovery of suppressor locus mutations for RB. Working initially with

Knudson and retinoblastoma, SH Moolgavkar and several colleagues have developed a two-stage model in which partially transformed cells may have a growth advantage, and this model fits a wide array of whole-population cancer data reasonably closely (Moolgavkar and Venzon, 1979; Moolgavkar, 1991), as well as dose-response data on smoking and lung cancer (Moolgavkar et al., 1989), and a variety of experimental data on exposure, such as to ionizing radiation (Moolgavkar et al., 1990).

The Weibull function is typically derived as a mathematical approximation to a more complete model. While a similar result has been derived from various starting points, one is as follows. Suppose a tumor arises when b events have occurred, each occurring with rate a per year. Here, I avoid referring to these as mutations and mutation rates, although this is usually the assumption of authors using such models whether or not so stated. Under this model, the probability of disease by time t, $P(t)$ is

$$P(t) = (1 - e^{-at})^{b} \qquad [1.1]$$

and from this, by the definition of the hazard (Elandt-Johnson and Johnson, 1980)

$$h(t) = \frac{(\mathrm{d}P(t)/\mathrm{d}t)}{(1 - P(t))}. \qquad [1.2]$$

The exact expression is algebraically messy, but tractable by computer. If one uses the approximation $\exp(-at) \approx 1 - at$ in [1.1], then [1.2] has an approximate Weibull form.

In 1978 Moolgavkar (see Moolgavkar, 1991) noted that this approximation may lead to inaccurate specifications of the hazard. Ranajit Chakraborty, Ryk Ward, and I independently noticed that the approximation was not good in another sense. Based on data from Waterhouse et al. (1976), the "unapproximated" model [1.2] fits the data slightly better than the approximated model (e.g., generates more of the curvature in the data), but more importantly, the estimated number of stages is very different from, and much more variable than the previously found 4–6, and is in the range 5–15 or more. These results were mentioned in Weiss and Chakraborty (1984) and were eventually published in detail as Chakraborty and Weiss (1989).

What we noticed was even more striking. For each population to which this model was fitted to the age pattern of some tumor, a pair of estimates results for the set of parameters (a, b). If the model is separately fitted to data for the same tumor from many populations (and here, we also fitted males and females separately) the scatter of the parameter estimates shows a highly constrained pattern. The points fitted across populations for the same tumor type lie closely along a straight line, as can be seen for colon cancer in Figure 1.10.

We also found that the parameters for many *different* tumor sites lie essentially on the same line (Figure 1.11). That is, within and largely across sites, one cancer hazard function can be transformed into another by a change along the single dimension of this "a–b" line, suggesting that common

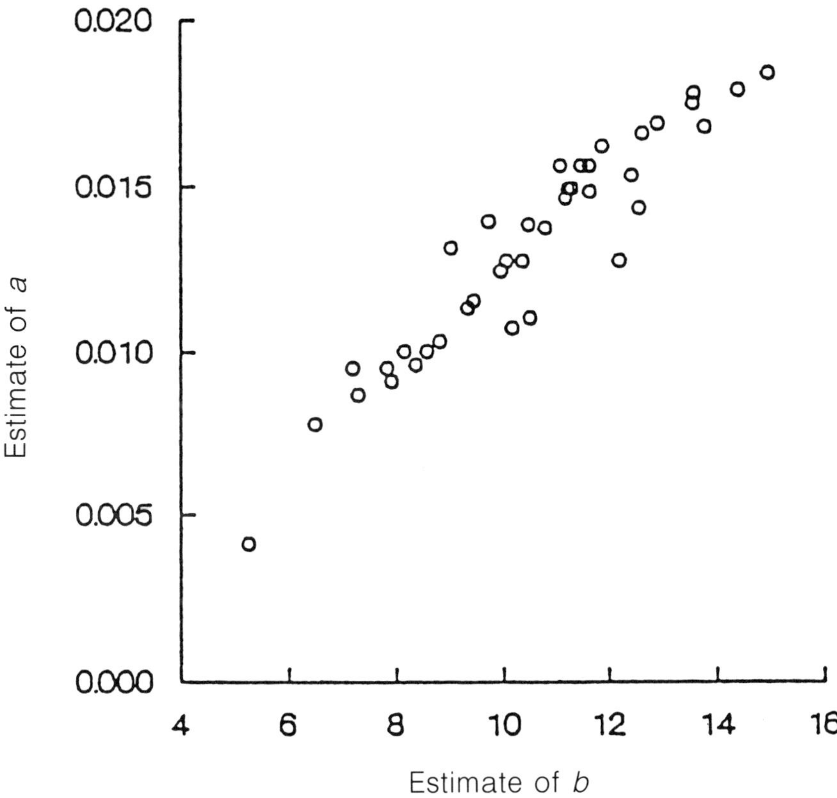

Figure 1.10. Scatter plot of sets of parameter values, a and b, for a multistage model fitted to 1970 US colon cancer hazard (incidence) function to show that across risk groups there is highly constrained variation in the shape of the hazard. Each point represents best-fitting point for a given population. Redrawn from subset of data given in Chakraborty and Weiss, 1989; original data from Waterhouse et al., 1976.

underlying cellular mechanisms are responsible for different types of tumors. This is consistent with evidence, some of it mentioned above, that diverse cancers fit a Weibull with similar slope parameter and that the same or homologous genes are involved in different tumors. This does not prove that different tumors are homologous, but is consistent with that idea, and various tests showed the pattern not to be an artifact of the method (Chakraborty and Weiss, 1989).

Since the value of b varies across populations and between males and females for the same tumor, the parameters of this model clearly cannot be interpreted in the genetic terms in which the model was formulated, a statement that applies *a forteriori* to the other, etiologically incompatible, models that fit the data comparably well. In fact, the estimate of b for FAP, in which one major event is inherited, is greater, not less, than in the general population (Chakraborty and Weiss, 1989; Weiss and Chakraborty, 1984; Weiss, 1990a).

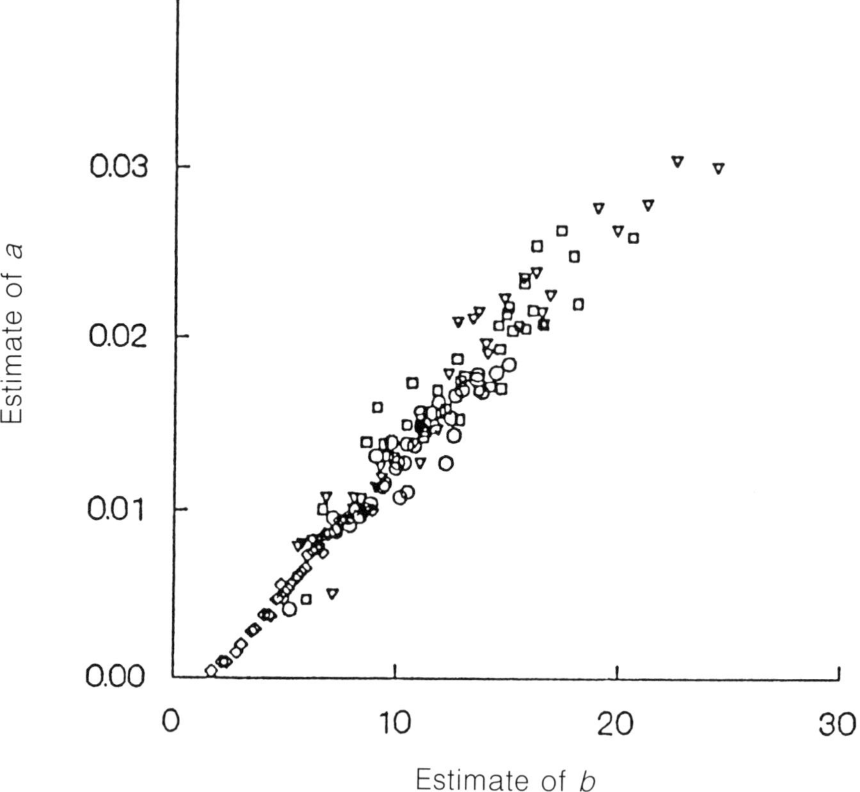

Figure 1.11. Data from selected sites like colon as shown in Figure 1.10, superimposed to show that, between sites, there is similarity in the constraint on the shape of the hazard function, suggesting that the same, or homologous, processes may be responsible. *Key*: diamonds: leukemia; circles: colon; downward-pointing triangles: lung; squares: stomach. For details, see Chakraborty and Weiss, 1989; Weiss, 1990a,b.

To be glib about it, the same cancer cannot require 2, 6, and 15 different mutations any more than Prester John can simultaneously live in three parts of Asia. A more definitive test must be found. Progress in molecular biology has prevented anyone from following Burch all the way into the mountain, although serious biostatisticians and epidemiologists continue to try to develop general models for carcinogenesis, presumably because of our conviction that there *is* something unitary about cancers. But what is it?

JERUSALEM

Jerusalem was the geographic as well as spiritual center of the medieval Christian world, to which important roads led. I represent this on the *mappamundi genetica*, by the confluence of ideas on genetic diversity. But since I can't claim holy status for this confluence, I simply call it *Nexus*.

The discovery of the cellularly recessive anti-oncogenes, and of a variety of cellularly dominant oncogenes seemed consistent with the fit of multistage models to cancer data. Whether the number of stages was two or six, for some years it seemed at least that we would be able to enumerate them. But as molecular assays have revealed the complexity and variability of cancer allelotypes, and we have understood the complexity of cell differentiation (e.g., as represented by Bishop's circuit diagram), the level of diversity that has been found suggests that a full-spectrum polygenic model may be closer to the truth (Weiss, 1990c), though this is not a consensus view (e.g., Moolgavkar, 1990).

Cancer phenotypes may be as diverse as cancer allelotypes. Extensive analyses of the biology of tumor cells in vivo as well as in vitro have shown that the array of cellular phenotypes, and biological pathways involved in cancer is large, complex, and even quasi-continuous, corresponding to Bishop's diagram, a network of effects including oncogene–anti-oncogene antagonism, repressors, locus complementary, diffusable factors, and a host of concentration-dependent determinants of the growth behavior of cells in tissues (e.g., Astrin and Costanzi, 1989; Franks and Teich, 1986; Klein, 1988; Herrlich and Ponta, 1989; Weinstein, 1988).

Cancer allelotypes appear to be related to these cancer phenotypes. The fraction of tested tumor loci that show LOH mutations is related to the tumor phenotype, specifically the cell type, response to treatment, progression, or prognosis (e.g., Ahuja et al., 1989; Chen et al., 1991; Devilee et al., 1989; Marx, 1990; Nigro et al., 1990; Sager, 1989; Scrable et al., 1990; Vogelstein et al., 1988, 1989; Weston et al., 1989). These mutations may contribute to the phenotype in an incremental, quasi-continuous way (e.g., Lane and Benchimol, 1990; Sager, 1989; Vogelstein et al., 1988), which an agricultural or evolutionary geneticist would refer to as additive polygenic effects. Even the classic major gene for colorectal cancer, the FAP region on chromosome 5, can experience a variety of mutations with lesser effects (Leppert et al., 1990). How can we make sense of this is regard to multistage models and the age pattern of cancer?

Oligogenic vs polygenic causation

We have two basic models of the genetic control of phenotypes, for qualitative and quantitative phenotypes, respectively, which correspond to the basic division in biological thinking that is as old as the dispute between Mendelians and Biometricians about how evolution works. For qualitative traits, we assume that some *specific* genotype(s) are responsible for the presence of the trait. Thinking of cancer in this way, we would model the disease as arising once a specific genotype has been produced in a dividing cell. This is what most of the stochastic waiting time models implicitly do, whether assuming two or six hits are required.

The other model is for quantitative traits like cholesterol level and blood pressure. Variation in these traits is produced by polygenic variation, that is, involving so large a number of loci, that until recent molecular biological methods came on the scene they could not individually be identified but only studied as aggregate effects. Such models have been the basis of animal and plant breeding for decades. A key feature that distinguishes qualitative from quantitative models is *allelic equivalency*: there are many genetic pathways to essentially the same phenotype, so that one cannot infer a genotype from the phenotype.

Given the spectrum of cellular phenotypes observed in cancer, can we profitably treat those phenotypes as a quantitative trait? Indeed, clinical oncologists and pathologists have warned epidemiologists not simply to regard cancer as present-or-absent but to recognize its variability, advice that has largely been ignored. The discovery of highly variable allelotypes suggests that the genetic architecture of cancer follows the same rules as other quantitative traits like cholesterol levels, that derive from the cardinal points on our genetic map: a panoply of generally unique mutations in each instance, occurring somatically as well as by inheritance, and involving members of multigene families.

If this is the case, we should be able to apply quantitative genetic models to cancer, in which we specify the probability of a tumor arising in a population of stem cells base on the aggregate effects of many loci with allelic equivalency. In a preliminary attempt to do that, let us assume that on some suitable phenotype scale we can score the behavior of a stem cell, and for reasons of practicality let us scale the inherited genotype to have value zero on that scale, assuming the individual to be conceived with a normal genotype compatible with embryogenesis. Now, further suppose that on this scale a cell becomes the progenitor of a clone that becomes clinically detectable as "cancer" only if its genetically controlled phenotype exceeds some approximate "threshold," say T.

At each of the many millions of mitoses among the stem cells of a given tissue, there is risk of somatic mutation all along the genome. Shown schematically in Figure 1.12, as an individual ages, genetic diversity (cellular genotypic variance) will gradually develop among these stem cells. If we can make some assumptions about the distribution of allelic effects of these mutations, relative to those of the original inherited genotype, and the rate at which mutations introduce new genetic variance per cell division, we can quantify the increase of genetic variance among the cells with age, and hence the probability that a cell will have a genotype that exceeds T.

To explore this kind of model, let $f(x|t)$ be the probability density that a stem cell in a given organ in a given individual has phenotype x, conditional on the individual being alive at time t (measured in cell division cycles assumed here to be approximately synchronous). Initially, we have $f(0.0|0) = 1.0$ and $f(x|0) = 0.0$ for $x \neq 0.0$. At time t, the probability that a *cell* exceeds the transformation threshold, T, is

$$\pi(t) = \int_T^\infty f(y|t)\,dy \qquad [1.3]$$

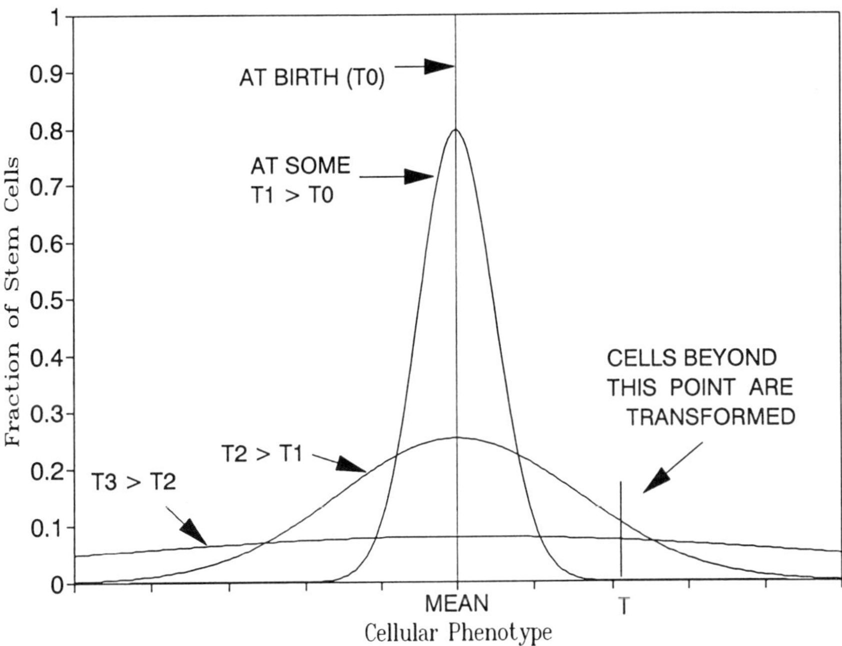

Figure 1.12. Schematic diagram of quantitative genetic threshold model illustrating how evolution in cells may occur. Distribution of cells is shown at birth and cell-division times $t = 1$, 10, and 100. At any time, the area of the cell phenotype distribution to the right of the threshold T represents the probability that a cell in an individual alive at time t exceeds the threshold and becomes transformed. This is represented by the shaded areas, which overlay each other in the figure.

and if there are N stem cells in the organ, the probability that in *individual* develops cancer in this cell cycle is

$$h_t = 1 - (1 - \pi(t))^N \approx e^{-N\pi(t)} \approx N\pi(t), \qquad [1.4]$$

where the approximations in [1.4] are numerically satisfactory for most human cancer situations, because $\pi(t) < N$ so $N\pi(t) < 1.0$. The probability h_t is closely related to the continuous-time hazard function $h(t)$ usually estimated from incidence data.

Ignoring any latency time to detection, the cell phenotype distribution is truncated at time t by the removal of individuals with at least one transformed cell. The new distribution is specified by the truncated, renormalized

$$\begin{aligned} f'(x|t) &= f(x|t) / \int_{-\infty}^{T} f(y|t)\,dy \qquad & x < T \\ &= 0.00 & x \geqslant T \end{aligned} \qquad [1.5]$$

After the next "round" of cell division and mutation, the new cell phenotype distribution will be determined by the effects of mutation. Assuming no correlation between the current cell phenotype and the effects of mutation, we can let the distribution of mutation effects, $\mu(z)$, be expressed as the probability of deviation by amount z from the current phenotype of a cell in which the mutations occur. At time $t+1$ we will have

$$f(x|t+1) = \int_{-\infty}^{T} f'(y|t)\mu(x-y)\mathrm{d}y. \qquad [1.6]$$

Then, h_{t+1} follows as in equation [1.4]. This specifies the evolving phenotype distribution among the dividing cells in a given organ as a recursion relationship relating mutations, their effects, and the collection of diagnosable phenotypes.

Numerically, we can make a first approximation to this simplified process by assuming that the distribution of additive allelic effects of new mutations is approximately exponential, so that the distribution of the consequent genotypic effects is approximately normal, relative to the original mean (e.g., $\mathrm{Nor}(0, \sigma^2)$. This simplification is numerically close to the available data on allelic effects (Connor, in prep; Connor and Weiss, in prep). For most human cancers h_t is quite small, implying that $\pi(t)$ is very small, and we can ignore the truncation effect of equation [1.5], so that the risk is approximately proportional to h_t. The process then is like a Brownian or Wiener diffusion process, in which the cell phenotype distribution at time t has mean 0 and variance $t\sigma^2$. Because we must translate $\pi(t)$ into h_t by way of the number of at-risk cells, this is not exactly like a simple process with an absorbing barrier at T, but is similar.

Figure 1.13 shows the fit of this simplified model to US age-specific cancer data for leukemia, colon, lung, and prostate cancers. These cover much of the range of sporadic cancer hazard functions. These fits are comparable in quality to those obtained with most multistage models. I include prostate cancer to illustrate that cancers that are unusual in other epidemiological regards are also unusual in regard to their fit to this model. Prostate cancer, for example, often remains benign, has long latency times, and is often undetedted before death.

With this model, cancer hazard functions vary approximately along a one-dimensional axis as our empirical parametric evidence suggests they seem to do (discussed above in regard to the (a, b) scatter plots); this dimension is the compound $t\sigma^2$ axis, a possibly satisfying result, since the complex of cell-division history and mutation intensity *ought* to calibrate cancer risk.

I do not wish to advocate still another model to explain cancer age patterns, another claim to have found Prester John. One of my major points is that one can't make strong etiologic inferences from one model when many others fit the same data. We need data with more discriminating power. Numerous complications were glossed over, too. A threshold is only an approximation; the model given here takes no account of variation in genotypes at birth, or varying intensities and histories of exposure to carcinogens, or even to the efficacy of tumor detection. I have assumed no epistasis (mutational effects independent of genotype of the cell in which the mutation occurs), and have modeled cell division as being synchronous. The last assumption precludes differential growth rates of preneoplastic cells, and we know this does occur in many tissues, a key feature of Moolgavkar's models (e.g., Moolgavkar, 1991), although not all tumors are necessarily of this type and the extra cell divisions in pre-malignant lesions may not be numerically important relative to the hazard function. Indeed, colorectal cancer has classically been viewed as frequently arising from adenomatous polyps, in FAP and even the general population (Cannon-Albright et al., 1988), but this impression may partly be an artifact of stage and ease of diagnosis (Shimoda et al., 1989).

What about the very high prevalence of certain LOH mutations found in colorectal cancers at loci like FAP, *ras*, and genes on chromosomes 17 and 18 (Vogelstein et al., 1988, 1989; Fearon et al., 1990)? Can we not view these as "causal" in the usual qualitative genetic sense—mappable directly from cellular genotype to phenotype? Perhaps most of the cancer allelotype is really just incidental mutations arising after the tumor has arisen, unrelated to the carcinogenic process itself. Perhaps there are so many stem cells that mutations in these loci will arise often enough solely to account for the observed tumors.

Until we know about the normal pattern of somatic mutation this is difficult to answer. But our ascertainment of mutations like p53 may be highly biased, leading to inflated estimates of their effects. Suppose that most p53 mutations have only minor phenotypic effects, even though occasionally having very large effects (e.g., in cells that have many other dangerous mutations). If one ascertains somatic cells only by way of tumors, a very distorted picture of the overall prevalence of p53 mutations will result. We have a parallel for this in FH. LDLR mutations have been thought to be essentially causal for FH because of their high prevalence in patients ascertained due to greatly elevated cholesterol. However, a recent study shows that a large fraction of LDLR mutations have only modest effects on cholesterol; ascertaining via FH misrepresents the prevalence of LDLR mutations, exaggerating their association with cholesterol variation in the general population (Roy et al., 1991).

It should not be surprising to find that cancer is complex at the cellular genetic level, and that the same complexities—of phenotype, of a variety of rare major genes, and so on—apply to cancer as apply to other phenotypes. After all, both a tumor and an organism are the amplification of a single cell. The general process suggested here would make cancer consistent with most other biological phenotypes.

I think a quantitative model for cancer is plausible, but its fit to the data may relate to another interesting issue. Many chronic diseases, including diseases which seem to have no somatic mutational component, fit the same models, even with similar parameter values. Why should this be so? One possible answer is that many chronic diseases arise in persons who are at high risk because of some physiologic risk factor (blood pressure, cholesterol, smoking history, obesity), which may be the cumulative result of long-term

A

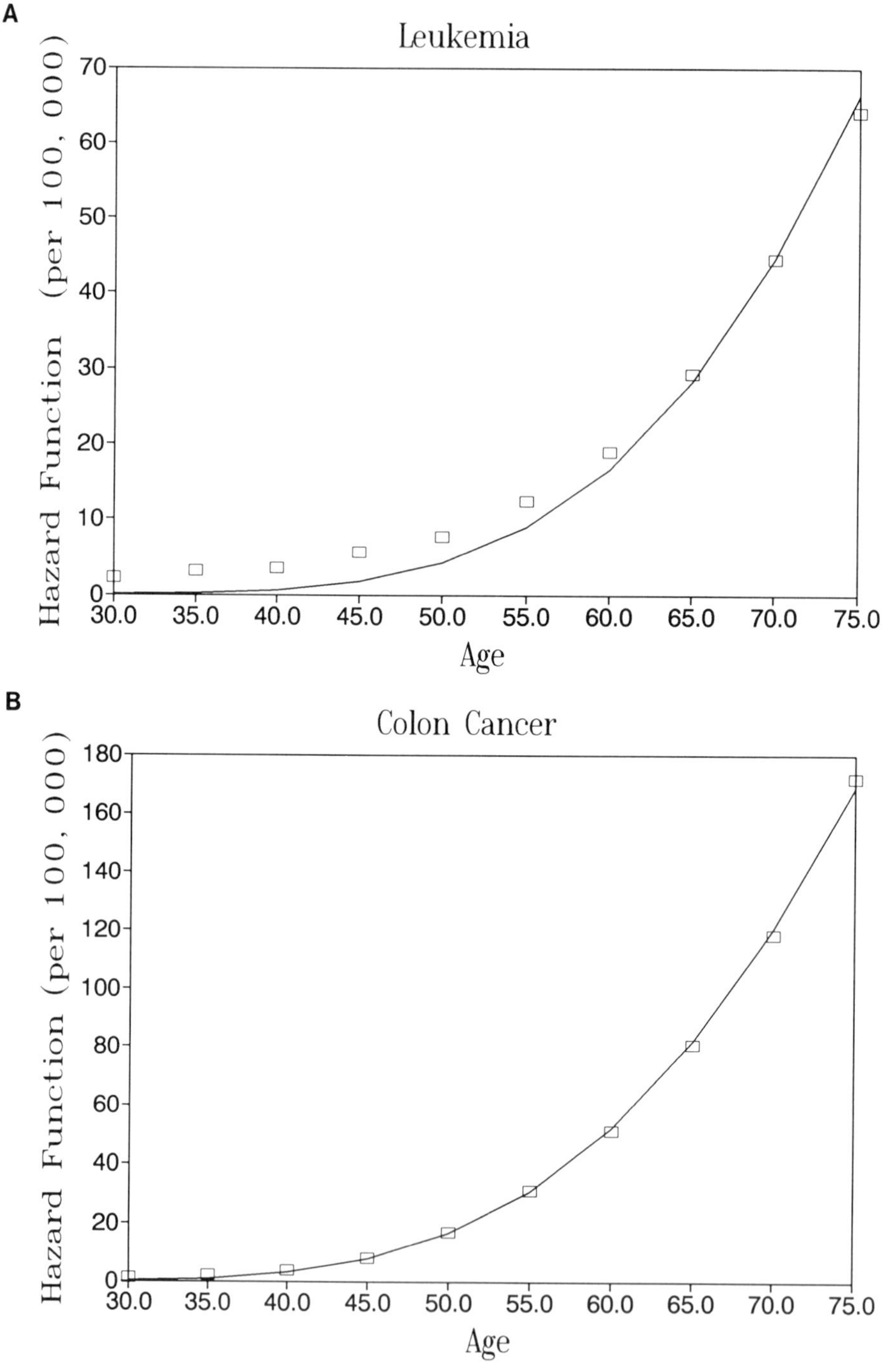

Leukemia

Hazard Function (per 100, 000)

70
60
50
40
30
20
10
0

30.0 35.0 40.0 45.0 50.0 55.0 60.0 65.0 70.0 75.0

Age

B

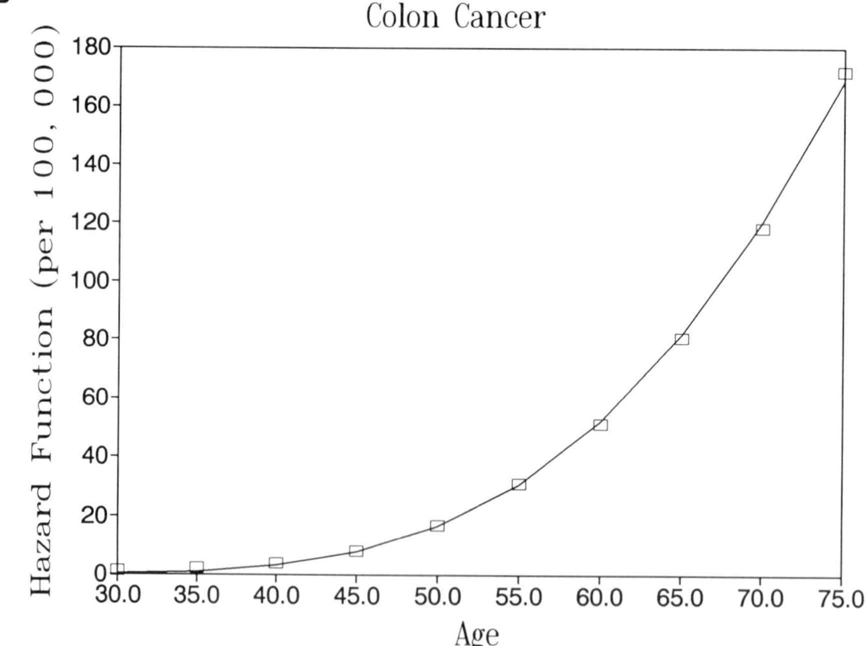

Figure 1.13a–d. Fit of simple quantitative genetic model to various cancers, US male, 1970. (A) leukemia, (B) colon, (C) lung, and (D) prostate. Data from Waterhouse et al., 1976.

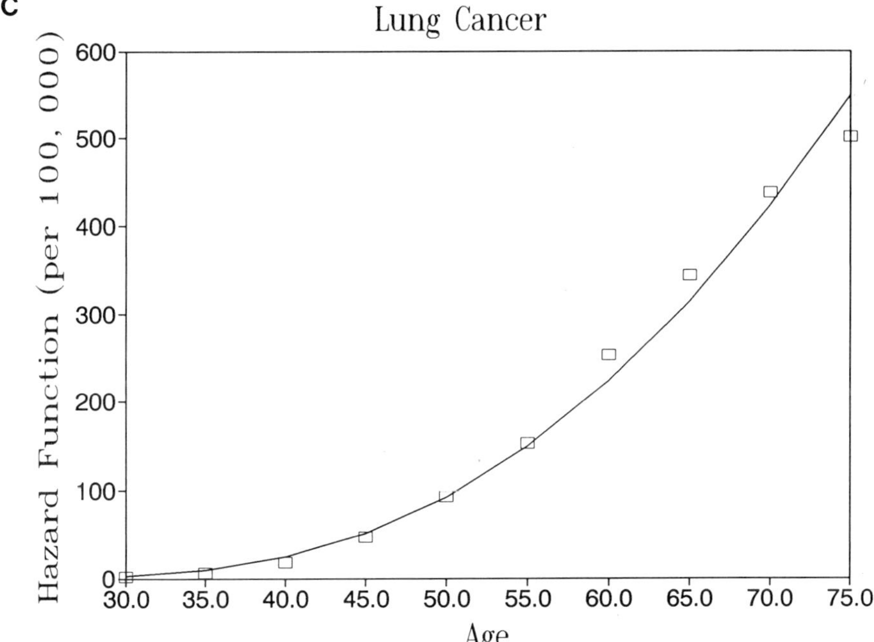

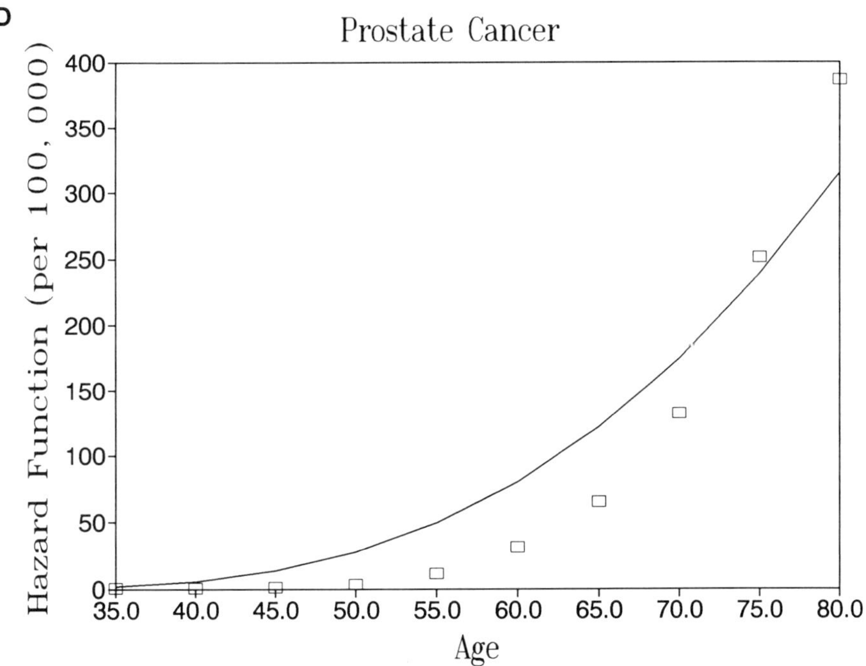

Figure 1.13 (*Continued*).

exposure to environmental or genetic factors. Exposures to such agents may often be approximately normally distributed in the population. As individuals age, the distribution of the effects of those exposures may "diffuse," or become more dispersed just as in our genetic cancer model. If the risk of disease, say, a heart attack, is roughly proportional to a person's position on such a risk-factor distribution, the hazard, as in the cancer model, would be proportional to the tail of that distribution. The diffusion model given here for cancer may reflect a generic kind of process that constitutes a kind of "null" model for senescence. If so, it is hopeless to infer a specific causal process from population hazard functions.

Whether cancer is a qualitative oligogenic, or quantitative polygenic phenomenon can have important consequences. Among them, it can be more difficult to undertake screening, therapeutic, counseling, and preventive measures for a highly heterogeneous, much less polygenic, trait than a simpler one (e.g., Yandell et al., 1990). And the quantitative model would suggest a much greater distribution of cancer susceptibility genotypes at birth than has been thought (in the past, we've only known of those born with a dramatically elevated susceptibility, such as due to FAP). The genes causing a qualitative genetic process can be enumerated, those causing a polygenic one cannot.

EAST: CONCLUSIONS

We have come full circle, and have returned to the East. Medieval maps replay Biblical history geographically as an allegory of knowledge. From city to hamlet, river to rivulet, and mountain to plain, maps reveal a few major features that are discrete and easily identifiable, grading into a large number of small, continuous subdivisions. The conceptual map of genetics shows that, at many levels, genetic traits also grade from the numbered to the numberless.

This sobering message, that has been raised elsewhere in this volume as well, may not imply that the genetics of complex traits cannot be understood, but perhaps only an impaired current way of understanding. In fact, we are quite accustomed to dealing with similar patterns of complexity. We find it every day in dealing with urban maps, for example, and no one would argue that we cannot understand the circulation of the blood just because, at some point, named arteries divide into unnamed arterioles and capillaries. If the observed level of diversity may make variants less *enumerable*, its Darwinian structure makes that same variation more *classifiable*. Membership in a gene family or genetic clade are new ways to turn disorder back into order. The processes of evolution, as we now understand them, represent a powerful organizing force for genetic variation.

A step into the unknown reaches of knowledge always reveals confusing complexity and new questions about where the path may be leading. But precedent shows that nature follows tractable rules which we must resolve from complex and imperfect data. Science rarely progresses best by assuming this cannot be done. In spite of the huge amount of genetic diversity revealed by our recent travels on the genetic map, and the complexity of causation and gene–environment interaction, I think that the new ways we will follow, our *via conceptionum*, will reveal interpretable structure, if we only know when to recognize it, and that we will not be led past the pillars of Hercules into an infinite, featureless ocean.

ACKNOWLEDGMENTS

This paper was written to honor Dr Schull's contributions to genetic epidemiology. Not the least of these is his early recognition of the importance of genetic variation in the epidemiology of complex chronic diseases (Neel and Schull, 1954; Neel, Shaw, and Schull, 1965). Being in his Center provided me, by osmosis, with most of my own career research interests. I thank him for providing me the working atmosphere in which to try, at least, to make thoughtful contributions in this area. He is thus ultimately responsible for this chapter, although any proximal errors must be blamed on me.

Henry Harpending helped me formulate the diffusion model. Anne Buchanan provided vital assistance in purging the version of this talk presented at the Montana conference of as much pomposity as she could take me out of. Other reviewers did their best.

REFERENCES

Ahuja H, Bar-Eli M, Advani SH, Benchmol S, Cline MJ (1989) Alterations in the p53 gene and the clonal evolution of the blast crisis of chronic myelocytic leukemia. Proceedings of the National Academy of Sciences, USA: 86:6783–6787.

Astrin SM, Costanzi C (1989) The molecular genetics of colon cancer. *Seminars in Oncology* 16:138–147.

Bagrow L (1964) *History of Cartography*. Revised by RA Skelton. London, CA Watts.

Baker SJ, Fearon ER, Nigro JM, Hamilton ST, Presisinger AC, Jessup JM, van Tuinen P, Ledbetter DH, Barker DF, Nakamura Y, White R, Vogelstein B (1989) Chromosome 17 deletions and p53 gene mutations in colorectal carcinomas. *Science* 244:217–221.

Barton NH, Turelli M (1989) Evolutionary quantitative genetics: how little do we know? *Ann Rev Genet* 23:337–370.

Beavan WL, Philott, HW (1873) *Mediæval Geography: An Essay in Illustration of the Hereford Mappa Mundi*. London, E Stanford.

Bigner SH, Bark J, Burger PC, Mahaley MS, Bullard DE, Muhlbaier LH, Bigner DD (1988) Specific chromosomal abnormalities in malignant human gliomas. *Cancer Research* 88:405–411.

Bishop JM (1987) The molecular genetics of cancer. *Science* 235:205–311.

Boorstin DJ (1985) *The Discoverers*. New York, Vintage.

Browning R (1845) The Pied Piper of Hamelin: A Child's Story. In Browning R (1952) *Poems and Lyrics*. Mt Vernon, NY, Peter Pauper Press.

Burch PRJ (1965) Natural and radiation carcinogenesis in man. I. Theory of initiation phase. II. Natural leukaemogenesis: initiation. III. Radiation carcinogenesis. *Proc. Roy Soc (Lond.)* 162B:223–287.

Burch PRJ (1969) *An Inquiry Concerning Growth, Disease, and Ageing*. Toronto, University of Toronto Press.

Burnet FM (1965) Somatic mutation and chronic disease. *Br Med J* 1:338–342.

Campbell T (1981) *Early Maps*. New York, Abbeville Press.

Cannon-Albright L, Skolnick MH, Bishop T, Lee RG, Burt RW (1988) Common inheritance of susceptibility to colonic adenomatous polyps and associated colorectal cancers. *New Engl J Med* 319:533–536.

Carter BS, Ewing CM, Ward WS, Treiger BF, Aalders TW, Schlaken JA, Epstein JI, Isaacs WB (1990) Allelic loss of chromosomes 16q and 10q in human prostate cancer. *Proc Natl Acad Sci USA* 87:8751–8755.

Chakraborty R, Weiss KM (1989) Age-specific risks for cancer as determined by multi-stage models of carcinogenesis, In Krisnian T (ed) *Medical Statistics*. Bombay, Himalayan Publishing House, pp 64–91.

Chen L-C, Neubauer A, Kurisu W, Waldman FM, Ljung B-M, Goodson W, Goldman Es, Moore D, Balazs M, Liu E, Mayall BH, Smith HS (1991) Loss of heterozygosity on the short arm of chromosome 17 is associated with high proliferative capacity and DNA aneuploidy in primary human breast cancer. *Proc Natl Acad Sci USA* 88:3847–3851.

Connor AC (in prep) PhD Dissertation, Penn State University

Connor AC, Weiss KM (in prep a) Region sharing in clonal chromosomes after strong selection.

Connor AC, Weiss KM (in prep b) The distribution of allelic effects.

Cook P, Doll R, Fellingham SA (1969) A mathematical model for the age distribution of cancer in man. *International Journal of Cancer* 4:93–112

Cortesao A (1969–71) *History of Portuguese Cartography*, 2 vols. Coimbra, Junta de Inv do Ultramar-Lisboa. (cited from Harley and Woodbury, 1987).

Daiger SP, Chakraborty R, Reed L, Fekete G, Schuler D, Berenssi G, Nasz I, Brdicka R, Kamaryt J, Pijackova A, Moore, S, Sullivan S, Woo SLC (1989) Polymorphic DNA haplotypes at the phenylalanine hydroxylase (PAH) locus in European families with phenylketonuria (PKU). *American Journal of Human Genetics* 45:310–318.

Devilee P, Van Den Borek M, Kuipers-Dijkshoorn N et al. (1989) At least four different chromosomal regions are involved in loss of heterozygosity in human breast carcinoma. *Genomics* 5:554–560.

Erlandsson R, Boldog F, Sumegi J, Klein G (1988) Do human renal cell carcinomas arise by a double-loss mechanism? *Cancer Genet Cytogenet* 36:197–202.

Fearon ER, Cho KR, Nigro JM, Kern SE, Simons JW, Ruppert JM, Hamilton SR, Preisinger AC, Thomas G, Kinzler KW, Vogelstein B (1990) Identification of a chromosome 18q gene that is altered in colorectal cancers. *Science* 247:49–56.

Fountain J, Bale SJ, Housman, DE, Dracopoli NC (1990) Genetics of melanoma. *Cancer Surveys* 9:645–671.

Franks LM, Teich N (eds) (1986) *Introduction to the Cellular and Molecular Biology of Cancer*. Oxford, Oxford University Press.

Friend SH, Dryja TP, Weinberg RA (1988) Oncogens and tumor-suppressing genes. *New Engl J Med* 318:618–622.

Gale JS (1990) *Theoretical Population Genetics.* London, Unwin Hyman.

Hanks SK, Quinn Am, Hunter T (1988) The protein kinase family: conserved features and deduced phylogeny of the catalytic domains. *Science* 241:42–52.

Harley JB, Woodward D (eds) (1987) *The History of Cartography, Vol 1. Cartography in Prehistoric, Ancient, and Medieval Europe and the Mediterranean.* Chicago, University of Chicago Press.

Hartl D, Clark AG (1989) *Principles of Population Genetics (2nd ed).* Sunderland, MA Sinauer Press.

Herrlich P, Ponta H (1989) "Nuclear" oncogenes convert extracellular stimuli into changes in the genetic program. *Trends in Genetics* 5:112–116.

Hollstein M, Sidransky D, Vogelstein B, Harris CC (1991) p53 Mutations in human cancers. *Sciences* 253:49–53.

Humphries SE, Talmud PJ, Kessling AM (1987) Use of DNA polymorphisms of the apolipoprotein genes to study the role of genetic variation n the determination of serum lipid levels. In Bock G, Collins GM (eds) (1987) *Molecular Approaches to Human Polygenic Disease.* Chichester, UK, John Wiley & Sons, pp 128–144.

Kimura M, Crow JF (1964) The number of alleles that can be maintained in a finite population. *Genetics* 49:725–738.

Kinzler KW, Nilbert MC, Su L-K, and 18 others (1991) Identification of FAP locus genes from chromosome 6q21. *Science* 253:661–665.

Klein G (1988) Oncogenes and tumor suppressor genes. *Rev. in Oncol* 1:427–437.

Knudson AG (1971) Mutation and cancer: statistical study of retinoblastoma. *Proc Natl Acad Sci, USA* 68:820–834.

Lande R (1981) The minimum number of genes contributing to quantitative variation between ane within populations. *Genetics* 99:541–553.

Lande R (1987) Quantitative genetics and evolutionary theory. In Weir B, Goodman MM, Eisen EJ, Namkoong G (eds) *Proceedings of the Second International Conference on Quantitative Genetics.* Sunderland, MA, Sinauer Associates, pp 71–84.

Lane DP, Benchimol S (1990) p53: Oncogene or anti-oncogene? *Genes & Develop* 4:1–8.

Leppart M, Burt R, Hughes JP, Samowitz W, Namamura Y, Woodward S, Gardner E, Lalouel J-M, White R (1990) Genetic analysis of an inherited predisposition to colon cancer in a family with a variable number of adenomatous polyps. *New England Journal of Medicine* 322:904–908.

Levine AJ, Momand J, Finaly CA (1991) The p53 tumour suppressor gene. *Nature* 351:453–456.

Lewin B (1990) Genes IV. Cambridge, MA, Cell Press.

Li W-H, Graur D (1991) *Fundamentals of Molecular Evolution.* Sunderland, MA, Sinauer Associates.

Manton KG, Malker H, Malker B (1986) Comparison of temporal changes in US and Swedish lung cancer, 1950–1951 to 1981–1982. *J Natl Cancer Soc* 77:665–675.

Manton KG, Stallard E (1984) *Recent Trends in Mortality Analysis.* Orlando, FL, Academic Press.

Manton KG, Stallard E (1988) *Chronic Disease Modelling.* New York, Oxford University Press.

Marx J (1990) Many gene changes found in cancer. *Science* 246:1386–1390.

Moir AL (1979) *The World Map in Hereford Cathedral.* Hereford, Friends of Hereford Cathedral.

Moolgavkar SH (1990) Cancer models. *Epidemiology* 1:419–410.

Moolgavkar SH (1991) Stochastic models of carcinogenesis. In Rao CR, Chakraborty R (eds) *Handbook of Statistics, Volume 8: Statistical Methods in Biological and Medical Sciences.* Amesterdam, Elsevier Science, pp. 373–393.

Moolgavkar SH, Cross FT, Luebeck G, Dagle GE (1990) A two-mutation model for radon-induced lung tumors in rats. *Rad Res* 121:28–37.

Moolgavkar SH, Dewanji A, Luebeck G (1989) Cigarette smoking and lung cancer: reanalysis of the British doctors' data. *J Natl Cancer Inst* 81:415–420.

Moolgavkar SH, Knudson AG (1981) Mutation and cancer: a model for human carcinogenesis. *J Natl Cancer Inst* 66:1037–1052.

Moolgavkar SH, Venzon DJ (1979) Two-event models for carcinogenesis: incidence curves for childhood and adult tumors. *Math. Biosci* 47:55–77.

Morita R, Ishikawa J, Tsutsumi M, Hikiji K, Tsukada Y, Kamidono S, Madea S, Nakamura Y (1991) Allelotype of renal cell carcinoma. *Cancer Research* 51:820–823.

Neel JV, Shaw MW, Schull WJ (eds) (1965) *Genetics and the epidemiology of chronic diseases.* US Public Health Service Publication 1163.

Neel JV, Schull WJ (1954) *Human Heredity. Chicago,* University of Chicago Press.

Nei M (1987) Molecular Evolutionary Genetics. New York, Columbia University Press.

Nigro JM, Baker SJ, Preisinger AC, and 13 others (1990) Mutations in the *p53* gene occur in diverse human tumor types. *Nature* 342:705–708.

Noble JN (1981) *The Mapmakers.* New York, AA Knopf.

Ohno S (1969) The role of gene duplication in vertebrate evolution. In Bittar ED, Bittar N (eds) (1969) *The Biological Basis of Medicine,* Vol 4, Chapter 4, London, Academic Press, pp 109–132.

Ohno S (1970) *Evolution by Gene Duplication.* New York; Springer-Verlag.

Ohno S, Wolf U, Atkin NB (1968) Evolution from fish to mammals by gene duplication. *Hereditas* 59:169–187.

Peterson AH, Damon S, Hewitt JD, Zamir D, Rabinowitch HD, Lincoln SE, Lander ES, Tanksley SD (1991) Mendelian factors underlying quantitative traits in tomato: comparison across species, generations, and environments. *Genetics* 127:181–197.

Peterson AH, Lander ES, Hewitt JD, Peterson S, Loncoln SE, Tanksley SD (1988) Resolution of quantitative traits into Mendelian factors by using a complete linkage map of restriction fragment length polymorphisms. *Nature* 335: 721–726.

Rey F, Berthelon M, Caillaud C, Lyonnet S, Abadie V, Blandin-Savoja F, Feingold J, Saudubray J-M, Frezal J, Munnich A, Rey J (1988) Clinical and molecular heterogeneity of phenylalanine hydroxylase deficiences in France. *American Journal of Human Genetics* 43:914–921.

Roberts L (1990) CF screening delayed for awhile, perhaps forever. *Science* 247:1296–1297.

Roy M, Sing CF, Betard C, Davignon J (1991) Impact of a >10 kb deletion in the low density lipoprotein receptor (LDLR) gene on means, variances and correlations of measures of lipid metabolism in French-Canadians. *Am J Human Genet* 49 (suppl):515.

Sager R (1989) Tumor suppresor genes: The puzzle and the promise. *Science* 246:1406–1412.

Scrable JK, Sapienza C, Cavenee WL (1990) Genetic and epigenetic losses of heterozygosity in cancer predisposition and progression. *Adv Cancer Res* 54:25–62.

Shimoda T, Ikegami M, Fujisaki J, Matsui T, Aizawa S, Ishikawa E (1989) Early colorectal carcinoma with special reference to its development *de novo*. *Cancer* 64:1138–1146.

Sing CF, Boerwinkle E (1987) Genetic architecture of inter-individual variability in apolipoprotein, lipoprotein and lipid phenotypes. In Bock G, Collins GM (eds) *Molecular Approaches to Human Polygenic Disease*. CIBA Foundation Symposium, No. 130. Chichester, England, Wiley, pp 99–121.

Sing CF, Moll PP (1989) Genetics of variability of CHD risk. *Int J Epidemiol* 18 (suppl 1):S183–S195.

Templeton AR, Boerwinkle E, Sing CF (1987) A cladistic analysis of phenotypic associations with haplotypes inferred from restriction endonuclease mapping. I. Basic theory and an analysis of alcohol dehydrogenase activity in *Drosophila*. *Genetics* 117:343–351.

Templeton AR, Sing CF, Kessling A, Humphries S (1988) A cladistic analysis of phenotype associations with haplotypes inferred from restriction endonuclease mapping. II. The analysis of natural populations. *Genetics* 120:1145–1154.

Turelli M (1987) Population genetic models for polygenic variation and evolution. In Weir B, Goodman MM, Eisen EJ, Namkoong G (eds) *Proceedings of the Second International Conference on Quantitative Genetics*. Sunderland, MA, Sinauer Associates, pp 601–618.

Waterhouse J, Muir C, Correa P, Powell J (1976) *Cancer Incidence in Five Continents, Vol III*. Lyons, France, International Agency for Research on Cancer.

Vogelstein B, Fearon ER, Hamilton SR, Kern SE, Preisinger AC, Lpeppert M, Nakamura Y, White R, Smits AMM, Bos JL (1988) Genetic alterations during colorectal-tumor development. *New England Journal of Medicine* 319:525–532.

Vogelstein B, Fearon ER, Kern SE, Hamilton SR, Preisinger AC, Nakamura Y, White R (1989) Allelotype of colorectal carcinomas. *Science* 244:207–211.

Weatherall DJ (1991) *The New Genetics and Clinical Practice*, 4th ed. Oxford, Oxford University Press.

Weeks SC, Weiss KM, Connor AC (in prep) Actuarial consequences of quantitative variation in biological risk factors.

Weinstein IB (1988) The origins of human cancer: molecular mechanisms of carcinogenesis and their implications for cancer prevention and treatment—twenty-seventh G.H.A. Clowes Memorial Award Lecture. *Cancer Res* 48:4135–4143.

Weiss KM (1990a) Biology, homology, and epidemiology. In Adams J, Hermalin A, Smouse PE (eds) *Convergent Issues in Genetics and Demography*. New York, Oxford University Press, pp 189–206.

Weiss KM (1990b) Duplication with variation: metameric logic in evolution from genes to morphology. *Yearbook of Physical Anthropology* 33:1–23.

Weiss KM (1990c) Cancer models and cancer genetics. *Epidemiology* 1:486–90.

Weiss KM (1993) *Genetic Variation and Human Disease: Problems and Evolutionary Approaches*. Cambridge, England, Cambridge University Press, in press.

Weiss KM, Buchanan AV, Valdez R, Moore JH, Campbell J (1992) Amerindians and the price of modernization. In Smith M, Bilsborough A, Schell L, Watts E (eds) *Urban Health and Ecology in the Third World.* Cambridge, England, Cambridge University Press, in press.

Weiss KM, Chakraborty R (1984) Multistage risk models and the age pattern in familial polyposis coli. *Cancer Investigation* 2:443–448.

Weiss KM, Connor AC, Weeks SC (in prep) The distribution of fitness in relation to genetic variation for quantitative phenotypes.

Weston A, Willey JC, Modali R, Sugimura H, McDowell EM, Resau J, Light B, Haugen A, Mann DL, Trump BF, Harris CC (1989) Differential DNA sequence deletions from chromosomes 3, 11, 13, and 17 in squamous-cell carcinoma, large-cell carcinoma, and adenocarcinoma of the human lung. *Proceedings of the National Academy of Sciences, USA* 86:5099–5103.

Whittemore A, Keller J (1978) Quantitative theories of carcinogenesis. *SIAM Review* 20:1–30.

Wilford JN (1981), *The Mapmakers.* New York, AA Knopf.

Wright S (1931) Evolution in Mendelian Populations. *Genetics* 16:97–159.

Wright S (1986) *Evolution and the Genetics of Populations: Vol 1, Genetics and Biometric Foundations.* Chicago, University of Chicago Press.

Yandell DW, Campbell TA, Dayton SH, Petersen R, Walton D, Little JB, McConkie-Rosell A, Bucklye EG, Dryja TP (1990) Oncogenic point mutations in the human retinoblastoma gene: their application to genetic counseling. *New Engl J Med* 321:1689–1695.

2

Impact of Genetic, Somatic, and Epigenetic Variation on Phenotype

ARAVINDA CHAKRAVARTI

It is a truism that for a complex organism such as the human, the pattern of inheritance for most phenotypes is complex. In McKusick's (1988) catalog of 4,344 uncommon traits, only 2,208 (51%) has a recognizable inheritance pattern. Of the latter, probably only 20% were single-gene Mendelian entities, the majority (31%) being autosomal dominant phenotypes with variable expressivity and reduced penetrance. The major chronic diseases in humans demonstrate familial aggregation and have a genetic component in their etiologies. However, these disorders also demonstrate a complex pattern of inheritance, and consequently are synonymously referred to as complex disorders. This designation arises from the fact that for chronic diseases, no single Mendelian pattern of inheritance can explain all observed types of familial aggregation.

Complex patterns of inheritance may be due to three causes:

(a) reduction in penetrance;
(b) segregation of multiple, independent, monogenic traits;
(c) segregation of multiple, interacting genes together with environmental effects.

Of these, model (c) is the most general scenario, of which (a) and (b) are special cases as explained in the following. Thus, interaction between a single gene and a common environment may lead to reduced penetrance, but this reduction is random and independent of the genetic background of the individual. On the other hand, interaction between a single gene and its genetic modifiers may also lead to reduced penetrance, but this reduction is nonrandom and dependent on the genetic background of the individual. Finally, when only two or three genes interact to produce a trait, there is a substantial probability that in some families only one locus will segregate, giving the impression of single-locus inheritance. However, by chance segragation, different single genes will segregate in different families producing multiple monogenic traits.

The classical tools for elucidating the mode(s) of inheritance for any phenotype have been the methods of segregation analysis as applied to family data (Morton, 1982). When multiple genes are involved, segregation analysis of phenotypes is per se inefficient, unless supplemented by simultaneous linkage analysis (Lander and Botstein, 1986; MacLean, Morton, and Yee, 1984). For the latter highly polymorphic genetic markers are a prerequisite. Even then, aside from the consideration of efficiency, not all relevant genes can be identified. This rather obvious yet important fact stems from the distinction between genes with an effect on the phenotype and the subset of genes that segregate within families and which can potentially be identified by genetic methods.

One of the primary tenets of genetics is that different phenotypes are caused by differences in gene sequences. However, such genetic differences may not be in the germ-line, but be somatic. Clearly, somatic alterations per se will not lead to familial aggregation, but may be an important contributor to a phenotype (see Ferrell, Chapter 3, this volume). Similarly, epigenetic modification of a phenotype, such as genomic imprinting, is also possible (Sapienza, 1989). Variation in somatic and epigenetic processes, be they random or under genetic control, has a significant impact on the phenotype and has many of the features of complex inheritance.

IMPACT OF GENETIC VARIATION ON PHENOTYPE

The patterns of inheritance due to rare recessive and dominant genes in humans are well understood and documented (Stern, 1973). When penetrance is complete, yet delayed, recognition of the mode of inheritance is not difficult. However, if not all individuals with genotypes that could lead to affection are indeed affected then, depending on the degree of "penetrance," recognition of the inheritance pattern may be exceedingly difficult. There are two solutions to this dilemma depending on whether the phenotype being studied is rare or common. When rare, single-locus inheritance is likely so that reduced penetrance is a real possibility. Such reduced penetrance may be due to random, stochastic factors or due to modification of the susceptible genotype(s) by other genes. This aspect was first recognized by Haldane (1941) who also suggested methods for distinguishing these possibilities. When the phenotype is common, the identification of individuals who carry the susceptible genotypes and yet are unaffected is difficult, since individuals marrying into the family may carry these genotypes as well. The observation that many individuals in the line of descent are unaffected and yet must transmit some susceptibility gene(s) suggests multiple genes and/or the environment is involved.

The model (c) mentioned previously covers this possibility. To explore the consequences of such a scenario we study below a simple genetic model. Consider a trait determined by the simultaneous homozygosity for recessive alleles at k independent loci. When k is large, such a trait has been termed "emergenic" (Li, 1987). Let x be the frequency for the recessive allele at each locus. Thus, the incidence of the trait is

$$I = x^{2k}. \qquad [2.1]$$

The probability of observing a genotype with homozygosity for the recessive allele at j *specific* loci and heterozygosity for the recessive allele at the remaining $k - j$ loci is

$$P_j = x^{2j}[2x(1-x)]^{k-j}$$

where $j = 0, 1, \ldots, k$. Consider now the collection of all genotypes that could produce gametes bearing only recessive alleles. The frequency of this class, discounting individuals with the affected trait genotype is,

$$S_k = \sum_{j=0}^{k-1} \binom{k}{j} P_j = x^k\{(2-x)^k - x^k\}. \qquad [2.2]$$

Thus, the total frequency of matings where neither parent is affected but which could produce affected offspring is S_k^2. Among these matings anywhere from 1 to k loci segregate. The frequency of matings with only single locus segregants is clearly

$$T_k = k\{2x^{2k-1}(1-x)\}^2$$

so that the proportion of families that demonstrate single locus inheritance is

$$\phi_k = T_k/S_k^2 = k\left\{\frac{2x^{k-1}(1-x)}{(2-x)^k - x^k}\right\}^2. \qquad [2.3]$$

Furthermore, among the parental genotypes considered, the contribution of a parent with j homozygous loci to the risk of bearing affected offspring is 2^{j-k}. Thus, the total frequency of affected offspring from these matings is,

$$R_k = \left\{\sum_{j=0}^{k-1} \binom{k}{J} P_j 2^{j-k}\right\} \Big/ S_k^2 = \left\{\frac{1-x^k}{(2-x)^k - x^k}\right\}^2. \qquad [2.4]$$

Among these, the relative frequency of affected offspring that arise from single locus segregants is

$$\rho_k = k\{x^{2k-1}(1-x)\}^2/R_k S_k^2 = k\left\{\frac{x^{k-1}(1-x)}{1-x^2}\right\}^2. \qquad [2.5]$$

In the following I calculate, using equations [2.1]–[2.5], numerical values of x, ϕ_k, R_k and ρ_k given an incidence of 1% and $k = 1, 2, \ldots, 5$. The results are

Table 2.1. Characteristics of single- and multilocus recessive models as a function of the number of loci (k) involved. The average gene frequency (x), proportion of families that demonstrate single-locus segregation (ϕ_k), the frequency of affecteds from these families (R_k), and this frequency as a proportion of all affecteds (ρ_k) are presented.

k	x	ϕ_k	R_k	ρ_k
1	0.10	1.000	0.250	1.000
2	0.32	0.050	0.108	0.115
3	0.46	0.013	0.065	0.049
4	0.56	0.006	0.047	0.030
5	0.63	0.003	0.037	0.021

provided in Table 2.1, which illustrates several specific features of recessive interacting loci.

First, for a fixed incidence, the gene frequency (x) for the recessive allele at each component locus increases in frequency as the number of component loci increases. Thus, although the specific trait is uncommon, the genes contributing to it are not. Second, the proportion of single locus segregants (ϕ_k) decreases sharply as k increases so that even for a two-locus trait it is only 5%. This value can, however, be considerably higher if the recessive allele frequency varies considerably between the component loci. In the above analysis we have not considered the problem of ascertainment of cases. In any genetic study, this proportion may be higher since the recurrence risk is greatest in these families; therefore, a large majority of *multicase* families will arise from this class. Third, the overall frequency of affection from segregating families decreases as k increases but not as rapidly as ϕ_k. Fourth, as a consequence of the rapid decline in single-locus segregants (ϕ_k) but a more-or-less linear decline in overall risk (R_k) as k increases, the contribution of single-locus segregants to overall risk (ρ_k) declines rapidly. Thus, this implies that even though multicase families may mostly arise from single- or two-locus segregants, the large majority of affected individuals arise out of multi-locus segregating families. Because these latter families have a low recurrence risk, such cases will appear as isolated or sporadic.

The practical advantage should be obvious. Selection of multicase families enriches for segregation at a few loci; however, these segregants will be at multiple, independent loci. This heterogeneity is an immediate consequence of interacting genes in an outbred, randomly mating organism.

A related question is: how many loci are involved in a trait? A priori, a large number, perhaps even a 100 genes, may be involved in the development of a trait. However, among these loci only those genes that are polymorphic in the population are relevant to genetic analyses. The number of such genes then depends on the extant genetic variability in the population under study. An approximate answer may be obtained by the following argument. Observe that, at least for recessive interacting genes, the component genes are highly polymorphic, even if the trait incidence is only 1%, with frequencies between 30% and 60%. If previous studies of protein polymorphisms are any indication of the variability in gene sequences, then the frequency of loci with the above characteristic is 20% to 30% (Nei, 1987). Thus, even for a phenotype controlled by 100 genes, only 20 to 30 would be variable in the human population. The results of the previous section suggest, however, that such traits would fail to show familial aggregation. Thus, two to five variable component genes are probably responsible for many complex phenotypes. This conversely implies that these traits are controlled by 5 to 25 genes. Further studies of genetic variation in different classes of genes should enable a more accurate answer to be obtained.

IMPACT OF SOMATIC VARIATION ON PHENOTYPE

A primary feature of many complex disorders is the occurrence of an excess of isolated cases. The classical explanation for this observation is the occurrence of new mutations or phenocopies. Genetically, both of these events are mutations, but occur in the germ-line or somatic cells respectively. As evidenced by the increasing body of data on human cancer many phenocopies, as for colon cancer (Bodmer et al., 1987), are due to somatic changes at the same locus involved in hereditary tumors. Thus, many phenocopies are, in fact, genocopies.

Knudson (1971) first suggested that somatic events play a crucial role even in the development of tumors in individuals with a hereditary predisposition. In this model, called the two-hit model, hereditary and non-hereditary cases are distinguished by the inheritance of the mutant gene and the occurrence of an early somatic mutation, respectively. However, the development of a tumor requires a second somatic event at the same locus, in both hereditary and non-hereditary cases. Knudson's model, first presented for explaining the complex pattern of inheritance in retinoblastoma, has been amply confirmed by experimental evidence (Cavanee et al., 1983). However, this evidence also suggests that a diversity of somatic events can occur, namely, point mutations, mitotic recombination, mitotic non-disjunction, and gene conversion.

Knudson's model postulated that the first event would be a specific mutation. The current evidence suggests, however, that the second event does not need to be an event specific to the locus under study. In fact, mitotic recombination, mitotic non-disjunction, and gene conversion are all normal processes of somatic cells that merely serve to reduce the target locus to homozygosity or hemizygosity. Little is known about the *de novo* rates of these processes or whether particular chromosomes or chromosomal regions are less prone or more prone to these processes. However, if there is variability in these processes then the result would be a change in the age at onset of the tumor.

A further point should be noted. Whereas the normal consequence of somatic mutation is to increase the variability within a somatic cell, the normal consequence of mitotic recombination, mitotic non-disjunction, and gene conversion is to decrease the variability within a somatic cell. Thus, given a primary deleterious mutation, its reduction to homozygosity may be virtually assured. Rates of cell division in various tissues, and consequently the rates of somatic changes, perhaps only determine the time course of reduction to homozygosity, and thus of penetrance.

Knudson's model has generally been discussed in the context of cancer, where a particular somatic cell with homozygosity at a specific locus has a growth advantage. However, the model is far more general in that the final outcome may be proliferation or ablation (see Ferrell, chapter 3, this volume). Thus, the two-hit model may be a general model for cancer, cardiovascular disease, autoimmune disorders, and congenital malformations. Of these, congenital malformations could be due to ablation of a specific cell type. The events would, however, have to occur early in development when the number of cells of a particular type are few. Recently, Kurnit et al. (1987) have proposed early stochastic events as contributing to the genesis of several congenital malformations, events that decrease the number of cells involved in the development of a particular tissue. These authors demonstrate that the inheritance pattern arising out of this process appears multifactorial even though controlled by a single gene. The above model would also produce outcomes similar to Kurnit's model.

IMPACT OF EPIGENETIC VARIATION ON PHENOTYPE

In Knudson's model, the first and second events are due to inherited differences at a specific locus in the germ-line and in somatic cells, respectively. A novel mechanism to explain the first hit is genome imprinting (Sapienza, 1989). Specific regions of mammalian genomes are differentially imprinted by being transmitted through the male or female germ-line (Cattanach and Kirk, 1985). A specific feature of imprinted genes is allele-inactivation, and this itself could be the first hit, as demonstrated for Wilm's tumor and embryonal rhabdomyosarcoma (Schroeder et al., 1987). Because the second event may be inevitable, this model makes the unrealistic prediction that all individuals would have the tumor. That this is not so is due to erasure of the imprint in a certain proportion of somatic cells. Sapienza (1989) suggests a model in which variation in the genes that control imprinting may produce a variety of phenotypes in somatic cells carrying a loss-of-function allele. In this scenario each somatic cell harbors a loss-of-function allele, whereas the other allele may or may not be completely inactivated. Because the process affects varying numbers of somatic cells, the inheritance of the mutant gene may be dominant, a dominant with variable expressivity or a recessive. A crucial aspect of the model is the variation observed in somatic cells which consequently affects the phenotype. Sapienza (1989) also argues that the imprinting genes may be unlinked to the imprinted gene. Thus, genetic mapping of the mutant phenotype may not identify the target-imprinted locus, but rather the imprinting loci.

ENVOI

The complexity of patterns of inheritance need not imply a complex genetic system. Many human disorders are probably due to few segregating genes but possibly subject to various somatic and epigenetic processes.

ACKNOWLEDGMENT

This work was supported by NIH grant HD00344 and Research Career Development Award HD00774.

REFERENCES

Bodmer WF, Bailey CJ, Bodmer J, et al. (1987) Localization of the gene for familial adenomatous polyposis on chromosome 5. *Nature* 328:614–618.

Cattanach BM, Kirk M (1985) Differential activity of maternally and paternally derived chromosome regions in mice. *Nature* 315:496–498.

Cavanee WK, Dryja TP, Phillips RA, et al. (1983) Expression of recessive alleles by chromosomal mechanisms in retinoblastoma. *Nature* 305:779–784.

Haldane JBS (1941) The relative importance of principal and modifying genes in determining some human diseases. *J Genet* 41:149–157.

Knudson AG (1971) Mutation and cancer: statistical study of retinoblastoma. *Proc Natl Acad Sci USA* 68:820–823.

Kurnit DM, Layton WM, Matthysee S (1987) Genetics, chance, and morphogenesis. *Am J Hum Genet* 41:979–995.

Lander ES, Botstein D (1986) Mapping complex genetic traits in humans: New methods using a complete RFLP linkage map. *Cold Spring Harb Symp Quant Biol* LI:49–62.

Li CC (1987) A genetical model for emergenesis: in memory of Laurence H. Snyder, 1901–86. Am J Hum Genet 41:517–523.

McKusick VA (1988) *Mendelian Inheritance in Man: Catalogs of Autosomal Dominant, Autosomal Recessive, and X-Linked Phenotypes* (8th ed). Baltimore, Johns Hopkins University Press.

MacLean CJ, Morton NE, Yee S (1984) Combined analysis of genetic segregation and linkage under an oligogenic model. *Computers Biomed Res* 17:471–480.

Morton NE (1982) *An Outline of Genetic Epidemiology*. Basel, S Karger.

Nei M (1987) *Molecular Evolutionary Genetics*. New York, Columbia University Press.

Sapienza C (1989) Genome imprinting and dominance modification. *Ann. NY Acad Sci* 564:24–38.

Stern C (1973) *Principles of Human Genetics*. San Francisco, J Freeman Co.

Schroeder WT, Chao LY, Dao DD, et al. (1987) Nonrandom loss of materal chromosome 11 in Wilms tumors. *Am J Hum Genet* 40:413–420.

II
INDIVIDUAL VARIABILITY

3

Impact of Genetic Variation in Individuals: Clonal Phenotypes Other than Cancer

ROBERT E. FERRELL

The idea of somatic mutations affecting the phenotype of an individual, but not being transmitted to offspring has been most thoroughly explored with respect to cancer. The recognition that the majority of tumors were clonal in origin contributed to acceptance of the view that cancer is a genetic disease at the cell level. A major goal of cancer genetics has been and continues to be the identification of the primary genetic events involved in cellular transformation, and to understand how these changes contribute to the malignant phenotype. That the pathway to transformation involves several genetic events is generally accepted and has been elegantly demonstrated in the molecular verification of Knudson's two-hit hypothesis in retinoblastoma and Wilm's tumor. The uninhibited clonal expansion of a somatically altered cell in retinoblastoma, as well as other tumors, simplifies the genetic analysis by providing a relatively large amount of tissue for biochemical and molecular studies. This led to the recognition that while susceptibility to retinoblastoma is inherited in an autosomal dominant pattern, the retinoblastoma phenotype is recessive at the cellular level. These studies also revealed that normal somatic recombination is a major mechanism by which cellular homozygosity occurs in retinoblastoma and Wilm's tumor, and that somatic events leading to homozygosity or hemizygosity for a mutant allele could be a mechanism for the phenotypic expression of any recessive mutation at the single-cell level.

Note. Since the presentation of this discussion, the International Commission for Protection Against Environmental Mutagens and Carcinogens has convened a subcommittee to specifically consider the subject of the possible involvement of somatic mutations in the development of atherosclerotic plaques. The full report of that subcommittee and related working papers was published in *Mutation Research*, Volume 239, 1990. They concluded that there was substantial evidence of the occurrence of somatic mutations in DNA from atherosclerotic plaques, and that further studies are needed to determine whether these mutations have an etiologic role in the development of atherosclerotic plaques. If such a role is deemed likely, the source of the mutagens, exogenous or endogenous, causing such changes and the relationship between these mutagens and known epidemiologic risk factors could be examined in appropriate animal models.

Table 3.1. Outcomes of somatic genetic events that may lead to phenotypic expression in individuals.

Cell death.
Gross clonal expansion due to a fundamental alteration in control of proliferation.
Limited clonal expansion due to normal expansion of a genetically altered stem cell or prolonged survival of cells of a mutant clone.
Clonal or focal alteration in expression of a normal gene product or a mutant form of an "appropriate" gene product.
Clonal or focal expression of an "inappropriate" gene product.

Note: In this context, an appropriate gene product refers to one normally expressed by the target cell involved in a clonal event, while an inappropriate gene product refers to one that is not normally expressed by the target cell.

Although a breakdown in the normal control of cell proliferation is only one of several possible outcomes of a somatic mutation, human geneticists have paid relatively less attention to other somatic phenotypes than to cancer. The ease of detection of the product of uncontrolled cell proliferation, a tumor, and the morbidity and mortality associated with cancer justify the effort that has gone into the study of cancer genetics. However, in the context of the impact of genetic variation in individuals, we consider here clonal phenotypes other than cancer that might result from somatic genetic events. I will describe several examples supported by experimental data and suggest others that have no experimental support as a starting point of the discussion of the impact of genetic variation in individuals.

A few of the possible outcomes of somatic mutation are listed in Table 3.1. Of these, mutations that alter the control of cellular proliferation and lead to clonal expansion of a particular cell, leading to a tumor, have been most intensively studied.

Somatic mutation in a stem cell population may lead to the expansion of a phenotypically altered cell population during normal stem cell differentiation without the development of malignancy. The size of the mutant cell pool and the specific cell lineages affected will depend on how early in differentiation the mutation occurs. Clonal phenotypes that arise by this mechanism are frequently recognized in hematopoiesis.

THE TN SYNDROME

The Tn Syndrome is an acquired disorder characterized by the exposure of a normally sequestered antigenic determinant on the surface of lymphocytes, erythrocytes, platelets, and granulocytes. Tn syndrome occurs in apparently healthy individuals but is frequently associated with mild anemia, leukopenia, and/or thrombocytopenia. Tn activation results from exposure of a normally hidden N-acetylgalactosamine residue found on cell surface glycoproteins which may then react with natural antibodies normally present in the serum of all individuals. Biochemically, Tn syndrome is the result of clonal loss of a 3-β-D galactosyltransferase in Tn positive cells. Tn patients have two types of blood cells Tn(+) and Tn(−) and in vitro studies by Vainchenker et al. (1982) have shown that hematopoietic colonies from Tn patients are composed exclusively of Tn(+) or Tn(−) cells. Thus, Tn syndrome is a clonal disease due to somatic loss of a specific glycosyltransferase. The presence of Tn(+) and Tn(−) clones in both erythroid and granulocytic colonies suggests that the Tn mutation occurs in a pluripotent stem cell of the bone marrow.

The precise genetics events that lead to expression of the Tn phenotype are not known, but features of the disease are suggestive of a cellular recessive mechanism. Tn is strictly clonal in that each erythroid clone is either Tn(+) or Tn(−). Intermediate colonies are not seen. This suggests the complete loss of T transferase activity in Tn(+) clones. In many single-gene metabolic disorders only a severe deficiency of the relevant enzyme is associated with a clinical phenotype. Thus, the Tn phenotype could arise by somatic mutation in a stem cell of an individual who is heterozygous for a null allele at the T transferase locus leading to homozygosity or hemizygosity for a null mutation as has been observed in the case of the RB locus in retinoblastoma. Quantitative genetic studies of T-transferase in families of Tn probands and molecular analysis of Tn(+) and Tn(−) clones to detect loss of heterozygosity in a specific chromosomal region could provide a direct test of this hypothesis.

A clonal nonmalignant phenotype due to somatic mutation occurring in a primordial stem cell population is seen in Tn Syndrome. Mutations occurring later in the differentiation pathway of affecting a genetic region which is active in only one derived cell population may lead to phenotypic expression in a single cellular compartment. Sideroblastic anemia appears to involve a gene region primarily involved in controlling heme synthesis. Phenotypic expression is restricted to the red cell series and is characterized by the deposition of iron within mitochondria. Dacie et al. (1959) first suspected that sideroblastic anemia was a clonal disorder based on the observation of two distinct cell populations in peripheral blood smears of patients. The clonal nature of this disorder was subsequently confirmed by studies of G6PD expression in female patients heterozygous for this X-linked enzyme. Several of the enzymes involved in heme synthesis have been mapped and should provide candidate regions for study of the molecular events leading to sideroblastic anemia. Sideroblastic anemia is one of a group of hematologic disorders classified as primary acquired myelodyspastic syndromes. All are clonal in origin and are not classed at malignant, although several of these disorders have an increased probability of evolving to frank leukemia.

DISEASES OF IMMUNOGLOBULIN IN PRODUCTION

A broad spectrum of diseases involve the excess production of immunoglobulins or their component light or heavy chains. Some are benign and Kyle and Lust (1989) have suggested the name "monoclonal gammopathies of undetermined significance," because the natural history of these conditions has not been fully explored and it is not possible to predict the course of the disease at the time of diagnosis. This condition is characterized by the presence of monoclonal protein in patients without evidence of multiple myeloma, amyloidosis, or macroglobulinemia. The M protein present in patients consists of two heavy chains of the same class and subclass and two light chains of the same class. This condition overlaps with a group of monoclonal immunogloubulin deposition diseases which ultimately lead to amyloidosis. The amyloid deposit present in different patients may consist of light chains of either the kappa or lambda type, but are more frequently of the latter, or may contain both heavy and light chains and are typically either IgG or IgA. The clinical phenotype of each patient will differ with respect to the characteristics of the deposit, the presence or absence of amyloid P component, and the distribution of deposits. Evidence that these are clonal disorders comes from the finding that deposits in an individual patient are of a single isotope. More recently, monoclonality has been demonstrated by the presence of identical heavy and light chain gene rearrangements in immunoglobulin producing cells.

CLONAL SOMATIC MUTATION IN ATHEROSCLEROSIS.

The hematopoietic system, characterized by the elaboration of multiple cell lineages from a self-renewing stem cell population, seems to provide a unique opportunity for the detection of clonal somatic mutations. However, there are other systems where clonal phenotypes have been postulated to be important. One such area of intense investigation is cardiovascular disease and the role of genetic variation in determining cardiovascular disease (CVD) risk. Most studies have focused on the quantitative genetics of levels of lipoprotein lipids which are primary risk factors for CVD rather than on the pathological lesion involved, namely the autherosclerotic plaque. These lesions are focal and are characterized by three fundamental phenomena: (1) proliferation of intimal smooth muscle cells of the aorta, (2) deposition of intracellular and extracellular lipid, and (3) the accumulation of extracellular matrix components. The presumed primary event, the focal proliferation of aortic endothelial cells, has been proposed to be initiated by a somatic genetic event occurring in a single cell. The evidence of the clonal nature of anthero-sclerotic plaques is summarized in Table 3.2.

Benditt and Benditt (1973) noted that the cells of spontaneous athero-sclerotic lesions differ from cells of the normal arterial wall and from cells populating the site of arterial wall repair following mechanical injury. This led them to propose that either the cells of atherosclerotic plaques were not derived from the population of cells of the normal arterial wall, or they are transformed cells. They predicted that if atherosclerotic plaques arose by transformation of a single normal cell of the aortic intima, they would be monoclonal. They examined cells from multiple atherosclerotic plaques from

Table 3.2. Findings related to the Clonal hypothesis in atherosclerosis

Finding	Source
Eighty percent of atheromatous plaques appear monoclonal.	Benditt and Benditt 1973
Ninety percent of fibrous plaques are monoclonal.	Pearson et al., 1975
No evidence of clonal selection in atherosclerotic plaques.	Pearson et al., 1978a; 1978b
Known chemical carcinogens cause fibromuscular plaques in the abdominal aortas of chickens.	Albert et al., 1977; Penn et al., 1981
Isolation of DNA segment with dominant transforming activity from atherosclerotic plaque DNA.	Penn et al., 1986

females heterozygous for the X-linked enzyme glucose-6-phosphate dehydrogenase and demonstrated that the majority of atherosclerotic plaques are monoclonal in origin. Pearson et al. (1975) confirmed these findings and went on to present evidence that the monoclonal nature of plaques was not due to selection for cells of particular enzyme type within the developing plaque and that the monoclonal character of atherosclerotic lesions appeared early in the development of the lesion (Pearson et al. 1978a, 1978b; Pearson et al., 1983).

The analogy between initiation of atherosclerotic plaques and the initiation of tumors has been extended in a variety of directions. The finding that infection of chickens with Marek disease virus led to the production of arterial lesions resembling human atherosclerosis led Benditt et al. (1983) to screen aortic specimens from patients with atherosclerosis with probes to herpes simplex virus, Epstein-Barr virus, and cytomegalovirus. Evidence of expression of herpes simplex virus but not Epstein-Barr virus or cytomegalovirus in these cells led them to conclude that viral infection might be an etiologic factor in the development of atherosclerosis. The finding that chemical initiation–promotion sequences used in experimental carcinogenesis could lead to focal smooth muscle cell proliferation is also consistent with a mutational, therefore clonal, origin of atherosclerotic plaques (Majesky et al., 1985; Penn and Snyder, 1988). The ultimate extension of this analogy came with the finding of Penn et al. (1986) that DNA isolated from human coronary artery plaques possessed dominant transforming activity in the NIH 3T3 cell transfection assay, and that secondary transformants from these cells were capable of producing tumors in nude mice.

These observations raise the question of whether those of us interested in the role of genes in determining atherosclerosis risk must consider both heritable genetic variation at loci affecting lipoprotein–lipid levels as well as somatic genetic events which might be involved in the initiation of atherosclerotic lesions. They also suggest a number of testable hypotheses related to the possible mutational origin of atherosclerotic plaques. Is there loss of heterozygosity for consistent chromosomal regions in atherosclerotic plaques, as has been demonstrated for a variety of human tumors? Do the oxidized derivatives of LDL-particles which have been shown to be particularly atherogenic also possess initiation and/or promotion activity in chemical mutagenesis assays? Use of the polymerase chain reaction will allow the sensitive and specific identification of somatic genetic changes that have occurred in atherosclerotic lesions and the detection of viral DNA sequences which may be present in the normal aorta and in atherosclerotic plaques from an individual.

SOMATIC EVENTS IN NON-PROLIFERATING CELLS

Most of the clonal phenotypes I have used as examples involve events in cells that retain some proliferative capacity or whose proliferative capacity is restored by somatic mutation. Are somatic mutations in terminally differentiated cells that do not activate cellular proliferation likely to have an impact on individual phenotype? That somatic genetic events can alter the cellular phenotype of non-dividing cells is suggested by the studies of Van Leeuwen et al. (1989) in the Brattleboro rat. In 1961 it was discovered that diabetes insipidus in the Brattleboro rat was due to lack of vasopressin synthesis in the hypothalamus of these animals. The vasopressin gene in the Brattleboro rat has suffered a single nucleotide mutation that causes a frame shift which replaces the C-terminal portion of the vasopressin precursor protein with a sequence that fails to undergo proper glycosylation. Affected animals are homozygous for this mutation. However, a small number of solitary hypothalamic neurons express apparently wild-type vasopressin gene products that are processed appropriately and these neurons are histologically identical to those seen in the heterozygous carriers of this mutation. The vasopressin gene is part of a gene family which includes another polypeptide hormone, oxytocin. This family arose by gene duplication and the vasopressin and oxytocin genes are closely linked, homologous, and similar in nucleotide sequence. It has been proposed that the appearance of heterozygous neurons in homozygous individuals is the result of a somatic, intrachromosomal gene conversion event within the vasopressin/oxytocin gene cluster. Mismatch repair or actual strand conversion between exons of the vasopressin and oxytocin genes would lead to restoration of the reading frame and synthesis of normal vasopressin. Because the enzyme systems involved are probably those involved in normal DNA repair, this event is independent of mitotic cell division. With the development of techniques that allow the analysis of genes from single cells, this proposed mechanism is testable at the molecular level.

In humans, a candidate for such a clonal somatic disease in Albright's syndrome. Albright's syndrome is characterized by polyostotic fibrous dysplasia and is commonly associated with sexual precocity in females and with goiter and/or hyperthyroidism and a variety of rarer endocrine associations (Albright et al., 1937). Hall and Warrick (1972) proposed that the endocrine sequelae of Albright's syndrome were the result of a hypothalamic abnormality causing the overproduction of a variety of releasing hormones leading to pituitary overactivity and increased function of the target organs.

The pituitary changes in Albright's syndrome are consistent with prolonged stimulation and activity, especially of the basophils responsible for the production of thyroid stimulating hormone, luteinizing hormone, and follicle stimulating hormone. However, there is no evidence of a pituitary tumor.

Cell death

I have deferred to the end what may on the surface seem to be the most trivial phenotypic outcome of a somatic mutation, cell death. The death of a single cell, or multiple cells, in a mature somatic tissue probably has little impact on an individual's phenotype. However, the death of a single cell in a small committed cell population during early embryogenesis could have substantial impact on the ultimate phenotype of an individual. If you allow me to extend the definition of cell death to include not only the metabolic death of a cell but ablation of a cell's program of migration, cell–cell interaction, and ultimate differentiation, then the impact of genetic variation in individuals may be substantial. The exciting work of Schull and his colleagues (1989) on development disorders, particularly mental retardation, in "in utero" exposed individuals in Hiroshima–Nagasaki are revealing the significant impact that somatic events may have on developmental disorders in humans. Although the number of individuals exposed to ionizing radiation in utero at particular periods of development is small, the results from magnetic resonance imaging of the brains of individuals exposed at 8 to 9 gestational weeks are consistent with the failure of neurons to migrate to their appropriate functional sites. Somatic mutations in genes coding for neural cell adhesion molecules or their receptors are candidates for mediating such effects.

IMPACT OF SOMATIC GENETIC VARIATION

The impact of somatic mutations due to environmental exposure to genotoxic agents is accepted to be large, particularly in terms of morbidity and mortality due to cancer. What is the potential impact of somatic variation due to genetic mechanisms that lead to homozygosity or hemizygosity for mutant alleles that may exist at polymorphic frequency in the population? If the most conservation estimates given by Neel (chapter 4, this volume) of the frequency of functional null mutations in the general population are correct, then the potential for generation of cellular homozygosity for null mutation by somatic genetic mechanisms is substantial. The frequent loss of heterozygosity for specific chromosomal regions in tumors, due to recombination, may be an indication of the potential magnitude of this impact. If even a small fraction of these null mutations occur in genes that are critical to normal development at an early stage, then the impact of clonal somatic genetic events in determining phenotypic variation in the population may be substantial.

ACKNOWLEDGMENTS

I wish to thank Ken Weiss, Jack Schull, and Aravinda Chakravarti for numerous stimulating discussions of somatic genetic events, some of which appear in mutated form in this chapter.

REFERENCES

Albert RE, Vanderlaan M, Burns F, Nishizumi M (1977) Effects of carcinogens on chicken atherosclerosis. *Cancer Res* 37:2232–2235.

Albright F, Butler AM, Hampton AD, Smith P (1937) Syndrome characterized by osteitis fibrosa disseminata, areas of pigmentation and endocrine dysfunction, with precocious puberty in females: report of five cases. *New Eng J Med* 216:727–746.

Benditt EP, Benditt JM (1973) Evidence for a monoclonal origin of human atherosclerotic plaques. *Proc Nat Acad Sci* 70:1753–1756.

Benditt EP, Barrett T, McDougall JK (1983) Viruses in the etiology of atherosclerosis. *Proc Nat Acad Sci* 80:6386–6389.

Dacie JV, Smith MD, White JC, Mollin DL (1959) Refractory normoblastic anaemia: a clinical and haematological study of 7 cases. *Br J Haematol* 5:56–82.

Hall R, Warrick C (1972) Hypersecretion of hypothalamic releasing hormones: a possible explanation of the endocrine manifestations of polyostotic fibrous dysplasia (Albright's syndrome) *Lancet* i:1313–1316.

Kyle RA, Lust JA (1989) Monoclonal gammopathies of undetermined significance. *Sem Hematol* 26:176–200.

Majesky MW, Reidy MA, Benditt EP, Juchan MR (1985) Focal smooth muscle proliferation in the aortic intima produced by an inidiation–promotion sequence. *Proc Nat Acad Sci* 82:3450–3454.

Pearson TA, Dillman JM, Solez K, Heptinstall RH (1978a) Clonal characteristics in layers of human atherosclerotic plaques. *Am J Pathol* 93:93–102.

Pearson TA, Dillman JM, Solez K, Heptinstall RH (1978b) Clonal markers in the study of the origin and growth of human atherosclerotic lesions. *Circulation Res* 43:10–18.

Pearson TA, Dillman JM, Heptinstall RH (1983) The clonal characteristics of human aortic intima. *Am J Pathol* 113:33–40.

Pearson TA, Wang A, Solez K, Heptinstall RH (1975) Clonal characteristics of fibrous plaques and fatty streaks from human aortas. *Am J Pathol* 81:379–387.

Penn A, Batastini G, Albert R (1981) Age-dependent changes in prevalence, size, and proliferation of arterial lesions in cockerels. II. Carcinogen-associated lesions. *Artery* 9:382–393.

Penn A, Garte JJ, Warren L, Nesta D, Mindich B (1986) Transforming gene in human atherosclerotic plaque DNA. *Proc Nat Acad Sci* 83:7951–7955.

Penn A, Snyder C (1988) Atherosclerotic plaque development is "promoted" by polynuclear aromatic hydrocarbons. *Carcinogenesis* 9:2185–2189.

Schull WJ, Otake M, Yoshimanu H (1989) Radiation-related damage to the developing human brain. In Bauerstock KF, Stather JW (eds) *Low Dose Radiation: Biological Bases of Risk Assessment*. New York, Taylor and Francis, pp 28–41.

Vainchenker W, Testa U, Deschamps JF, Henri A, Titeux M, Brenton-Gorris J, Rochant H, Lee D, Cartron J-P (1982) Clonal expression of the Tn antigen in erythroid and granulocyte colonies and its application to determination of the clonality of the human megakaryocyte colony assay. *J Clin Invest* 69: 1081–1091.

VanLeeuwen F, VanderBeek E, Seger M, Burbach P, Ivell R (1989) Age-related development of a heterozygous phenotype in solitary neurons of the homozygous Brattleboro rat. *Proc Nat Acad Sci* 86:6417–6420.

4

Human Consanguinity Effects Revisited: Why is the Measurable Impact of Inbreeding So Small?

JAMES V. NEEL

Between the years 1958 and 1964, Jack Schull was the senior partner in a series of joint studies in Japan on consanguinity effects, a series that provides the most comprehensive view of the results of human inbreeding in a defined population to this date. The results have been presented and analyzed in a book and a series of papers published between 1965 and 1972. At that time, studies on inbreeding effects were seen as one of the two or three strongest approaches to understanding the amount and significance of the genetic variation present in the human genome. Given the theme of this symposium, it seemed particularly appropriate to review these studies in light of some of the developments in biochemical, molecular, and population genetics since this last analysis. Not only do they recall one of Jack's very significant contributions to the genetic literature, but, considered in conjunction with more recent developments, they will pose some searching issues for those interested in evaluating the biological significance of the avalanche of genetic variation now upon us. In brief, I will first suggest that there is an apparent gap between the results of inbreeding and what we now know about human mutation rates and the amount of variation at the protein and DNA level, and will then raise some questions concerning ways to bridge this gap.

CONSANGUINITY STUDIES IN JAPAN

During the years 1946 to 1947, in the course of the planning for the studies on the genetic effects of the atomic bombs, it became apparent that by Western standards, consanguineous marriages were unusually frequent in Japan. This situation presented a potential complication in the studies, since if because of the internal structure of the cities, consanguineous marriages were more common in one area than another, such an unequal distribution could be a source of bias. A pivotal aspect of the planning of the A-bomb studies was the finding that in those difficult times for Japan, pregnant women, if they registered the fact at the completion of the fifth lunar month of pregnancy, were entitled to access to a variety of rationed items. Registration for ration purposes was found to be very complete. With the approval of the municipal authorities, we engrafted a questionnaire relevant to the upcoming studies on A-bomb effects on this civil procedure; the questionnaire included an item concerning the consanguinity of the parents, ultimately determined to involve 6.0% of the Hiroshima pregnancy terminations and 8.0% of those in Nagasaki. We therefore had the basis for a *prospective* study on a *large* sample of children born to consanguineous unions, a study which should be *free of many of the ascertainment biases* of the past, and, of course, an appropriate set of controls in the children born under similar circumstances to non-consanguineous parents. In 1956 we took the initial steps toward a rather comprehensive, bi-national collaborative study of the sample so ascertained, the sample consisting of 4305 terminations to related parents and 4817 matched control terminations; in neither city were the offspring of parents exposed to significant amounts of radiation included in the study.

Table 4.1 summarizes some of the salient findings of that study, in terms of the average offspring of a first-cousin marriage in Hiroshima and Nagasaki (Schull and Neel, 1965). On average, there is excess homozygosity at 6.25% of the genetic loci of such a child. Nevertheless, cumulative mortality (including stillbirths) up to an average life expectancy of 98 months in Hiroshima and 105 months in Nagasaki was increased only 1.5% (9.0% vs 10.5%). The cumulative (since birth) frequency of one or more major medical problems in surviving children was 8.5% in the controls and 11.7% in the offspring of first cousins. Because the children incorporated into this study had for the most part been ascertained at the completion of their fifth lunar month of gestation, this particular study did not rule out the possibility of major consanguinity effects earlier in gestation. Incidentally, in view of all that has been written about the effect of inbreeding on central nervous system function, we note the relatively small effects of inbreeding on neuromuscular performance, psychometric scores, or school performance—and the consistency between these indicators (Table 4.2).

Table 4.1. Some salient findings in the study of inbreeding effects in Hiroshima and Nagasaki, after Schull and Neel, 1965.

Characteristic	Control children	Offspring of first cousins	Percentage difference from control
Cumulative mortality	0.090	0.105*	17.2
One or more major defects	0.085	0.117**	37.5
One or more minor defects	0.079	0.098*	24.1

*$0.05 > p > 0.01$
**$0.01 > p > 0.001$

Table 4.2. The effect of inbreeding on neuromuscular tests, intelligence, and school performance.[a]

Characteristic	Control children	Offspring of first cousins	Percent change with inbreeding*
Dynamometer grip (rt, kg)	14.04	13.71**	2.1
Tapping rate (10 sec, rt)	28.02	27.62**	1.1
Psychometrics (WISC)			
Verbal	58.67	55.34**	4.7
Performance	57.37	54.94**	3.6
School performance (average grade)			
Language	3.09	2.95**	3.2
Mathematics	3.21	3.04**	4.0

[a]Results are presented for Hiroshima males only; findings in Nagasaki and in females were similar.
*Corrected for differences in socioeconomic status.
**$0.01 > p > 0.001$

Although these studies provided excellent data on the outcome of consanguineous marriages, they did not touch upon an aspect of consanguinity that studies of domestic animals had suggested might be quite important, namely, the reproductive performance of the woman who was the product of consanguinity (albeit married to an unrelated man). In an effort to extend the findings to this aspect of consanguinity, a further study, again involving numerous Japanese collaborators, was carried out on Hirado, an island with some 41,000 inhabitants off the southwestern tip of the Japanese archipelago, on which fertility was high prior to WWII (summary in Schull and Neel, 1972). This study included an island-wide survey intended to obtain reproductive histories on (1) consanguineous marriages, (2) non-consanguineous marriages where either the mother or father was the product of a cousin marriage but unrelated to her/his marital partner, and (3) suitable controls. A total of 10,530 marriages were ascertained for which one or both spouses were alive and residing on Hirado. The frequency of consanguineous marriage was 15% for the total island, and another 9% of mothers were known to be the product of consanguinity.

Total prereproductive-age mortality, including recognized early fetal losses (net zygotic loss), among the offspring of first-cousin marriages was estimated to be 6.4% greater than among the controls (39.1% vs 32.7%) (Schull et al., 1970). These percentages, indicating a higher risk than was observed in Hiroshima and Nagasaki, probably reflect both the greater portion of the life span covered and the harsher life on Hirado. Furthermore, the live-born offspring of a woman who was the product of a first-cousin marriage (but not married to a relative) were estimated to have an increased risk of death before the age of reproduction of 2.9%. There was thus an important "delayed" consanguinity effect. Total pregnancies and total live births were significantly *increased* with consanguinity, but "net fertility" was not, once allowance was made for the role of socioeconomic factors. The absence of an increase in net fertility could be accounted for by the increased risk of death for the children resulting from a consanguineous marriage, i.e., there was reproductive compensation with respect to children lost early. In a subsequent paper, Schull (1973) examined the important question of birth intervals in consanguineous and non-consanguineous marriages. The data are complicated. For both types of marriage, in the event of an early loss (stillbirth or neonatal death), the interval to the next pregnancy was shorter than if the child had survived. Because such losses were somewhat higher in consanguineous marriages, this introduced a source of bias into a comparison of pregnancy intervals for the two types of marriage. Be that as it may, for all consanguineous marriages, the average interval between the first and second pregnancies (arbitrarily selected as the most appropriate interval to study), was 32.8 months, whereas for non-consanguineous, it was 32.3 months. Variances are not given, but although it would be of interest to explore with modern modeling techniques the possibilities for hidden loss, the impression was that hidden losses (which should extend the interval between recognized conceptions) were small, a conclusion reached more formally in a later paper by Rossmann and Schull (1974).

At the time of these studies, it was a matter of some contention whether the outcome of consanguineous marriage suggested that the genetic variation exhibited by populations was predominantly maintained by mutation pressure or by heterozygote advantage. Morton et al. (1956) had argued that the ratio of (1) the mortality predicted at full inbreeding ($F = 1$) from the regression of indicator on degree of consanguinity, termed B, to (2) the mortality in the outbred (control) population, termed A, provided a decisive insight into the question. The uncertainties introduced in the extrapolation from inbreeding coefficients usually not exceeding 0.06 to a coefficient of 1.0 were obviously tremendous, as was the bias introduced by the fact that the intercept (A) included death from all causes, only a fraction of which was genetic. Nevertheless, based primarily on the data of Sutter and Tabah (1953) from France, which yielded a B/A ratio of 18.5, they had concluded that the "genetic load" revealed by inbreeding was primarily mutational in origin. Jack and I argued that although the high B/A ratios characterizing specific, rare deleterious recessive traits were compatible with maintenance of the phenotype in question by mutation pressure, in a study such as our own, the lower B/A ratios for survival, namely, 6.1 in Hiroshima, 1.1 in Nagasaki, and 6.7 in the more comprehensive data of Hirado, were best explained by segregation involving a preponderance of balanced genetic systems of various types, i.e., a segregational load. (Other contemporary studies, reviewed in Schull and Neel [1965], had yielded B/A ratios similar to those we reported.) I note in this connection that in the simplest case, a balanced polymorphism leads to a B/A ratio of 2:1, but B/A ratios as high as 10:1 can readily be explained by a population structure in which the preponderance of the inbreeding effect results from a segregational load (Neel and Schull, 1962).

Not surprisingly, we felt that the Japanese data, because of the frequency of consanguineous marriage and manner of its ascertainment, were more appropriate than the French data in reaching general conclusions on the effects of inbreeding. Furthermore, we had corrected our data insofar as possible for socioeconomic differentials between consanguinity classes.

While these studies were in progress, and as the true complexity of the allelic structure of populations became more apparent, a number of investigators explored the concept of truncation selection as a more effective mode of elimination of deleterious genes than the independence of gene action implied in the Morton-Crow-Muller formulation (King, 1967; Milkman, 1967; Sved et al., 1967). While truncation selection is a convenient concept, evidence for its action is very limited; the Japanese inbreeding data revealed no systematic trend toward the nonlinearity in the inbreeding regressions which would indicate positive synergism of recessive traits (i.e., truncation selection). However, the limited range in the observed inbreeding coefficients did not permit a strong test for synergism. Other investigators explored the role of linkage in creating "super" genes which would function as a unit (Franklin and Lewontin, 1970; Wills et al., 1970). The result was a variety of calculations indicating that several thousand balanced polymorphisms could be maintained by selection even in an animal with the mortality structure of humans, although the success of the argument rests on foetal losses of the order of 50% (cf. Schull and Neel, 1972). Total mutation rates were unknown; Jack and I were regarded as overly generous when in our 1954 text we suggested that the average gamete might carry one new mutation. The formulations just mentioned could accommodate such a rate. While no formal consensus was ever reached, I believe that as the debate died down, most objective participants, which Jack and I certainly were, felt that the inbreeding depression was probably as much due to balanced polymorphic systems as to mutational pressure, and that it was impossible to be more precise. Note that in stating that conclusion, I have eschewed speaking of the inbreeding depressions as a "load," a term that many have argued is pejorative when applied to a process as basic and diverse as mutaton (cf. esp. Wallace, 1989).

But whether we refer to the outcome of inbreeding as depression or load, it is real enough, and next I propose to direct your attention to a series of developments during the past 25 years which are relevant to the interpretation of inbreeding effects. Some of these developments were in progress during these earlier studies, but their full impact is only now apparent. They are all interrelated, and the order in which they are considered is somewhat arbitrary. Most of these developments are now common genetic knowledge and need not be referenced in detail.

At this juncture, I would like to draw a distinction between what I will term the *proximal* and the *distal* (or ultimate) causes of an inbreeding effect. The proximal cause is the array of genetic variation currently present in the genome. The distal, ultimate cause is the mutation pressure that is responsible for this variation. It is convenient to consider the distal cause first.

MUTATION

The anatomy of the genetic material

Any contemporary treatment of mutation really has to begin with some consideration of the nature of the genetic material. As details concerning the organization of the 3×10^9 nucleotides in a human gamete have emerged, surprise has followed surprise. Let me mention, with no attempt at full documentation, some of these surprises. The first of these surprises was the amount of *apparently* superfluous DNA in our genome. As the first detailed studies of gene structure appeared, it seemed that about 5%—at most, 10%—of the DNA should be sufficient to encode the genetic information necessary to the development and functioning of a human being. The remaining 90% to 95% was thought to be composed of very abundant or semi-abundant repetitive sequences, retroviral footprints, pseudogenes, and a relatively high proportion of sequences that fit into none of these classifications. As it became clear how much larger some genes were than the hemoglobin prototype of some 2 kb—the HPRT locus of some 44 kb (Melton et al., 1984; Patel et al., 1984), the von Willebrand factor locus of approximately 175 kb (Collins et al., 1987), the retinoblastoma gene of 200 kb (Hong et al., 1989), the cystic fibrosis gene of about 250 kb (Rommens et al., 1989), and the monstrous dystrophin locus of about 2000 kb (Koenig et al., 1987), with mutations hundreds of bp 5′ and 3′ to the coding sequences having effects on gene expression, the amount of irrelevant DNA has been shrinking steadily. Nevertheless, there is still a substantial amount of the DNA—perhaps 80%—to which it is difficult to assign function, but let us not yet write it off as junk.

The second of these surprises regarding genome structure is the business of introns and exons. The discovery of "genes in pieces," as Gilbert termed it in 1978, was immediately followed by bountiful speculation concerning the justification for this organization. An obvious possibility was that the exons often encoded for functional domains, the potential building blocks for new proteins. In fact, several examples are now known where this reassortment of domains seems to have occurred, such as the gene for coagulation factor IX, which apparently results from the combination of a Gla domain (encoding for multiple [12] gamma-carboxylated glutamic acids), two epidermal growth factor domains, an activation peptide domain, and a catalytic domain such as is present in the serine proteases (Yoshitake et al., 1985)! Imagine the unsuccessful experiments of nature—mutations—which had to have preceded this successful structure. There is, moreover, a price for maintaining the structure that permits this domain shuffling. The splice site junctions, which of course include intron DNA, must be conserved, and, indeed, intron length seems also conserved. Mutation at these sites, or mutation in introns which creates new splice sites, can wreak havoc with a protein. Furthermore, a priori it would seem metabolically inefficient for the organism to devote so much energy to the elaboration of intron transcripts, most of which must immediately be degraded. In short, there has to be a cost to this structure, a cost that would not occur if the gene were continuous, a cost met each generation, and a cost no one has yet calculated. With only two introns, as in the β and α globin genes that dominated the early studies, the issue did not emerge in full force, but as the complexity of some gene structures emerged—the factor VIII gene with 25 introns (Gitschier et al., 1984), the $\alpha 1$(I) collagen gene with approximately 50 introns (Chu et al., 1982; Myers et al., 1982), the dystrophin gene with some 60 introns (Koenig et al., 1987)—the potential costs are soaring much as the military budget did under Reagan. Evolution does not look ahead. It would seem that the occasional reward to the organism from a structure that facilitates the emergence of new proteins has to be quite high, especially if the new mutation is to take hold in the population structure to be described shortly. I am led to postulate that if they are to persist, mutations from the recombination of parts of existing genes should confer large selective advantages, larger than we are accustomed to considering. I come back to this thought later.

The third surprise is the frequency of gene families, presumably resulting from repeated duplication of a successful DNA configuration, followed by divergence of this configuration into genes whose products have different specificities or which may even become ghost genes. The α and β gene families of hemoglobin are perhaps the best studied examples of this, both with seven members. There are other more extensive (and chromosomally scattered) gene families, however, such as the carbonic anhydrases (reviewed in Tashian, 1989), the serine proteases (reviewed in Hewett-Emmet et al., 1981), the coagulation factors (reviewed in Kurachi and Chen, 1987), and, of course, the special cases of the rRNA, the tRNA, the immunoglobulin, and the histocompatibility gene families, concerning which whole books have been written. Pseudogenes constitute a high proportion of some gene families. As we will see, this aspect of the genome, while reflecting our evolutionary past, and providing much of the basis for an evolutionary future, also creates problems for the accurate replication of the genome, problems the magnitude of which we are just now beginning to realize.

The biochemical basis for mutation

Equally as exciting as the foregoing insights have been the advances in our understanding of how the structure we have just mentioned determines the nature of spontaneous mutation. For expository purposes, three types of mutation can be recognized. The first consists of mutations involving gain or loss of entire chromosomes or major parts of chromosomes, translocations between chromosomes, and inversions within chromosomes. These have been recognized for years, but a major surprise was their frequency in humans, as revealed through studies of early abortuses. Since 50% of foetal losses during the first trimester are characterized by chromosomal abnormalities so gross as to be detected with the light microscope (cf. Evans, 1984), the true proportion must be even higher. Genetic defects of this type will not accumulate and be subject to inbreeding effects, but to the extent they account for early foetal loss, then proportionately fewer early foetal loses can be attributed to homozygosity for recessive genes.

The second type of mutation, at the opposite end of the spectrum, is the true point mutation, presumably resulting for the most part from misrepair of a spontaneous depurinization or depyrimidinization or misrepair of a chemical adduct. An often-quoted calculation with respect to spontaneous depurinization is that the probability of occurrence at any specific site in mammalian DNA is 3×10^{-11}/sec. This is a remarkable stability for any specific site, but given the number of nucleotides in a human cell, the nucleus should lose by spontaneous hydrolysis about 10,000 purines (and 500 pyrimidines) during a 20-hour generation period (Lindahl, 1977). Failure to repair each of these precisely can lead to a mutation. In *E. coli*, the DNA polymerase responsible for "proofreading" the replicated strand detaches many correctly positioned nucleotides during this process (Fersht et al., 1982). We presume the same for humans. These must be correctly replaced. In addition to these accidents, there is endogenous DNA damage from various sources, the amount of which is only now being appreciated. For example, Ames (1989) has calculated that oxidative damage to DNA results in about 10^3 oxidized thymine adducts per day for each of the human body's 6×10^{13} cells and estimates that the total numbers of all types of oxidative hits of DNA per cell per day in humans is about 10^4. The amount of repair necessary to offset the potential for damage to DNA from the natural pesticides of common foods is only now becoming apparent (Ames, 1989). We can only guess, but I suggest that correct repair of all these lesions must be better than 99‰, to account for the mutation rates we will consider.

But the greatest advances in understanding mutation stem from recognition of a large class of events whose occurrence is facilitated by the anatomy of the genome as just discussed plus previously unappreciated nuances of chromosomal behavior. Unequal intragenic sister chromatid exchange or homologous chromosome exchanges, in gonial or meiotic cells, now seem documented as causes of mutation in humans. Reciprocal intragenic exchanges can also result in mutation if the genes involved in the exchange have differing nucleotide substitutions. Mutations within splice sites, or creating new splice sites which introduce extraneous material into the messenger RNA, are well known. Gene conversion, formerly thought to occur only in lower eukaryotes, now seems a characteristic of human DNA. For instance, in the α-globin cluster, there are three functional genes: ζ, $\alpha 2$, and $\alpha 1$; three nonfunctional genes, $\Psi\zeta$, $\Psi\alpha$, and Θ; and one whose functioning is uncertain. The case has been made for at least three instances of gene conversion in this cluster. Proudfoot and colleagues (1980, 1982) suggested that the close homology between the $\Psi\zeta$ and ζ genes and the increased homology between the $\Psi\alpha$ and α genes relative to their flanking DNA was best explained by gene conversion. Further evidence for gene conversion may be found in the close correspondence between the Z-blocks containing the $\alpha 2$ and $\alpha 1$ genes (Michelson and Orkin, 1983) and the 36-bp repeated element between $\Psi\zeta$

and ζ (Goodburn et al., 1983). With respect to the β-globin cluster, evidence has been presented for a major conversion involving the $^{G}\gamma$ and $^{A}\gamma$ genes (Slightom et al., 1980). For both the α-globin and β-globin clusters, closely related genes are involved; limits to the divergence between genes consistent with conversion are yet to be defined. Gene conversion carries with it the possibility of delayed repair of mutational damage, but such repair would require a sensing mechanism for "normality."

Mutation may also result from recombination between non-allelic members of gene families, the best known example being the hemoglobin Lepore's that result from recombination between the β and δ genes of the β-globin gene family, genes which are some 5 kb apart. Honig and Adams (1986) list seven different examples of hemoglobins resulting from crossing over between the β and δ loci (the hemoglobin Lepore's and anti-Lepore's). These two loci may be unusually prone to such events, since the coding sequences for the β and δ chains, of 423 and 438 bp respectively, differ in only 31 nucleotides, a similarity undoubtedly favoring such exchanges. Other examples of such events in gene families have been summarized by Sinnott et al. (1990). If this behavior characterizes gene families, what is the total frequency each generation of mutations of this type, mutations that are clearly deleterious? Thus the evolutionary strategy of developing tandem variations of a proven functional DNA configuration carries with it the risk of mispairing and inappropriate crossing-over.

The frequency with which gene families have been documented in the genetic material suggest that primary duplications of entire loci are not infrequent mutational events, although the (indirect) evidence for such duplication is available only for the case where there are already duplicated loci (e.g., the α hemoglobins [Goossens et al., 1980; Zimmer et al., 1980] and the color vision genes [reviewed in Motulsky, 1988]). Finally, the frequency of retroviral footprints in the human genome—estimated to constitute 5% of the total primate genome—raises questions as to the role of retroviruses in a type of gene transfer that could result in mutagenesis; the evidence for retroviral-mediated *inter*specific gene transmission followed by germ-line integration introduces a new dimension into evolution (reviewed in Benveniste, 1987).

The frequency of mutation

There has thus been this rush of insight in recent years into the many ways the integrity of functional DNA may be compromised. Our conceptual basis for mutation today bears little resemblance to that of 35 years ago, when each gene was a unique identity on a string of such entities. The extent to which the potential for mishap we have just discussed is realized constitutes the mutation rate. Advances in estimating the frequency of mutation have come much more slowly than insights into how it can occur. The first human mutation rates to be measured with any accuracy were those resulting in certain dominantly inherited phenotypes. It was generally recognized by those of us who practiced this art that while these rates had heuristic value, in documenting the contribution of one type of mutation to disease, they were the basis for only limited generalizations. After all, one usually undertook a systematic study of a specific mutation rate only if several documented examples of that mutation were already known. To restate the obvious, one did not study the rate with which mutation produced a non-existent phenotype. There was also the bias introduced by the possibility that the phenotype in question could result from mutation at several different loci.

The advent of the biochemical era in human genetics introduced in principle a less biased approach to the study of germ-line human mutation rates. Now the choice of indicator—in this case, protein—was not dictated by circumstances, but became the investigator's prerogative. Unfortunately, new mutations of proteins were not—unlike the dominant mutations previously studied—ascertained through the medical triage system, but had to be dug out by large-scale surveys. The most convenient technique for accomplishing this objective was electrophoresis, which only detected mutations characterized by the substitution of a differently charged amino acid in a protein, these usually the result of a single nucleotide substitution. Processing the number of specimens necessary to generate a mutation rate was both laborious and expensive. Fortunately, the interest in the genetic effects of radiation helped generate the necessary funding. In order to present data collected under uniform conditions, I shall restrict myself to data generated in Ann Arbor and in a parallel program of the Radiation Effects Research Foundation in Hiroshima. At this writing there have been 743, 628 tests for spontaneous mutation resulting in an electrophoretic variant, involving some 32 polypeptides; three putative mutations have been detected, a rate of 0.4×10^{-5}/locus/generation (Neel et al., 1988a,b). Electrophoresis is thought to detect between 1/3 and 1/2 of all amino acid substitutions; correction for this fact yields a mutation rate of approximately 1×10^{-5}/locus/generation. If there were an average of 10^3 exon nucleotides at risk at the average locus, this finding would correspond to a nucleotide rate of 1×10^{-8} per generation.

Estimates of this nature have also been approached indirectly, based on the results of surveys of presumably isolated, stable, long-endogamous populations for rare (non-polymorphic) electrophoretic variants of a battery of erythrocyte enzymes and serum transport proteins. We have pursued this approach with material from some of the least acculturated Amerindian tribes of Central and South America. The estimate has evolved as the database changed, but now yields for electromorphs a rate of 1.4×10^{-5}/locus/generation (Neel et al., 1986). By a different approach, based not only on the presence but also on the numbers in which the rare variants occur, Neel and Thompson (1978) have estimated the rate to be 0.7×10^{-5}/locus/generation.

Chakraborty and I have explored still another indirect approach to the estimation of mutation rates resulting in electromorphs (Chakraborty and Neel, 1989). This approach manipulates data on the total number of alleles per locus across loci and across tribes to develop maximum likelihood estimates of individual locus mutation rates and tribal N_e. Now the average mutation rate for electromorphs was estimated to be 1.1×10^{-5}/locus/generation. However, the concomitant estimates of N_e were, after two tribes known to have expanded their numbers recently were excluded, some two times the estimated number of reproductive-age adults. Two obvious explanations for this discrepancy are a recent decrease in population size and unappreciated intertribal exchanges. Both of these explanations would imply that the figure of 1.4×10^{-5}/locus/generation cited above was an overestimate.

The average of all three of these estimates, all based on tribal populations, is 1.1×10^{-5}/locus/generation for electromorphs, almost three times the value obtained in civilized populations by the direct method, using essentially the same loci. Given the sources of error in both the direct and indirect estimators, we conclude there is "satisfactory" agreement between the results of the two approaches. At the very least, the data do not suggest an increase in mutation rates with the coming of industrialization. The Chakraborty–Neel approach, as noted, generates estimates of mutation at individual loci. For the loci exhibiting variants, these mutation rates ranged from 0.25×10^{-5} (several loci) to 3.48×10^{-5} (PGM1). Inclusion in that estimate of loci which were monomorphic in this study but were associated with variants in other studies would only extend (downwards) that range in mutation rates. Given Benzer's classic demonstration (1961) of a mutational hot spot in the rII region of phage T4, locus differences in mutability in higher eukaryotes are not surprising, but it will be important to define the range and, ultimately, the basis of such differences.

Flushed with this new knowledge, several years ago we succumbed to the temptation to speculate on the implications of these data for total gametic and zygotic mutation rates (Neel et al., 1986). On the assumption that the direct estimate of the nucleotide rate we had derived was typical of the entire genome, the total nucleotide rate would be $3 \times 10^9 \times 1 \times 10^{-8}$, or 30 mutations/gamete. Variants characterized by lack of enzyme activity, which are predominantly the results of gene inactivation or deletion, have about the same frequency in populations as the rare electromorphs which are the basis of most direct and indirect estimates of mutation rate. Thus far, efforts to measure the mutation rate for variants of this type directly have been very limited: no spontaneous mutations have been encountered in 74,025 locus tests (Mohrenweiser, 1981; Neel et al., 1988a). Because the frequency of null variants is about the same as rare electrophoretic variants, despite the heavier negative selection to which they must be subject, we will assume that the mutation rate per locus for this type of variant is at a minimum as high as for electromorphs. The mutation rate per locus now becomes 2×10^{-5}/locus/generation or, when applied to the entire haploid genome, some 60 mutations resulting in nucleotide substitutions and small insertions/deletions/rearrangements leading to gene inactivation. This estimate does not include allowance for gene duplication events, nor for the results of unequal crossing over in the highly repetitive DNA.

I now believe there may have been significant bias in this estimate of total nucleotide mutation rates. The two-dimensional polyacrylamide gel electrophoresis technique is revealing substantially less genetic variation in the more abundant proteins visualized in lysates of such cell types as fibroblasts, liver and brain cells, and lymphocytes, than revealed by one-dimensional electrophoresis, as applied predominantly to erythrocyte enzymes and serum transport proteins (reviewed in Neel, 1990). While this may be due in part to technical factors, I am now convinced that as measured by the heterozygosity index, these proteins are on average less than half as variable as the proteins previously studied with one-dimensional electrophoresis. As this fact began to emerge, several investigators (Edwards and Hopkinson, 1980; McConkey et al., 1979) suggested that these proteins, many of which could be presumed to have a structural role in the cell, were less free to vary than, for example, enzymatic proteins. Elsewhere I have presented reasons for believing that while a role for stabilizing selection cannot be excluded, this explanation is insufficient by itself and that serious consideration must be given to the possibility of lower average mutation rates at the loci encoding for these proteins (Neel, 1990). At this time point it appears as if a suitable adjustment could reduce the estimate of total mutation rates by one half. Even with this downward modification, I still suggest that the total gamete mutation rate is much higher than envisioned a generation ago, when the Muller-Haldane equation of one mutation = one eventual genetic death held sway.

There is a completely different approach to the estimation of mutation rates, stemming from "neutral substitution" theory. Kimura (1968) has shown that under this theory, the mutation rate can be estimated from the rate of amino acid substitutions in proteins, or nucleotide substitutions in DNA. Using this approach, Nei (1987) estimated from the rate of nucleotide substitutions in mammalian pseudogenes that the mutation rate could be 4.7×10^{-9}/nucleotide/site/year. Li et al. (1987), also analyzing pseudogene evolution, estimate a somewhat lower rate (1.7×10^{-9}/site/year). In humans, with a 25-year generation time, Nei's estimate would correspond to a nucleotide rate of 1.2×10^{-7}/generation, unless, as seems likely (cf. esp. Li et al., 1987), human rates have been down-regulated (see below). Nei has further suggested that the total nucleotide mutation rate for the mammalian α globin exons is 2.0×10^{-6}/year. On the basis of an average generation time for the human lineage during the last million years of about 20 years, he estimates a total mutation rate for the coding region of the human α-globin gene of 4.0×10^{-5} per generation. It is noteworthy that this estimate is within a factor of two of the rate we obtained by the indirect approach, after correction for silent amino-acid substitutions. We note, however, that neither estimate includes allowance for nucleotide substitutions with deleterious effects or the other types of mutation mentioned earlier.

Although the data are not as hard as is desirable, the relative similarity in the mutation rates of drosophila, mice, and humans on a generational basis has for some time argued for a considerable genetic control of mutation rates (reviewed in Neel, 1983). Incidentally, that apparent similarity argues against the constancy of mutation rates per year across species which entered into some of the above calculations. The constancy hypothesis requires that on a generational basis, human rates are some 10 times greater than mouse rates. The possibility, mentioned above, of differences in locus mutability related to class of protein encoded raises an intriguing line of thought. At present the prevailing viewpoint is that differences in the rate of nucleotide substitutions in proteins in evolutionary time is predominantly a reflection of differential selective pressures working on constant mutation rates. In fact, would it not be more efficient if selection, once a protein functioning within narrow tolerances had evolved, led to down-regulation of the rate of mutation at the locus encoding that protein? This would certainly decrease the cost of stabilizing selection. Note how very important this renders the choice of protein (or DNA) probes for directly studying mutation rates.

None of the foregoing takes into consideration the potential role of a variety of "mutator genes," the term encompassing such diverse phenomena as transposable elements, retroviral derivatives, or repair enzyme deficiencies. Although the action of such genes is not yet clearly identified for humans, it seems unlikely that our species is exempt from phenomena so wide-spread in plants and other animals.

In leaving this section, I will express surprise that average mutation rates, whether they be the upper or lower of our estimates, are as low as these estimates suggest. It is difficult for me to visualize so cluttered a factory as our genome functioning as efficiently as seems to be the case, especially considering the regular need for precise meiotic pairing. It is now clear that all the different possibilities for error inherent in the complex structure of DNA do occur. That the realized error rate is so low, as measured by the mutation rate, suggests incredibly efficient repair systems possibly, given gene conversion, extending across generations. It is to these repair systems we probably have to look for the chief explanation of species and locus differences in mutation rates.

The nonequilibrium of civilized/industrialized/national populations

I must now make what will appear to be a digression but really is not. Until perhaps 10,000 years ago—roughly the last 0.1% of humankind's evolution—all humankind lived in bands, hordes, or tribes. During this last 10,000 years, a steady process of amalgamation has been underway, and now there are very few populations left with any real semblance to the population structure that obtained while we were evolving. Clearly the number of generations that have elapsed since this amalgamation began is inadequate for a new genetic equilibrium for the entire species; the allele profiles of tribal populations are so strikingly different from those of these recently conglomerate, civilized populations that any manipulation of the latter in the expectation of undestanding the forces of evolution can only lead to confusion. Dr. Chakraborty (Chapter 12) will deal in some detail with the precise impact of amalgamation on allele frequencies, leaving me free to discuss briefly the population structure that resulted in these tribal allele profiles.

Our group has by now done field work among some 20 Amerindian tribes of Central and South America, but the Yanomama of northern Brazil and southern Venezuela received much more attention that any other tribe (reviewed in Neel, 1978). We have been criticized for presenting swidden horticulturists like the Yanomama as an approximation to early human population structure; my response is, give me adequate data from a more appropriate group, and I'll be happy to use them. Our use of these data included an effort to develop life tables and fertility schedules (Neel and Weiss, 1975). At the time of study, the growth rate for the Yanomama was estimated to be between 0.5% and 1.0% annually. We believe tribes tended to alternate between such periods of high growth and periods of lower growth or actual decline, but the figures that follow must be viewed in the light of such an expanding population. Infant mortality rates from natural causes were about 25%, lower than in most agricultural populations throughout much of human history. Infanticide was practiced, involving not only deformed children or children thought to be conceived extramaritally, but in addition, about 25% of all females. When natural causes of death during childhood were added in, survivorship to age 15 was about 50%.

The Yanomama, like most tribal peoples, are polygamous, with headmen in particular enjoying the multiple wives. Inbreeding beyond that resulting from small village size and a preference for village endogamy is encouraged by a prescriptive bilateral cross-cousin marriage system. Female infertility is very rare. The annual number of births per fertile female is on the basis of several different approaches, estimated to be 0.25. This child-spacing is achieved by intercourse taboos, prolonged lactation, and crude abortion practices. Even with these measures, the mean number of full-term births for females who reached at least age 15 is estimated to be 5.7, and the mean family size for females surviving to menopause is approximately 8.2, but only 35% of all females who reached age 15 live to age 50. Primarily because of polygyny, headmen (who earn their status) father twice as many children as non-headmen.

With this population growth rate, a newly founded village will in the course of several decades often exceed the size optimal to the economy, and undergoes a fission, the fission often accelerated by intravillage rivalries for leadership. These splits are along lineal lines, in consequences of which the resulting villages are far more different genetically than would result from the randomized draw envisioned by classical drift theory genetics (Smouse et al., 1981). This process, repeated again and again, plus the breeding structure just defined, leads to marked genetic microdifferentiation between the villages of a tribe. These villages, and presumably before them, equally differentiated bands or hordes, are the basic units of competition between human populations. Social structure tends to maximize the genetic differences between competing population units. Interestingly, there are numerous genetic clines within the Yanomama (Ward and Neel, 1976), a finding confirmed by spatial autocorrelation analysis (Sokal et al., 1986). We believe these clines—usually taken as a hallmark of selection—can to a large degree be explained by the vagaries of the fissioning process. Should one of these "fission" pseudopods become the nidus for a new tribe, a major step in tribal genetic differentiation could be taken at the outset.

The potential genetic consequences of this demography are far reaching and have been explored in multiple ways. To supplement the shallow time depth of the Yanomama, we developed a population simulation program patterned after the parameters just mentioned (MacCluer et al., 1971). For the cohort of 86 males aged 0 to 9 years in the four villages that formed the basis of the simulation, the projected number of grandchildren ranged from 0 (for 47 of the males) to 32 (for one) (Neel, 1980). On the assumption that the population structure we observed had in its essentials obtained since the advent of humans to the New World, the probability of identity by descent for a pair of randomly designated alleles in a village was estimated to be an incredible 0.5 (Spielman et al., 1977). This inbreeding should limit the accumulation of recessive deleterious alleles. The small effective population sizes that result from this mortality, fertility, and inbreeding structure led in the above-mentioned population simulation to an average survival time (t_0) for a mutant allele introduced in a newborn, of 2.9 generations, in contrast to the 11.7 predicted by the Kimura diffusion model based on an idealized, stable population of the same size (Li and Neel, 1974; Li et al., 1978). Even assigning a mutant allele introduced into a young adult a selective advantage of 0.05 (rather than zero) only increased the probability of survival after 16 generations from 8.6% to 13.0% (Thompson and Neel, 1978). However, the same stochastic process that results in the rapid loss of most mutations can result in a rapid increase in numbers for the surviving mutations, including mutations with slightly deleterious effects.

Two overriding impressions from these studies were the intensity of inbreeding and the strong potential role of the stochastic process in determining both survival of a mutation, whether deleterious or beneficial, and the population allele frequencies. Now, there is a useful dichotomy in thinking of the mutations responsible for evolution as mutations of minute effect and mutations of major effect (cf. Nei, 1988). Along with Kimura (1968), Nei (1975, 1988), and Wilson (1975), I prefer to emphasize the role of genes of major effect, on the negative grounds that given early human population structure, most mutations with small effects would be lost in the stochasticism. Assuming evolution to be a continuing process, however, endorsement of this concept implies the existence of transient polymorphisms with rather pronounced inbreeding effects. In a formal sense this stochasticism can be met in population theory by adjusting N_e, but given our demonstration (Li and Neel, 1974) that the allele variance N_e, the inbreeding N_e, and the mutation survival N_e are all different, which N_e to choose?

Given mutation rates of the kinds and frequencies the facts suggest, plus the structure of tribal populations, the gene pool during human evolution appears in a much more dynamic light than previously. As a young geneticist, I visualized a stately progression of the gene pool through time and space; now, it is clearly to be seen as a much more active arena, an arena that if not sometimes chaotic certainly deserves to be termed a bit helter-skelter. Perhaps the best term is "organized chaos." What most needs emphasis in the present context is that in view of the relatively recent advent of "civilization" to most human populations, the contemporary populations on which inbreeding studies have been performed can in no way be seen to be in "equilibrium," any more than the populations from which they were derived, a fact with profound implications for the interpretation of inbreeding effects.

An aside on the genetic effects of the A-bombs

I cannot, at this point, resist the temptation to slip in a brief speculation concerning the outcome of the studies of the genetic effects of the atomic bombs. As you know, in recent years Jack Schull and I, together with our Japanese colleagues, have suggested that humans are less sensitive to the genetic effects of radiation than projected from the murine studies (Neel et al., 1990). Now, much of our early thinking about the genetic risks of radiation was formulated when the germ-line and the genes that comprise it were visualized as relatively isolated from the buffeting of the soma which is the lot of living creatures. Radiation was seen as unique in its ability to "get at" the genes. Now that we understand just how dynamic that gene pool is, radiation can be viewed philosophically as just one more factor perturbing the gene, its effects to be dealt with by the same homeostatic mechanisms evolved to handle other threats. I do not mean to denigrate the importance of controlling radiation exposures; I do mean to supply some philosophical perspective.

THE ACCUMULATION FACTOR

Let us turn now to the proximal aspect of inbreeding effects, the extent to which genetic variation has accumulated in the geome. With each well-studied locus yielding its own set of surprises, it is dangerous to generalize. Nevertheless, we can begin to ask, given the surprising genomic structure which was so briefly reviewed and the preliminary data on the frequency with which it mutates, what is the genomic accumulation factor, representing the balance between mutation and selection? I insist, however, that this "balance" still reflects humankind's tribal days as much as our current population structure. We are in transition. Most of the attention in recent years has been directed toward the common enzyme and protein polymorphisms, so useful as descriptors of populations. Many investigators have contributed to the finding that in civilized populations the index of heterozygosity (HI) for electrophoretically detectable amino acid polymorphisms is about 6% to 7%; correction for silent amino acid substitutions should increase that index to about 15% (Neel, 1984). Rare electrophoretic variants, for which the question of a balanced polymorphism does not arise, contribute about 0.2% to that HI (Neel et al., 1988a). Of particular interest in the present context are three types of variants whose frequency is not measured by electrophoresis.

The first type of variant is characterized by absence of gene function, a type of variant most easily studied in erythrocyte enzymes. Recognized by activity assays, such variants result either from absence of gene product or elaboration of an inactive gene product. Their frequency (as heterozygotes) in a battery of nine enzymes studied in an urban, largely Caucasian population was 2.5/1000 determinations (Mohrenweiser, 1981). Their frequency in two urban Japanese populations, in a largely overlapping set of 11 enzymes, was 2.4/1000 determinations (Satoh et al., 1983). The allele frequency for the loci studied would be approximately 0.001.

The second type of variant is characterized by "medium sized" insertion/deletion/rearrangements (I/D/R) of DNA. These are well known as a cause of genetic disease, but only recently has there been a population survey for their frequency with reference to a series of "representative" probes (Mohrenweiser et al., 1989). In this survey, 18 different probes, ranging in size from 1 kb to 52.7 kb were employed; in a total of 9950 screened DNA segments, 31 variants were identified, i.e., 1 such variant per 321 determinations. Variants involving a change in size of less than $\sim 2\%$ of any of these fragments, i.e. $< \sim 40$ bp, would probably not be detected.

The third type of variant, to which so many laboratories have contributed, is the restriction fragment length polymorphism (RFLP), usually resulting from the occurrence of alternative states with respect to specific nucleotides. It has been estimated that there is on average about 1 variable site for each 200 nucleotides, with 1 heterozygote per 1000 nucleotides (Ewens et al., 1981). These estimates are still based on very limited data, largely derived from functional genes, but extrapolated to the total genome would correspond to some 1.5×10^7 variable sites, with heterozygosity at 3×10^6 of them.

INBREEDING EXPECTATIONS IN LIGHT OF THE FOREGOING

Let us now return to the inbreeding results described earlier in the light of the data just mentioned concerning mutation rates and the variation at the protein and gene levels, most of which have become available since our original studies on inbreeding were completed. Extensions to the total genome on the basis of the limited data available are precarious, and must be regarded with great skepticism, but three oversimplified calculations (all ignoring linkage) are illustrative of the potentialities in the situation.

(1) If there are 50,000 functional genes, with respect to each of which there is a 0.001 probability of a null, then at conception the zygote resulting from a first cousin marriage ($F = 1/16$) should be homozygous for an additional $50{,}000 \times 1/16 \times q(1-q) = 3.12$ such alleles, where $q = 0.001$. (I ignore the q^2 term.)

(2) With respect to the I/D/R DNA variants, the probes employed were

not complete genes, and the calculation must proceed somewhat differently. An estimated total of 77,400,000 bp's were surveyed (from 130 persons), 31 variants were identified, or one such variant per 2,496,373 bp. Extrapolation to the entire genome from this limited sampling is perilous, but the data suggest the presence of approximately $(3 \times 10^9)/(2.5 \times 10^6) \cong 1.2 \times 10^3$ such variants per haploid genome. The conceptus of a first-cousin marriage would be homozygous at an additional $(1.2 \times 10^3) \times 1/16 = 75$ such sites, 7.5 to 15.0 of which might be in functional genes (on the basis of 10% to 20% of the genome constituting "genes").

(3) The calculation with respect to RFLPs is more difficult because of a lack of population survey data, but if Ewens' estimate of heterozygosity for 1 RFLP per every 1000 nucleotides is correct, then the excess homozygosity at RFLP sites in a first-cousin marriage would be $1 \times 10^{-3} \times 1/16 \times 3 \times 10^9$, or 1.9×10^5 sites. If these sites were randomly distributed throughout the genome, then between 1.9×10^4 and 3.4×10^4 of them might be in functional genes. The evidence in fact suggests that these sites are not randomly distributed in the genome, being relatively less common in exons (Neel, 1984), but even a factor of two difference between exons and the remaining DNA with respect to the frequency of these sites would still imply excess homozygosity for some 10^4 of these RFLPs in the functional genes of the zygote resulting from a first-cousin marriage. Recall that these RFLPs are qualitatively the same as the DNA variants responsible for the biochemical polymorphisms, including hemoglobins S and C, the stop codons which may be produced through mutation, and the variants creating new splice sites. If only 1 in 100 of these RFLPs in genes has a deleterious effect when homozygous, the expectation is for 100 such additional homozygotes in the conceptus of a first-cousin marriage.

BACK TO THE INTERPRETATION OF INBREEDING EFFECTS

How do all of these developments and calculations bear on the interpretation of the inbreeding effects we observed in Japan? In many respects, the issues have only grown more complex in recent years. I must confess, coming back to the problem after an interval of almost 20 years, a sense of wonder that inbreeding effects are as small as they appear to be. Twenty years ago, suitable explanations seemed at hand for inbreeding results—now I am not so sure. As the full complexity of the genome began to emerge, the calculations mentioned earlier, of how many polymorphisms could be maintained under certain assumptions regarding the operation of selection, plus Kimura's bold thesis (cf. Kimura, 1983) of the selective neutrality of much of the biochemical variation being discovered, gave population genetics a brief respite from strict accountability for inbreeding results. Now that a fuller picture of the structure of the genome and its variability is emerging, that respite is over: we are groping for new formulations. Six years before we began to plan for these consanguinity studies, Muller (1950) in his much-quoted paper on "Our Load of Mutations," had envisioned humankind as precariously balanced on a razor's edge dividing mutational input and species deterioration. Dobzhansky (1955), although not denying the deleterious genetic impact of radiation, saw the genotype as much more resilient because of its balanced polymorphisms, a concept substantially advanced in this generation by Wallace (esp. 1987) and Carson (esp. 1989). As our own inbreeding results emerged, we felt, as mentioned earlier, that they reflected a great deal of homeostasis in the system, a homeostasis that would not be plumbed by studies of the action of radiation on specific genetic loci. The recent insights into the frequency and nature of mutation and the amount of variation in the genome now begin to strain the conventional explanations of inbreeding results which have been more or less satisfactory for the past 20 years. These thoughts lead me to ask what questions if answered would substantially advance our understanding of where all the mutations have gone. Below are six somewhat interrelated queries. The first three questions concern the distal (mutational) causes of an inbreeding effect, the last three the proximal (observed variation) causes.

(1) Have we been badly led astray in our projections of total gamete mutation rates by the studies on erythrocyte enzymes and serum proteins? If total mutation rates were really two or three times less than suggested by the studies to date, inbreeding effects would be expected to be correspondingly decreased.

(2) Is it possible that deleterious mutations are largely eliminated in the heterozygous state, so that there are relatively few classically "recessive" deleterious genes to be brought to light by inbreeding?

(3) How far in error are we in introducing into interpretations of the outcome of inbreeding the concept of genetic equilibrium, a balance between mutational input and selectional elimination? Jack Schull and I recognized 25 years ago that this was not a proper assumption for any civilized population, but only in recent years has it become clear how far off base that assumption could be. The relaxation of inbreeding in recent years plus the amalgamation of long-separated populations implies that modern populations are a "mutational sink," the results of inbreeding not fully reflecting the mutation–selection balance. It could lead to serious error, to argue that inbreeding effects fully measure the mutational input we have postulated. Is there any way to measure the potential magnitude of the sink effect?

But even if we agree that inbreeding effects do not reflect our present view of mutational input because total mutation rates are lower than we recently postulated, or heterozygote selection plays a prominent role in removing "recessive" genes, or that civilized populations are a mutation sink, inbreeding results should still reflect the variation in the gene pools that we now recognize. With respect to the proximal aspect of inbreeding effects, there would appear to be three, somewhat overlapping questions to be addressed with reference to why the measured inbreeding effects are so small.

(1) Early foetal loss is undoubtedly an important avenue for the elimination of deleterious mutant genes and perhaps represents the principal arena for

the action of truncation selection. As noted earlier, there is no evidence from the Hirado studies that this type of selection is substantially increased with inbreeding, but, on the other hand, these data are not sufficiently refined to define the exact role for this phenomenon. Can a better data set for addressing this question be developed?

(2) The genetic material may be much more loosely organized than the perceptions of the past. This thought is inherent in the Kimura thesis that many amino acid substitutions in proteins which occur during evolution have no selective significance. The thought also follows from the extent to which protein structure can be altered through "cassette" mutagenesis without apparently compromising protein function (cf. Horwitz et al., 1989). The frequency with which we encountered I/D/R variants in probes of "working" genes (Mohrenweiser et al., 1989) also suggests more redundancy in the gene pool than previously seemed likely. In this connection, I note that with respect to the genes encoding for the 30 proteins included in our survey for biochemical variants, the effects of homozygosity for null states has been established for 12 of the autosomal loci; for 4 of these, homozygosity is not accompanied by obvious disease (cf. Neel et al., 1988c). Just what is the degree of redundancy in the genome? Redundancy implies a high degree of "forgiveness" for mutations involving redundant loci, and forces a reconsideration of the proportion of all mutations that are neutral.

(3) Finally, we return to the issue of the balanced homeostatic properties of the genome. If there is even a quasi-equilibrium between mutational input and segregational output, one interpretation of the inbreeding results is that these abundant mutations are being absorbed into balanced polymorphic systems. With respect to new formulations Kondrashov (1988) may have taken a significant step forward, developing the thesis that as long as the total mutation load, (v), defined as the ratio of the deleterious mutation rate/genome/gamete (u) to the standard deviation of mean genome contamination (σ), is in the neighborhood of 1, the total mutation rate is appropriate to the species. "Genome contamination" may be roughly equated to total heterozygosity. However, this formulation relies heavily on truncation selection to hold σ within sustainable bounds. There is an urgent need for data on the occurrence and magnitude of truncation selection.

It would have been intellectually very exciting to return to these studies of consanguinity effects of some 20 years ago with decisive new insights. Unfortunately, while substantial progress has been made in defining the questions raised by the relatively small impact of human inbreeding on morbidity and mortality, I do not feel we are yet in a position to deal decisively with most of these questions. What does seem to be emerging is a family of interrelated potential explanations of inbreeding effects, all bearing on what has now become the overarching question of our genetic times: how is the newly discovered, really incredible complexity of our genome maintained at the cellular, organismic, and population levels? This question will not be answered by the further complexity which will be identified by the Human Genome Project, but only intensified. Sorting these explanations out, especially in the face of the relatively recent, dramatic changes in the structure of human populations which I touched upon so briefly, is surely one of the primary challenges facing molecular and population genetics today.

REFERENCES

Ames BN (1989) Mutagenesis and carcinogenesis: endogenous and exogenous factors. *Env Mol Mutag 14* (suppl 16):66–77.

Benveniste RE (1987) The contributions of retroviruses to the study of mammalian evolution. In MacIntyre RJ (ed) *Molecular Evolutionary Genetics*. New York, Plenum Press, pp 359–417.

Benzer S (1961) On the topography of the genetic time structure. *Proc Natl Acad Sci USA* 47:403–415.

Carson HL (1989) Genetic imbalance, realigned selection, and the origin of species. In Giddings LV, Kaneshiro KY, Anderson WW (eds) *Genetics, Speciation and the Founder Principle*. Oxford, Oxford University Press, pp xvii and 373.

Chakraborty R, Neel JV (1989) Description and validation of a method for simultaneous estimation of effective population size and mutation rate from human population data. *Proc Natl Acad Sci USA* 86:9407–9411.

Chu ML, Myers JC, Bernard MP, Ding JF, Rameriz F (1982) Cloning and characterization of five overlapping cDNAs specific for the human pro α (I) collagen chain. *Nucleic Acids Res* 10:5929–5934.

Collins CJ, Underdahl JP, Levine RB, Ravera CP, Morin MJ, Dombalagian MJ, Ricca G, Livingston DM, Lynch DC (1987) Molecular cloning of the human gene for von Willebrand factor and identification of the transcription initiation site. *Proc Natl Acad Sci USA* 84:4393–4397.

Dobzhansky TH (1955) A review of some fundamental concepts and problems of population genetics. *Cold Spring Harbor Symp Quant Biol* 20:1–15.

Edwards Y, Hopkinson DA (1980) Are abundant proteins less variable? *Nature* 284:511–512.

Evans HJ (1984) Genetic damage and cancer. In Bishop JM, Rowley JD, Greaves M (eds) *Genes and Cancer*. New York, Alan R Liss, pp 3–18.

Ewens WJ, Spielman RS, Harris H (1981) Estimation of genetic variation at the DNA level from restriction endonuclease data. *Proc Natl Acad Sci USA* 78:3748–3750.

Fersht AR, Krill-Jones JW, Tsui WC (1982) Kinetic basis of spontaneous mutation. Misinsertion frequencies, proofreading specificities and cost of proofreading by DNA polymerases of *Escherichia coli*. *J Mol Biol* 156:37–51.

Franklin I, Lewontin RC (1970) Is the gene the unit of selection? *Genetics* 65:707–734.

Gilbert W (1978) Why genes in pieces? *Nature* 271:501.

Gitschier J, Wood WI, Goralka TM, Wiou KL, Chen EY, Eaton DH, Vehar GA, Capon DJ, Lawn RM (1984) Characterization of the human factor VIII gene. *Nature* 312:326–330.

Goodburn SEY, Higgs DR, Clegg JB, Weatherall DJ (1983) Molecular basis of length polymorphism in the human zeta-globin gene complex. *Proc Natl Acad Sci USA* 80:5022–5026.

Goossens M, Dozy AM, Embury SH, Zachariades Z, Hadjiminas MG, Stamatoyannopoulos G, Kan YW (1980) Triplicated α-globin loci in humans. *Proc Natl Acad Sci USA* 77:518–521.

Hewett-Emmet D, Czdusniak J, Goodman M (1981) The evolutionary relationships of the enzymes involved in blood coagulation and hemostasis. *Am NY Acad Sci* 370:511–527.

Hong FD, Huang HJS, To H, Young LJS, Oro A, Bookstein R, Lee EY-HP, Lee W-H (1989) Structure of the human retinoblastoma gene. *Proc Natl Acad Sci USA* 86:5502–5506.

Honig GR, Adams JG III (1986) *Human Hemoglobin Genetics*. New York, Springer-Verlag, pp xv and 452.

Horwitz MSZ, Dube DK, Loeb LA (1989) Selection of new biological activities from random nucleotide sequences: evolutionary and practical considerations. *Genome* 31:112–117.

Kimura M (1968) Evolutionary rate at the molecular level. *Nature. Lond* 217:624–626.

Kimura M (1983) *The Neutral Theory of Molecular Evolution*. Cambridge, Cambridge University Press, pp xv and 367.

King JL (1967) Continuously distributed factors affecting fitness. *Genetics* 55:483–492.

Koenig M, Hoffman EP, Bertelson CJ, Monaco AP, Feener C, Kunkel LM (1987) Complete cloning of the Duchenne muscular dystrophy (DMD) cDNA and preliminary genomic organization of the DMD gene in normal and affected individuals. *Cell* 50:509–517.

Kondrashov AS (1988) Deleterious mutations and the evolution of sexual reproduction. *Nature* 336:435–440.

Kurachi K, Chen S (1987) Human genes for factor IX and other vitamin K dependent blood proteins. In Wessler S, Becker CG, Nemerson Y (eds) *The New Dimensions of Warfarin Prophylaxis*. New York, Plenum Publ Co., pp 67–81.

Li FHF, Neel JV (1974) A simulation of the fate of a mutant gene of neutral selective value in a primitive population. In Dyke JW, MacCluer JW (eds) *Computer Simulation in Human Populations*, Academic Press, New York, pp 221–240.

Li FHF, Neel JV, Rothman ED (1978) A second study of the survival of a neutral mutant in a simulated Amerindian population. *Amer Nat* 112:83–96.

Li WH, Tanimura M, Sharp PM (1987) An evaluation of the molecular clock hypothesis using mammalian DNA sequences. *J Mol Evol* 25:330–342.

Lindahl T (1977) DNA repair enzymes acting in spontaneous lesions in DNA. In Nichols WW, Murphy DG (eds) *DNA Repair Processes*. Miami, Symposia Specialists, pp 225–240.

MacCluer JW, Neel JV, Chagnon NA (1971) Demographic structure of a primitive population: a simulation. *Am J Phys Anthro* 35:193–208.

McConkey EH, Taylor BJ, Phan D (1979) Human heterozygosity: a new estimate. *Proc Natl Acad Sci USA* 76:6500–6504.

Melton DW, Konecki DS, Brennand J, Caskey CT (1984) Structure, expression, and mutation of the hypoxanthine phosphoribosyltransferase gene. *Proc Natl Acad Sci USA* 81:2147–2151.

Michelson AM, Orkin SH (1983) Boundaries of gene conversion within the duplicated human alpha-globin genes. Concerted evolution by segmental recombination. *J Biol Chem* 258:15245–15254.

Milkmàn RD (1967) Heterosis as a major cause of heterozygosity in nature. *Genetics* 55:493–495.

Mohrenweiser HW (1981) Frequency of enzyme deficiency variants in erythrocytes of newborn infants. *Proc Natl Acad Sci USA* 78:5046–5050.

Mohrenweiser HW, Larsen RD, Neel JV (1989) Development of molecular approaches to estimating germinal mutation rates. I. Detection of insertion/deletion/rearrangement variants in the human genome. *Mut Res* 212:241–252.

Morton NE, Crow JF, Muller HJ (1956) An estimate of the mutatonal damage in man from data on consanguineous marriages. *Proc Natl Acad Sci USA* 42:855–863.

Motulsky AG (1988) Normal and abnormal color-vision genes. *Am J Hum Genet* 42:405—407.

Muller HJ (1950) Our load of mutations. *Am J Hum Genet* 2:111–176.

Myers JC, Chu ML, Faro SH, Clark WJ, Prockop DJ, Rameriz F (1982) Human type I procollagen genes are located on different chromosomes. *Proc Natl Acad Sci USA* 79:6627–6630.

Neel JV (1978) The population structure of an Amerindian tribe, the Yanomama. *Annu Rev Genet* 12:365–413.

Neel JV (1980) On being headman. *Persp Biol Med* 23:277–294.

Neel JV (1983) Frequency of spontaneous and induced "point" mutations in higher eukaryotes. *J Hered* 74:2–15.

Neel JV (1984) A revised estimate of the amount of genetic variation in human proteins: implications for the distribution of DNA polymorphisms. *Am J Hum Genet* 36:1135–1148.

Neel JV (1990) Average locus differences in mutability related to protein "class": an hypothesis. *Proc Natl Acad Sci USA* 87:2062–2066.

Neel JV, Mohrenweiser HW, Gershowitz H (1988a) A pilot study of the use of placental cord blood samples in monitoring for mutational events. *Mut Res* 204:365–377.

Neel JV, Mohrenweiser HW, Rothman ED, Naidu JM (1986) A revised indirect estimate of mutation rates in Amerindians. *Am J Hum Genet* 38:649–666.

Neel JV, Satoh C, Goriki K, Asakawa J, Fujita M, Takahashi N, Kageoka T, Hazama R (1988b) Search for mutations altering protein charge and/or function in children of atomic bomb survivors: final report. *Am J Hum Genet* 42:663–676.

Neel JV, Satoh C, Goriki K, Fujita M, Takahashi N, Asakawa J, Hazama R (1986) The rate with which spontaneous mutation alters the electrophoretic mobility of polypeptides. *Proc Natl Acad Sci USA* 83:389–393.

Neel JV, Satoh C, Smouse P, Asakawa J, Takahashi N, Goriki K, Fujita M, Kageoka T, Hazama R (1988c) Protein variants in Hiroshima and Nagasaki: tales of two cities. *Am J Hum Genet* 43:870–893.

Neel JV, Schull WJ (1962) The effect of inbreeding on mortality and morbidity in two Japanese cities. *Proc Natl Acad Sci USA* 48:573–582.

Neel JV, Schull WJ, Awa AA, Satoh C, Kato H, Otake M, Yoshimoto Y (1990) The children of parents exposed to atomic bombs: estimates of the genetic doubling dose of radiation for humans. *Am J Hum Genet* 46:1053–1072.

Neel JV, Schull WJ, Awa AA, Satoh C, Otake M, Kato H, Yoshimoto Y (1989) Implications of the Hiroshima-Nagasaki genetic studies for the estimation of the human "doubling dose" of radiation. *Proc 16th Intern Congr of Genetics, Genome* 31:853–859.

Neel JV, Thompson EA (1978) The rate of mutation consistent with the number of private polymorphisms in Amerindian tribes. *Proc Natl Acad Sci USA* 75:1904–1908.

Neel JV, Weiss K (1975) The genetic structure of a tribal population, the Yanomama Indians. XII. Biodemographic studies. *Am J Phys Anthro* 42: 25–52.

Nei M (1975) *Molecular Population Genetics and Evolution.* Amsterdam, North Holland, pp xiii and 288.

Nei M (1987) *Molecular Evolutionary Genetics.* Amsterdam, North Holland, p 512.

Nei M (1988) Relative roles of mutation and selection in the m[illegible] of genetic variability. *Phil Trans Roy Soc London* (B) 319:615–629.
Patel PI, Nussbaum RL, Framson PE, Ledbetter D, Caskey CT, Chinault AC (1984) Organization of the HPRT gene and related sequences in the human genome. *Somat Cell Mol Genet* 10:483–493.
Proudfoot NJ, Gil A, Maniatis T (1982) The structure of the human zeta-globin gene and a closely linked, nearly identical pseudogene. *Cell* 31:553–563.
Proudfoot NJ, Maniatis T (1980) The structure of a human alpha-globin pseudogene and its relationship to alpha-globin gene duplication. *Cell* 21:537–544.
Rommens JM, Iannuzzi MC, Kerem B, Drumm ML, Melmer G, Dean M, Rozmahel R, Cole JL, Kennedy D, Hidaka N, et al. (1989) Identification of the cystic fibrosis gene: chromosome walking and jumping. *Science* 245:1059–1065.
Rossman DL, Schull WJ (1974) Recessive lethals and the birth interval. In Dyke B, MacCluer JW (eds) *Computer Simulation in Human Population Studies.* New York, Academic Press, pp 143–160.
Satoh C, Neel JV, Yamashita A, Goriki K, Fujita M, Hamilton H (1983) The frequency among Japanese of heterozygotes for deficiency variants of 11 enzymes. *Am J Hum Genet* 35:656–674.
Schull WJ (1973) Birth intervals, inbreeding, and reproductive compensation. *Hum Biol in Oceania* 2:1–16.
Schull WJ, Nagano H, Yamamoto M, Komatsu I (1970) The effects of parental consanguinity and inbreeding in Hirado, Japan. I. Stillbirths and prereproductive mortality. *Am J Hum Genet* 22:239–262.
Schull WJ, Neel JV (1965) *The Effects of Inbreeding on Japanese Children.* New York, Harper and Row, pp xii and 419.
Schull WJ, Neel JV (1972) The effects of parental consanguinity and inbreeding in Hirado, Japan. V. Summary and interpretation. *Am J Hum Genet* 24:425–453.
Sinnott P, Collier S, Costigan C, Dyer PA, Harris R, Strachan T (1990) Genesis by meiotic unequal crossover of a *de novo* deletion that contributes to steroid 21-hydroxylase deficiency. *Proc Natl Acad Sci USA* 87:2107–2111.
Slightom JL, Blechl AC, Smithies O (1980) Human fetal $^{G}\gamma$ and $^{A}\gamma$-globin genes: complete nucleotide sequences suggest that DNA can be exchanged between these duplicated genes. *Cell* 121:627–638.
Smouse PE, Vitzhum VJ, Neel JV (1981) The impact of random and lineal fission on the genetic divergence of small human groups: a case study among the Yanomama. *Genetics* 98:179–197.
Sokal RR, Smouse PE, Neel JV (1986) The genetic structure of a tribal population, the Yanomama Indians. XV. Patterns inferred by autocorrelation analysis. *Genetics* 114:259–287.
Spielman RS, Neel JV, Li FHF (1977) Inbreeding estimation from population data: models, procedures, and implications. *Genetics* 85:355–371.
Sutter J, Tabah L (1953) Structure de la mortalité dans les familles consanguines. *Population* 8:511–526.
Sved JA, Reed TE, Bodmer WF (1967) The number of balanced polymorphisms that can be maintained in a natural population. *Genetics* 55:469–481.
Tashian RE (1989) The carbonic anhydrases: widening perspective in their evolution, expression, and function. *BioEssays* 10:186–192.
Thompson EA, Neel JV (1978) Probability of a founder effect in a tribal population. *Proc Natl Acad Sci USA* 75:1442–1445.
Wallace B (1987) Fifty years of genetic load. *J Hered* 78:134–142.

Wallace B (1989) One selectionist's perspective. *Quart Rev Biol* 64:127–145.
Ward RH, Neel JV (1976) The genetic structure of a tribal population, the Yanomama Indians. XIV. Clines and their interpretation. *Genetics* 82:103–121.
Wills C, Crenshaw J, Vitale J (1970) A computer model allowing maintenance of large amounts of genetic variability in Mendelian populations. I. Assumptions and results for large populations. *Genetics* 64:107–123.
Wilson AC (1975) Evolutionary importance of gene regulation. *Stadler Symp* 7:117–134.
Yoshitake S, Schach BG, Foster DC, Davie EW, Kurachi K (1985) Nucleotide sequence of the gene for human factor IX (antihemophilic factor B). *Biochem* 24:3736–3750.
Zimmer EA, Martin SL, Beverley SM, Kan YW, Wilson AC (1980) Rapid duplication and loss of genes coding for the α chains of hemoglobin. *Proc Natl Acad Sci USA* 77:2158–2162.

5

Genetic Risks from Exposure to the Atomic Bombs: Hiroshima and Nagasaki

MASANORI OTAKE

Worldwide, tens, possibly hundreds of millions of individuals have been or are bing exposed to doses of ionizing radiation above the "natural" levels either as a consequence of (1) the diagnostic or therapeutic use of x-ray or radioactive materials, for example, radium or cobalt-60; (2) their occupations; (3) residing in geographic areas with "high" natural background or human-made radiation, such as in the vicinity of the Chernobyl accident; or (4) their presence in Hiroshima or Nagasaki at the time of the atomic bombings. Exposure in these various instances varies substantially, qualitatively, and quantitatively (UNSCEAR, 1982, 1986). It may be acute or chronic, of single quality or several, whole body or partial, prompted by illness and hence possibly confounded by health status, and so forth. The ubiquity of these exposures and the differences in their nature make the estimation of the genetic risks to human beings not only an important task, but a difficult one.

Studies of the potential genetic effects of exposure to the atomic bombing of Hiroshima and Nagasaki began under the auspices of the Atomic Bomb Casualty Commission (ABCC) in 1948 after more than a year of planning and preparation. In the presentation of the results of these studies, and those conducted subsequently under the aegis of the Radiation Effects Research Foundation (RERF), it is taken as a fact that some mutations were induced in the atomic bomb survivors, and, therefore, our remarks focus on the estimation of risk rather than tests of significance. This position is defensible not only on general principles, but also because of the chromosomal damage and somatic mutations observed in lymphocytes and red blood cells of the atomic bomb survivors (Awa et al., 1988a,b; Nakamura et al., 1987), as well as the increase in leukemia and other malignant neoplasms in survivors (Preston and Pierce, 1988; Shimizu et al., 1987), and the well-known correlation between carcinogenesis and mutagenesis. Reanalysis of the genetic data at this time was prompted by the recently available gonadal dose estimate based on the new DS86 dosimetry (Roesch, 1987).

The aim of this chapter is to generate an estimate of the effect that may be presumed to exist. The most convenient way to express this estimate is in terms of the doubling dose, that is, that dose of ionizing radiation that will produce a 100% increase over the spontaneous mutation rate.

MATERIALS AND METHODS

Study sample

Planning of a continuous surveillance of the children born in Hiroshima and Nagasaki subsequent to the atomic bombings was begun in 1946 and a full-scale program was initiated in 1948. The birth registration process that has been the basis of identifying the various study samples has been described in length elsewhere, and will not be repeated here (for details, see Neel and Schull, 1956; Schull et al., 1981a,b; Neel et al., 1990 Yoshimoto et al., 1990, 1991). However, we will describe briefly the five different, albeit interrelated samples that are involved.

The *first* sample, that based upon the clinical examination of pregnancy outcomes obtained in a surveillance of births in Hiroshima and Nagasaki between 1948 and 1953, consists of 70,073 children whose parents were assigned T65DR doses (Kerr, 1979; Kerr and Salomon, 1976; Milton and Shohoji, 1968). The new DS86 doses can be directly estimated or assigned, through an ad hoc procedure described below, to the majority of the parents of these children, specifically the parents of 69,706 of these 70,073 children (see Table 5.1 and DS86 dosimetry below). This sample has been used in the analyses of malformations, stillbirths, and early mortality, collectively termed "untoward pregnancy outcome" (UPO).

The *second* sample (Neel et al., 1990; Schull et al., 1981; Yoshimoto et al., 1991), the basis of the mortality surveillance, is composed of 50,689 live, single births occurring between 1 May 1946 and 31 December 1958, and two subsequent extensions, namely 13,128 livebirths, occurring between 1 January 1959 and 31 December 1980 (total of 63,817 children), and an additional 3,385 livebirths occurring between 1 January 1981 and 31 December 1985 (total of 67,202 children). It should be noted that in the selection of the mortality cohorts multiple births were excluded because their mortality rates are appreciably different from those of single births.

Table 5.1. Number of children of UPO, malformations, stillbirths, and early mortality.

Item	T65DR Sample	DS86 Sample	DS86 Sample Extended	Not Assigned
Subjects	70073	55303	69706	367
UPO	3521	2760	3498	23
Malformations	955	770	950	5
Stillbirths	1160	894	1148	12
Early mortality	1572	1230	1565	7

Note: These individually designated categories are overlapping, 134 for 2,760 UPO and 165 for 3,498 UPO. Neither formal nor ad hoc doses were assigned.

The *third* sample, on which the frequency of cytogenetic abnormalities is based, is composed of 8,322 children of proximally exposed parents (i.e., exposed within 2000 m of the hypocenter), and 7,976 children of distally exposed parents (exposed at or beyond 2500 m) who were examined during 1967 through 1985, including the members of an initial pilot study (Awa et al., 1988a,b).

The *fourth* sample, on which the biochemical findings rest, consists of 23,661 children, members of the mortality cohort previously described. Electrophoretic data are available on some 30 serum or erythrocyte proteins, representing 32 gene loci, for most of these individuals. Collectively, these data embrace results of 667,404 locus product tests of children of proximally exposed parents and 466,881 locus products of children of distally exposed parents (Neel et al., 1988), including a pilot study period from 1972 to 1975. Among these tests are 122,270 involving loss of enzyme activity (60,529 tests on the proximally exposed and 61,741 on the distally exposed).

The *fifth* and *last* sample (Schull et al., 1981b; Otake et al., 1990) concerns the sex ratio of 70,073 children born between 1948 and 1953, and another 72,902 born between 1954 and 1962.

DS86 dosimetry

When this reanalysis began, parental gonadal doses could be calculated on the basis of the DS86 dosimetry system for only 55,303 of the 70,073 children in the UPO study on whom parental T65DR doses were available. We call these 55,303 pregnancy terminations the DS86 sample. To avoid the loss of the 14,770 children whose parental DS86 doses had not then been calculated, an ad hoc estimation procedure was introduced. It involved dividing all parents, including those on whom a DS86 dose could be calculated and those on whom it could not, into two groups according to their distance from the hypocenter (2000 m or less, and 2000+ m), and the presence or absence of acute symptoms of radiation sickness, namely, epilation, subcutaneous bleeding, oropharyngeal lesions. We then proceeded on the principle that within a given distance interval the T65DR dose on an individual on whom a DS86 dose had not been calculated could be mutiplied by the ratio of the DS86 to the T65DR mean total kerma seen in the DS86 known dose group within the same distance interval to obtain an ad hoc dose. Accordingly, within the DS86 known dose group, the mean total kerma as well as its gamma and neutron components were computed for the two cities separately. To obtain the gonadal dose we have used the average transmission factors seen among members of the Life Span Study sample (Shimizu et al., 1987). These 69,706 individuals including the ad hoc dose group are here called the DS86 sample extended (Table 5.2). Neither formal nor ad hoc doses could be assigned to the parents of 367 pregnancy terminations; these have been excluded in all subsequent UPO analyses (Otake et al., 1990).

The same rule used in assigning ad hoc doses to parents with children in the initial clinical study who had a T65DR dose but no directly computed DS86 dose has been employed in the estimation of gonadal doses for those

Table 5.2. Distribution of joint parental DS86 gonadal dose equivalents per sievert (RBE = 20) in DS86 sample and DS86 sample extended by dose category.

Dose Category (Sv)	DS86 Sample			DS86 Sample Extended		
	Subjects	Gamma	Neutron	Subjects	Gamma	Neutron
2.50 +	164	3.00	0.70	285	2.96	0.69
1.00–2.49	702	1.32	0.18	916	1.31	0.18
0.50–0.99	1074	0.65	0.06	1404	0.65	0.06
0.10–0.49	2669	0.23	0.01	3318	0.23	0.01
0.01–0.09	5485	0.03	0.00	6486	0.03	0.00
< 0.01	45209	0.00	0.00	57297	0.00	0.00
Total	55303	0.05	0.00	69706	0.06	0.01

parents with a T65DR dose but no directly computed DS86 dose in all other genetic studies.

Statistical models

To estimate the increments of change in these indicators with changing dose, two dose-response models have been fitted to the occurrence of the indicators treated as binary response variables. The individual observations, y_i ($i = 1, \ldots, n$), are assumed to be independently and binomially distributed. Under these assumptions, the joint likelihood function of an independent binary array is

$$L_i = \prod_{i=1}^{n} P_i^{y_i}(1 - P_i)^{1-y_i} \qquad [5.1]$$

where y_i is 1 for an individual with any abnormality and 0 for others.

First, a standard linear regression model was fitted expressed by

$$P_i = \text{Constant} + \sum_{j=1}^{k} \beta_j X_{ij} + \beta\text{Dose} \qquad [5.2]$$

where P_i is the expected frequency of the event of interest in the ith individual, X_{ij} the background variables, and dose.

Second, mutations were assumed to follow a "one-hit" radio-biological model represented by

$$P_i = 1 - \exp - \left(\text{Constant} + \sum_{j=1}^{k} \beta_j X_{ij} + \beta\text{Dose} \right). \qquad [5.3]$$

If the powers higher than the first in the algebraic expansion of the logarithm in the one-hit model are negligible, the one-hit model is equivalent to a standard linear model. Because the estimates under the two models are very close, suggesting that the higher powers are negligible, only the estimates of the parameters for the standard linear model are given here. The reader interested in the actual estimates under the one-hit model is referred to Otake et al., 1990.

Insofar as a fully genetically determined indicator is concerned, the doubling dose (D_g) is defined by the ratio of the spontaneous rate of mutation, a, to the induced rate per unit dose, b (Neel, et al., 1990). However, if the indicator is not fully genetic, as is certainly true of UPOs, then the intercept, a, of the dose response relationship must be adjusted to reflect that component arising through spontaneous mutation in the previous generation. The magnitude of this component is imperfectly known. In Schull et al. (1981a,b), it was assumed to be 0.0025 (1/400), but more recent estimates suggest a better value lies in the range 0.0033 to 0.0053 (Neel et al., 1990). We use here the value 0.0043, or the midpoint of the suggested range.

The 95% lower limit of D_g in a one sided test is given by

$$LD_g = 0.0043/(b + (1.645 \times S_b)) \qquad [5.4]$$

where S_b denotes the standard error of the regression parameter b. In a previous report (Schull et al., 1981a, b), the estimate of the doubling dose was divided by 2 on the assumption of equal mutation contribution for both sexes. However, Neel et al. (1990) have argued that this was inappropriate and estimate the gametic doubling dose without dividing by 2. We follow the latter argument here.

In a 2×2 contingency table, estimates of the doubling dose and its 95% lower limit are derived from

$$D_g = P_c/(P_e - P_c)/M \text{ and } LD_g = D_g - (1.645 \times S_g) \qquad [5.5]$$

where M denotes the average joint (or combined) parental gonadal dose based on a neutron RBE of 20, P_e the mutation rate in children of proximally exposed parents, P_c the mutation rate in children of distally exposed parents, and S_g is the approximate standard error of the doubling dose.

Results

Clinical findings

The increments of change in the frequency of abnormality regressed on joint (combined) parental gonadal dose, assuming a neutron RBE of 20, were 0.42%/Sv for UPO, the values for the components of UPO being 0.10%/Sv for malformations, 0.15%/Sv for stillbirths, and 0.24%/Sv for early mortality in the DS86 sample; whereas in the DS86 extended sample, they were 0.26%/Sv for UPO, 0.10%/Sv for malformations, 0.09%/Sv for stillbirths and 0.13%/Sv for early mortality (Table 5.3). In a previous paper (Schull et al., 1981a,b), this increment, based on a neutron RBE of 5 and the T65DR joint parental gonadal dose, was 0.18%/Sv for UPO. The frequency of UPO for DS86 joint parental gonadal doses (RBE = 20) is 1.4 fold higher than that for T65DR gonadal doses (RBE = 5). The estimate of the doubling dose for UPO, and its 95% lower limit in the DS86 sample are 1.02 Sv and 0.43 Sv, respectively. The estimates derived from the DS86 extended sample are all slightly higher than these values, but not significantly so.

Table 5.3. Increments of decrements of change in the frequency of abnormalities per sievert of joint parental gonadal dose based on a neutron RBE of 20.

	DS86 Sample		DS86 Sample Extended	
Item	Reg estimate (Sv)	SE (Sv)	Reg estimate (Sv)	SE (Sv)
UPO	0.00422	0.00343	0.00264	0.00277
Malformations	0.00099	0.00184	0.00101	0.00154
Stillbirths	0.00151	0.00199	0.00092	0.00163
Early mortality	0.00237	0.00233	0.00128	0.00185

Note: *SE* denotes standard error of regression coefficient.

F_1 mortality study

The estimates of the regression coefficients obtained from the three different periods do not change much with increments ranging from 0.085%/Sv to 0.073%/Sv for all causes of death, assuming a neutron RBE of 5 in the T65DR gonadal dose, to 0.076%/Sv for F_1 morality, when the DS86 doses (RBE = 20) are used. In the F_1 mortality study, the estimates of the doubling dose and its 95% lower limit are 5.89 Sv and 1.43 Sv in the 1946 to 1980 cohort based on the T65DR dosimetry with an RBE of 5, respectively, and 5.66 Sv and 1.31 Sv in the 1946 to 1985 cohort based on the DS86 dosimetry with a neutron RBE of 20.

Cytogenetic and biochemical studies

The cytogenetic study was initiated in 1967, expanded in 1976 as a part of the Genetics Platform Research Program at the Radiation Effects Research Foundation, and was continued until 1985 (Awa et al., 1988a). The frequency of sex-chromosome aneuploids in the children of proximally exposed parents was found to be 16/5726 (0.28%), and that in children of distally exposed parents to be 12/5058 (0.24%) in the 1967 to 1979 study (Schull et al., 1981b). The regression estimates of sex-chromosome aneuploids, assuming an RBE of 20 in the DS86 gonadal dose, give a positive genetic risk of 0.044%/Sv ± 0.069%/Sv in the 1967 to 1985 study (Table 5.4). The positive regressions term is due to a clustering of affected children at the higher parental exposures.

Table 5.4. Cytogenetic and biochemical variant studies.

		Proximally Exposed		Distally Exposed	
Year	RBE (Dose)	Examined	Abnormal (%)	Examined	Abnormal (%)
Sex Chromosome Aneuploids					
1967–79[a]	5 (T65DR)	5,762	16 (0.278)	5,058	12 (0.237)
1967–85	20 (DS86)	8,322	19 (0.228)	7,976	24 (0.301)
Biochemical Variants					
1972–79	5 (T65DR)	289,868	1 (0.34 × 10)	208,196	0
1972–85	20 (DS86)	667,404	3 (0.45 × 10)	466,881	3(0.64 × 10)

[a]The sex chromosome aneuploids are based on the results in *Science* (Schull et al., 1981b).

The estimates of the doubling dose and its 95% lower limit for sex-chromosome aneuploids are 5.04 Sv and 1.76 Sv, respectively, from a 2 × 2 contingency table for the T65DR mean dose with an RBE of 5 in the former period, and 5.68 Sv and 1.58 Sv from the regression value for DS86 dose with an RBE of 20 in the later study period.

The frequency of electrophoretic or activity variants in the children of proximally and distally exposed parents has recently been reported to be 4/667,404 or 0.60×10^{-5} in the former group and 3/466,881 or 0.64×10^{-5} in the latter (Table 5.4). The 95% confidence intervals of these estimates are $0.2–1.5 \times 10^{-5}$ and $0.1–1.9 \times 10^{-5}$, respectively.

Sex ratio

In the early years of the genetics study, specifically 1948 to 1962, the expectation for sex ratio effects was set by the simple theory of sex-linked inheritance that then prevailed. However, the discovery of Lyonization and sex chromosome aneuploidy has greately complicated the interpretation of sex ratio changes, and it is arguable whether such changes can now be interpreted in a meaningful way. Accordingly, we merely recapitulate the earlier findings briefly and make no effort to compute a doubling dose.

As regards the frequency of male births when the father was not exposed but the mother was, no clear trends with maternal T65DR doses are disclosed by these data. Although the frequency of male births appeared to diminish with increasing maternal exposure in the years 1948 to 1953, male births actually increased in frequency with maternal exposure in 1954 to 1962 if the father was not exposed. This relative excess of male births in the children of exposed mothers is, of course, inconsistent with the argument that the sex ratio should decline with maternal exposure because of the induction of sex-linked lethal mutations.

DISCUSSION

Five genetic indicators of possible mutational damage arising from exposure to atomic bomb radiation have been examined, and previous publications and recent reports reviewed (Table 5.5). Any effort to estimate the doubling dose of radiation for mutation in humans must of necessity involve a series of debatable and sometimes tenuous assumptions. The validity of these assumptions can change with time, and hence the estimation of the doubling dose must be seen as a dynamic or continuing process. Moreover, because the mortality surveillance continues, further data will become available and these data, when coupled with future molecular biological studies, will undoubtedly result in some revision of the doubling dose estimates presented here.

Weighted average estimates of the doubling dose and the 95% lower limit for the three indicators of UPO, F_1 mortality and sex-chromosome aneu-

Table 5.5. Gametic doubling dose estimates per sievert (RBE = 20) and the lower 95% bound.

Year at Risk	RBE (Dose)	Gametic Doubling Dose (Sv)	95% Lower Bound (Sv)
UPO			
1948–53	5 (T65DR)	2.36	0.60
1948–53	20 (DS86)	1.02	0.43
F_1 Mortality Cohort			
1946–71	5 (T65DR)	5.06	0.99
1946–80	5 (T65DR)	5.89	1.43
1946–85	20 (DS86)	5.66	1.31
Sex Chromosome Aneuploids			
1967–79	5 (T65DR)	5.04[a]	1.76
1967–85	20(DS86)	5.68	1.58

[a]Note that the gametic doubling dose in a 2 × 2 tables was estimated by $D_g = (P_c/P_e - P_c)/(\text{Mean Dose})$.

ploids by joint parental DS86 gonadal dose per sievert with a neutron RBE of 20 become

$$D_g = (1.02\,W_1 + 5.66\,W_2 + 5.68\,W_3)/(W_1 + W_2 + W_3) = 3.55\,\text{Sv} \quad [5.6]$$

and

$$LD_g = (0.43\,W_1 + 1.31\,W_2 + 1.58\,W_3)/(W_1 + W_2 + W_3) = 0.94\,\text{Sv} \quad (5.7)$$

where the weights $W_i (i = 1, 2, 3)$ are based on the size of the sample from which each estimate was derived, i.e., $W_1 = 69{,}704$, $W_2 = 67{,}202$, and $W_3 = 16{,}298$.

A previous treatment of the doubling dose (Schull et al., 1981a, b) resulted in an arithmetic average gametic estimate of 3.12 Sv and a 95% lower limit of 0.90 Sv when the joint parental T65DR gonadal dose with a neutron RBE of 5 was used. Using weight $W_i = 1$ for joint parental DS86 gonadal dose with neutron RBE of 20, we now obtain 4.12 Sv for the estimate of the doubling dose and 1.11 Sv for its 95% lower limit. Another recent estimate of the doubling dose, incorporating two additional indicators and employing a somewhat different statistical approach, resulted in an estimate of between 1.7 and 2.2 Sv (Neel et al., 1990).

ACKNOWLEDGMENTS

The author wishes to express his appreciation to Drs James V. Neel (Department of Human Genetics, University of Michigan) and William J. Schull (formerly Permanent Director, RERF, and the Genetics Centers, Graduate School of Biomedical Sciences, University of Texas Health Science Center, Houston) for their valuable suggestions and comments, and to extend his thanks to Drs A.A. Awa, T. Honda, C. Satoh, and Y. Yoshimoto for their Contributions to the genetic studies upon which this presentation is based.

REFERENCES

Awa AA, Honda T, Neriishi S, Shimba H, Ohtaki K, Nakano M, Kodama Y, Itoh M, Hamilton HB (1988a) Cytogenetic study of the offspring of atomic bomb survivors, Hiroshima and Nagasaki. RERF Technical Report 21–88.

Awa AA, Ohtaki K, Itoh M, Honda T, Preston DL, McConney E (1988b) New dosimetry at Hiroshima and Nagasaki and its implications for risk estimates. Chromosome aberration data for A-bomb dosimetry reassessment. Proceedings No. 9, 185–202, NCRPM.

Committee of the Biological Effects of Ionizing Radiation (1990) *Health Effects of Exposure to Low Levels of Ionizing Radiation* (BEIR V). Washington DC, National Academy Press.

Kerr GD (1979) Organ dose estimates for the Japanese atomic bomb survivors. *Health Phys* 37:487–508.

Kerr GD, Salomon DD (1976) The epicenter of the Nagasaki weapon: a reanalysis of available data with recommended value. ORNL-TM-5139, Oak Ridge National Laboratory.

Milton RC, Shohoji T (1968) Tentative 1965 radiation dose estimation for atomic bomb survivors, Hiroshima and Nagasaki. ABCC Technical Report 1–68.

Nakamura N, Akiyama M, Kyoizumi S, Langlois, RG, Bigbee WL, Jensen RH, Bean MA (1987) Frequency of somatic cell mutations at the glycophorin A locus in erythrocytes of atomic bomb survivors. RERF Technical Report 1–87.

Neel JV, Lewis SE (1990) The comparative radiation genetics of humans and mice. *Annual Review of Genetics* 24:327–362.

Neel JV, Satoh C, Goriki K, Asakawa J, Fujita M, Takahashi N, Kageoka T, Hazama R (1988) Search for mutations altering protein charge and/or function in children of atomic bomb survivors: final report. *Am J Hum Genet* 42:663–676 (see also, RERF Technical Report 10–87, 1987).

Neel JV, Schull WJ (1956) *The effects of exposure to the atomic bombs on pregnancy termination in Hiroshima and Nagasaki.* Washington DC, NAS-NRC Publication No. 461.

Neel JV, Schull WJ, Awa AA, Satoh C, Kato H, Otake M, Yoshimoto Y (1990) The children of parents exposed to atomic bombs: estimates of the genetic doubling dose of radiation for humans. *Am J Hum Genet* 46:1053–1072.

Otake M, Schull WJ, Neel JV (1990) The effects of parental exposure to the atomic bombings of Hiroshima and Nagasaki on congenital malformations, stillbirths and early mortality among their children: a renanalysis. *Radiat Res* 122:1–11 (see also RERF Technical Report 13–89, 1989).

Preston DL, Pierce DA (1988) The effect of changes in dosimetry on cancer mortality risk estimates in the atomic bomb survivors. *Radiat Res* 114:437–466 (see also RERF Technical Report 9–87, 1987).

Roesch WC (1987) *US–Japan joint reassessment of atomic bomb radiation dosimetry in Hiroshima and Nagasaki. Final Report.* Vols. 1–2. Hiroshima, Radiation Effects Research Foundation.

Schull WJ, Neel JV, Otake M, Awa AA, Satoh C, Hamilton HB (1981a) *Hiroshima and Nagasaki: three and a half decades of genetic screening.* Tokyo, Proc III Int Conf Environmental Mutations.

Schull WJ, Otake M, Neel JV (1981b) Genetic effects of the atomic bombs: a reappraisal. *Science* 213:1220–1227 (see also RERF Technical Report 7–81).

Shimizu Y, Kato H, Schull WJ, Preston DL, Fujita S, Pierce DA (1987) Life Span Study 11. Part 1. Comparison of risk coefficients for site-specific cancer mortality based on the DS86 and T65DR shielded kerma and organ doses. RERF Technical Report 12–87.

UNSCEAR (1982) *Ionizing Radiation: Sources and Biological effects.* New York, United Nations.

UNSCEAR (1986) *Genetics and Somatic Effects of Ionizing Radiation.* New York, United Nations.

Yoshimoto Y, Neel JV, Schull WJ, Kato H, Soda M, Eto R, Mabuchi K (1990) The frequency of malignant tumors during the first two decades of life in the offspring of atomic bomb survivors. *Am J Hum Genet* 46:1041–52.

Yoshimoto Y, Schull WJ, Kato H, Neel JV (1991) Mortality among the offspring (F_1) of atomic bomb survivors, 1946–85. RERF Technical Report 1–91.

6

Genotype-by-Environment Interaction: It's a fact of life

ERIC BOERWINKLE
D. MICHAEL HALLMAN

It is a developing paradigm in medical practice and research that both genetic and environmental factors are key contributors to disease etiology and pathophysiology (Braunwald et al., 1987). Consideration of both contributing components, as well as other factors, is necessary for predicting those individuals at increased risk for disease and treating those already afflicted. However, simultaneous recognition of both is too often only rhetorical, with research efforts pursuing either genetic factors or environmental factors alone without consideration of the other. Human geneticists traditionally focus on the role of genes and familial factors influencing risk of disease, either ignoring environmental factors such as smoking or diet, or treating them as concomitants whose effects are to be removed by linear regression. Epidemiologists, on the other hand, describe in detail the contribution of environmental factors to risk of disease in the population, ignoring the role of genetic factors leading to inter-individual variation in response to environmental perturbations. It is our thesis that simultaneous consideration of both genes and environment, and their interactions, is necessary to appreciate their true contribution to the disease process and their utility in disease prevention and treatment. In this Chapter, we attempt to develop this thesis and present useful results from our own studies of genotype-by-environment and genotype-by-genotype interactions.

The chronic diseases of later life such as hypertension, cancer, diabetes, and coronary heart disease (CHD) are common in all Westernized populations. These diseases are the result of the interaction of numerous environmental and genetic factors. Disease liability is not attributable to a single factor such as a solitary foodstuff or a single genetic alteration. Rather, it is the result of interactions among numerous factors, each making a relatively small contribution to overall liability. As a result, the contribution to disease risk of variation at any single gene is likely small when compared to all other risk factors combined. Also, the genes affecting disease risk do not act alone;

they are constantly interacting with other genes and with environmental factors. Therefore, the effect of a gene measured in one environment may be very different than its effect in another. The common chronic diseases are also characterized by considerable pathophysiological and genetic heterogeneity (Sing et al., 1985).

The use of genetic information in predicting susceptibility to the common chronic diseases presents several unique problems and opportunities. Humankind is witnessing spectacular advances in the prevention and understanding of several single-gene disorders such as cystic fibrosis (Lemna et al., 1990; Riordan et al., 1989). It would be naive to suppose that we can simply apply the lessons established for using genetic information for the prevention of these rare inborn errors of metabolism to the common chronic diseases. Multifactorial etiologies will always confound prediction. As a direct consequence of the aforementioned heretogeneity, no single genetic defect will cause all or even a majority of cases. However, because of their immutability throughout life and from generation to generation, genes potentially represent useful predictors very early in life.

The inevitable interactions between genotypes and environments also will confound prediction of the common chronic diseases. Multiple genetic and environmental factors interacting will affect the occurrence of disease or the timing of onset. Once a genetic etiology has been assigned to a disease, however, there is a tendency to oversimplify the relationship between genotype and disease when making statements about risk or making predictions. For example, many people potentially afflicted with favism are not even aware that they carry the mutations (Luzzatto and Mehta, 1989). However, in this case there is no hesitation in predicting that an individual with the defect will show the disease if appropriately exposed. Similarly, sickle cell anemia is caused by a defect in a single gene; however, there is considerable heterogeneity in the onset of disease and its severity (Odenheimer et al., 1983). Where multiple genes and multiple environmental interactions are involved in disease etiology, definitive predictions of disease onset in an individual have little applicability. Clearly, genetic defects may predispose to disease (e.g., LDL-receptor mutations leading to familial hypercholesterolemia); however, the effect is subject to modification and amelioration by other genetic or environmental factors (e.g., Utermann et al., 1989). Having the genetic defect is not equivalent to having the disease. As examples of how complex interactions among genetic and environmental variation may affect risk factors for the common chronic diseases, we will consider the role of genes and environments in normal lipid variation.

THE ROLE OF GENES IN NORMAL LIPID VARIATION

Two basic lines of evidence document the role of genes in affecting lipid levels. First, several inborn errors of lipid metabolism have been reported and characterized. For example, Brown and Goldstein (1986, 1989) have described lesions in the low density lipoprotein (LDL) receptor gene that lead to elevated LDL levels and premature atherosclerosis. An insertion in the gene encoding lipoprotein lipase (LPL) causes a deficiency in LPL activity resulting in familial combined hyperlipidemia (Langlois et al., 1989). Brown et al. (1989) have reported that a mutation in the gene coding for cholesterol ester transfer protein leads to an increase in plasma HDL-cholesterol levels. The second line of evidence comes from quantitative genetic studies showing that a significant fraction of variability in lipid, lipoprotein, and apolipoprotein (apo) levels is attributable to polymorphic genetic variability. For example, Hamsten et al. (1986) estimate that the proportion of population variability in high density lipoprotein (HDL)-cholesterol and apo B levels attributable to genetic factors is 0.42 and 0.51, respectively. This value is 0.53 for total serum cholesterol levels (Hamsten et al., 1986) and 0.98 for plasma Lp(a) levels (Hasstedt and Williams, 1986). Of particular interest in the biometrical genetic studies of CHD risk factors has been the detection of statistical evidence for single unknown genes with large effects. Hasstedt et al. (1986) report the characteristics of such a "major gene" with a large effect on HDL-cholesterol (HDL-C) levels. Apo A-I is the major apolipoprotein in HDL, and Moll et al. (1989) report evidence supporting the existence of a gene with a large effect on plasma apo A-I levels. Evidence supporting the role of a single locus with a large effect on plasma lipoprotein(a) [Lp(a)] levels is well established (e.g. Hasstedt and Williams, 1986). Apolipoprotein (a), the unique protein component of Lp(a), is closely linked and homologous to the serum protease plasminogen (McLean et al., 1987; Murray et al., 1987). Linkage analyses using a protein polymorphism in plasminogen have determined that the major gene contributing to Lp(a) levels is likely the apolipoprotein(a) [apo(a)] structural gene (Drayna et al., 1989).

IDENTIFYING AND CHARACTERIZING POLYGENES: A MEASURED GENOTYPE APPROACH

Although traditional quantitative genetic studies indicate that genes are contributing, they do not yield information about the identity and role of specific gene loci. Knowledge concerning individual loci is necessary for determining which individuals carry specific mutations, and for studying the interaction between genes and environments. Advances in atherosclerosis research and molecular biology have identified numerous genes and gene products that may be contributing to cardiovascular disease. Plasma lipids do not circulate freely but rather are transported as lipoprotein particles containing specific proteins known as apolipoproteins. The lipoproteins, named by their density properties, are primarily formed at the intestinal wall or by the liver. During circulation, various enzymes alter the lipid and protein components of these particles. The lipoprotein particles may be taken up by cell surface receptors which recognize and bind the apo moieties. The genes encoding these apolipoproteins, enzymes, and receptors are candidates for those responsible for the underlying polygenetic and major gene effects discussed above.

The gene whose effects on normal lipid variation are best understood is apolipoprotein E. Apo E is a structural component of chylomicrons, VLDL lipoproteins, and a subset of HDL lipoproteins. Apo E is thought to have a major regulatory role in lipid metabolism as a ligand for apo E receptors. Human apo E is polymorphic (Utermann et al., 1977). In a sample of 223 unrelated individuals from Nancy, France, the relative frequencies of the ε2, ε3, and ε4 alleles were 0.13, 0.74, and 0.13, respectively (Boerwinkle et al., 1987). The average effects of the apo E alleles on total serum cholesterol levels were −0.52, 0.05, and 0.26 mmol/l for the ε2, ε3, and ε4 alleles, respectively. In this and most other populations (Davignon et al., 1988), the average effect of the ε2 allele is to significantly lower total cholesterol levels and the average effect of the ε4 allele is to significantly raise total cholesterol levels. Using family data, Boerwinkle and Sing (1987) directly estimated that the apo E polymorphism acounts for 12.5% of the overall polygenetic variance of total serum cholesterol levels. Boerwinkle and Utermann (1988) have investigated the effect of the apo E polymorphism on plasma apo E and total cholesterol levels in a sample of 563 blood bank donors from Marburg and Giessen, Germany. The average effects of the apo E alleles are shown in Table 6.1. The effects of the apo E polymorphism on plasma apo E levels were in the opposite direction from that for total cholesterol levels. On the basis of these and other results they formulate a model for the effect of the apo E polymorphism on lipid metabolism and predict that the effects of this gene will differ among populations differing in dietary fat intake. According to this model, differences among alleles in receptor binding, chylomicron clearance, and subsequent regulation of LDL receptors account for the observed effects of this polymorphism on total cholesterol levels and CHD risk (Boerwinkle and Utermann, 1988).

Laboratory methods have revealed a high degree of genetic variability in targeted candidate genes (e.g., Boerwinkle et al., 1988). Some of this genetic variability has a direct biochemical and physiological effect on the gene product and on lipid metabolism. An example of this type of effect is that of the apo E polymorphism on lipid metabolism, discussed previously. However, all variability does not directly affect the phenotype of interest. Marker loci with no effects may be associated in the population with genetic variability with direct physiological effects. Non-random association between alleles at two gene loci is linkage or gametic disequilibrium (Hedrick, 1987).

Table 6.1. Average effects of the apo E polymorphism.[a]

Phenotype	ε2	ε3	ε4	Variance[b]
Apolipoprotein E (mg/dL)	0.95	−0.04	−0.19	20.2%
Total cholesterol (mg/dL)	−14.2	−0.16	7.09	4.0%

[a] Boerwinkle and Utermann (1988). Table 6.1 is reproduced with permission from the *American Journal of Human Genetics* (1988; 42:104–112).

[b] Percentage of the total phenotypic variance attributable to the apo E polymorphism.

A marker with no direct effect on the phenotype may have a statistical association with lipid levels or a disease endpoint. The size of the association is a function of the size of the effect of the phenotypically important locus (PIL) and the disequilibrium between the two loci (Chakraborty et al., submitted). As the linkage disequilibrium decreases, the estimated contribution of the marker locus to the variance of the quantitative trait decreases until it is zero when the two loci are independent. The decrease can be described by the relationship given in equation [6.1],

$$\sigma^2_{\text{Marker}} = D^2 \cdot \sigma^2_{\text{PIL}} \qquad [6.1]$$

where σ^2_{Marker} is the variance of the quantitative trait, D is the correlation measure of linkage disequilibrium (Hendrick, 1987), and σ^2_{PIL} is the contribution of the phenotypically important locus to the variance of the quantitative trait.

Recently, there have been numerous reports of significant associations between restriction site polymorphisms and lipid levels. We do not review the results of these studies here, because several excellent reviews have been recently published (Cooper and Clayton, 1988; Humphries, 1988). Rather, we present results using one locus and discuss some of their ramifications. One such restriction site polymorphism is a silent mutation at codon 2488 of the apo B gene detectable by the restriction enzyme Xba I. Table 6.2 and 6.3 summarize results from several association studies using this polymorphism. The first report of a significant RFLP-CHD association was by Hegele et al. (1986), who found a higher frequency of the $X(-)$ allele in survivors of a myocardial infarction than in age- and sex-matched controls. Other studies have failed to detect this association or have found a significant association in the opposite direction (Talmud et al., 1987). Likewise, studies testing the equality of average total serum cholesterol levels among genotypes have reported conflicting results (Hegele et al., 1986; Talmud et al., 1987). The results summarized in Table 6.2 and 6.3 point out several trends worth noting. First, significant associations of DNA variation with risk factor levels and overt disease are being found and most of these associations are assumed to be attributable to disequilibrium between the RFLP and a nearby gene locus with direct physiological effects. When an association exists only because of linkage disequilibrium, the predictive value of a gene marker is

Table 6.2. Summary of three Apo B Xba I RFLP case-control studies.[a]

Reference	Disease	#cases/#controls	p_{case}	p_{control}	Significance
Hegele et al. (1986)	MI survivors	84/84	.64	.50	.01
Myant et al. (1989)	CAD(+)	124/146	.54	.47	NS
Talmud et al. (1987)	Type III	44/62	.32	.50	.01

[a]These studies were selected for review because each presented results from case-control and random samples.

Table 6.3. Summary of three Apo B Xba I RFLP association studies.[a] *A comparison of means among genotypes.*

Reference	Trait	*N*	*p*	Effect	Significance
Hegele et al. (1986)	Cholesterol	84	.50	X(−) < X(+)	NS
	LDL-C			X(−) < X(+)	NS
	Apo B			X(−) < X(+)	NS
Myant et al. (1989)	Cholesterol	146	.47	X(−) > X(+)	NS
	Triglycerides			X(−) > X(+)	NS
	LDL-C			X(−) = X(+)	NS
	Apo B			X(−) > X(+)	NS
	HDL-C			X(−) < X(+)	NS
	Apo AI			X(−) < X(+)	$p < .05$
Talmud et al. (1987)	Cholesterol	62	.50	X(−) < X(+)	$p < .05$
	Triglycerides			X(−) < X(+)	$p < .05$
	LDL-C			X(−) < X(+)	$p < .1$

[a] These studies were selected for review because each presented results from case-control and random samples.

less than that of the physiologically important gene mutation. Identifying gene variation associated with disease is not enough; we must work toward identifying gene variation with direct effects on the disease process. And second, observational studies of the same phenomenon may differ with respect to their conclusions. This is likely attributable to different populations of inference, stochastic variation in sampling, and different statistical methods.

GENETIC INTERACTIONS

The regulation of gene expression and the action of gene products are determined by the micro- and macro-environments of the organism and its cellular constituents. These environments are influenced by both the expression of other genes and by the ecology of the organism, such as its diet. Much has been discussed about the potential for, and effects of, genotype-by-environment interaction, but only a few studies have attempted to quantitate the role of genotype-by-environment interaction in humans (Eaves, 1984; Orr et al., 1981). Although these biometrical genetic studies are important because they provide evidence supporting interaction, they are complicated and generally do not reveal the biological nature of any significant interactions. Genotype-by-environment and genotype-by-genotype interaction studies are greatly enhanced by incorporation of measured genotype information. One example of this approach is given by studies of the apo E polymorphism.

Nestruck et al. (1987) have reported significant interaction between the apo E polymorphism and the cholesterol-lowering drug probucol. Probucol has both antioxidatant and lipid-lowering properties, and was previously shown to have a variable efficacy among patients (Cortese et al., 1982). Nestruck et al. (1987) report that probucol-treated type II hypercholesterolemic individuals with an ε4 allele showed a larger reduction in plasma cholesterol levels when compared with ε3/3 individuals. They hypothesize that the E4 isoform and probucol act synergistically to promote enhanced lipoprotein catabolism.

Gueguen et al. (1989) have investigated the interaction between the apo E polymorphism and longitudinal changes in several CHD risk factors in a sample of 158 nuclear families from Nancy, France. The estimated frequencies of apo $\varepsilon 2$, $\varepsilon 3$, and $\varepsilon 4$ alleles in this sample were .120, .764, and .116, respectively. There was no significant evidence for an effect of the apo E polymorphism on the longitudinal profile of any of the variables considered (Figure 6.1, Panel A). These result are consistent with other reports that have demonstrated consistency of apo E effects in both children and adults (Boerwinkle and Sing, 1987). There was a significant interaction between apo E effects and weight change on the longitudinal change of serum triglyceride and β-lipoprotein levels. Accompanying weight gain, individuals with an $\varepsilon 4$ allele showed a larger increase in triglyceride levels ($0.15 \pm .03$ mmol/1/kg) compared to individuals with no $\varepsilon 4$ allele (Figure 6.1, Panel B).

Tikkanen et al. (1990) and Xu et al. (in press) have investigated the effects of the apolipoprotein E polymorphism in a sample of adults from Finland who switched from their normal diet to a diet low in total and saturated fat. The frequencies of the apo E alleles in this sample, and their effects on lipid levels when the subjects were on their normal diets were similar to those previously reported in other studies from Finland (Ehnholm et al., 1986). In contrast, when the subjects switched to a diet lower in total and saturated fat, average total serum cholesterol levels were not statistically significantly different among apo E genotypes. These results support the proposition of Boerwinkle and Utermann (1988), that the response to changes in dietary cholesterol will differ among apo E types. A direct test of whether dietary response differs among apo E genotypes, however, did not detect significant differences (Xu et al., in press). The decrease in total serum cholesterol levels on the low total and saturated fat diet was not significantly different among apo E types (Figure 6.2).

Interaction effects are not limited to those between genes and the environment. Gene expression and action may also be influenced by other genes. Utermann et al. (1989) have reported a multiplicative interaction between the effects of the apo(a) protein polymorphism and heterozygous familial hypercholesterolemia. Lp(a) was originally detected as a variable antigenic property of LDL (Berg, 1963). Several studies have shown an association between elevated Lp(a) levels and atherosclerosis (e.g., Rhoads et al., 1986). Apo(a) exhibits a genetically determined size polymorphism with multiple alleles (Utermann et al., 1987). The underlying cause of this variation is hypothesized to be inter-individual variation in the number of Kringle 4 domains in the apo(a) molecule (McLean et al., 1987). Apo(a) size isoforms have a significant effect on plasma Lp(a) levels, explaining up to 40% of its population variation (Boerwinkle et al., 1989). Apo(a) isoforms also have a significant effect on Lp(a) levels in patients with heterozygous familial hypercholesterolemia (Utermann et al., 1989). In 102 patients with familial hypercholesterolemia, Lp(a) levels were found to be threefold higher than in

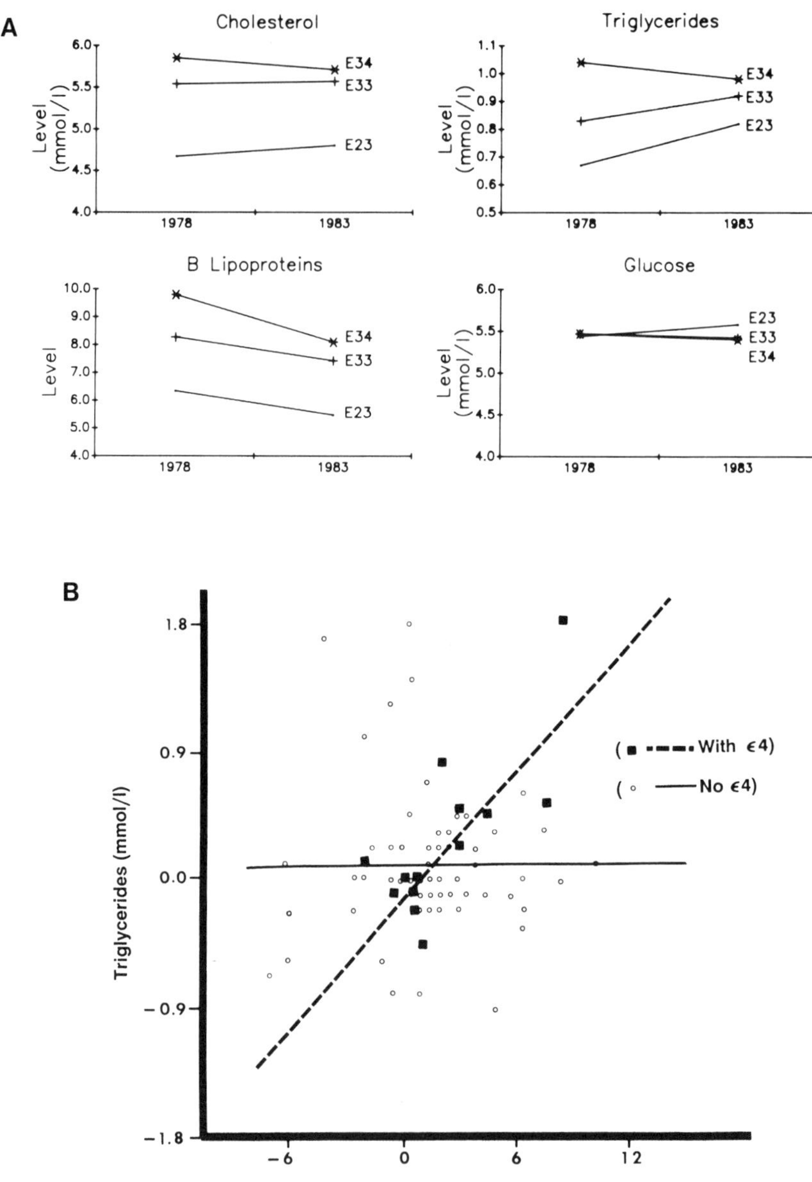

Figure 6.1. Longitudinal change over a 5-year period for plasma cholesterol, triglycerides, β-lipoprotein, and glucose levels for each of the three common apo E genotypes (A) and the longitudinal change of triglyceride levels as a function of weight change for individuals with one ε4 allele and for those with none (B). From Gueguen et al., 1989. Reprinted with permission from the *American Journal of Human Genetics.*

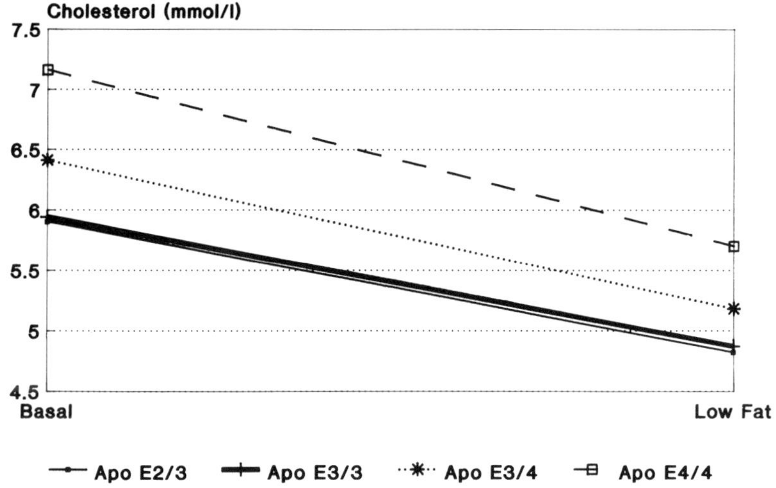

Figure 6.2. Average total serum cholesterol levels on each of two diets (a basal diet, and a diet low in total fat and saturated fat) for each of the common apo E genotypes in the Finnish dietary intervention study.

279 normal subjects. This increase could not be attributed to a high frequency of those Apo(a) types associated with elevated Lp(a) levels in the hypercholesterolemic subjects. The effects of variation in the Apo(a) and LDL-receptor genes on Lp(a) levels were multiplicative. As shown in Table 6.4, Lp(a) levels were elevated threefold for each of the apo(a) phenotypes examined. In addition, we have recently shown that plasma Lp(a) level is a significant predictor of CHD in patients with familial hypercholesterolemia (Seed et al., 1990). Apo(a) allele frequencies were significantly different between patients with and without CHD. Alleles associated with raised Lp(a) levels were more frequent in the diseased group, whereas alleles associated with lower Lp(a) levels were more frequent in the group without disease. Clearly, the effects

Table 6.4. Average Lp(a) levels for each of the common apo(a) types in a random sample and FH patients[a]

	Random Sample		FH patients	
Phenotypes	*n*	Mean (s.d.)	*n*	mean (s.d.)
S1	8	34.4 (20.7)	4	58.7 (32.0)
S2	50	24.5 (24.2)	24	62.8 (34.8)
S3	58	10.2 (9.7)	17	35.3 (31.6)
S4	89	5.7 (7.6)	28	23.4 (14.0)
null	16	0.4 (1.3)	5	2.0 (4.5)
S2/S3	18	27.9 (27.0)	5	45.4 (11.9)
S2/S4	8	34.1 (30.7)	8	61.4 (36.9)
S3/S4	27	8.8 (8.6)	5	40.2 (24.1)

[a]Utermann et al. (1989). Table 6.4 is reproduced with permission from the National Academy of Science (USA).

of variation in the LDL receptor and apo(a) genes combine in an interactive multiplicative manner affecting plasma Lp(a) levels and disease risk.

CONCLUSIONS

The above examples underscore the role of genotype-by-environment and genotype-by-genotype interaction in determining CHD risk factor variation. Knowledge of these interactions will impact the ability to implement the two basic aims of this research, prediction and etiology. Genotype-by-environment interaction will confound the ability to use gene information to identify individuals at increased risk for disease without detailed consideration of the environment. However, if individuals with a genetic susceptibility can be identified early, they may be counselled toward more healthy environments. On the other hand, by understanding more about these interactions, they can be exploited for the design and implementation of more efficacious intervention strategies. The light shed on chronic disease etiology by discovery of its genetic bases is likely to be of immediate value in suggesting effective interventions based on knowledge of genotype-by-environment interaction. Epidemiologists traditionally teach that the identification of a risk factor for a disease represents an opportunity for intervention, and possibly prevention (Lilienfeld and Lilienfeld, 1980; Mausner and Kramer, 1985). However, for the foreseeable future, we will not be intervening in or preventing disease by altering those genes influencing the common chronic diseases. The challenge for the future is to identify those genotypes that are susceptible to specific environmental influences, such as diet or drugs, so that appropriate actions can be taken.

ACKNOWLEDGMENTS

This work was carried out in part through support from HL-40613 to E.B. The authors would like to thank Professors Gerard Siest and Gerd Utermann for their patience, support, and encouragement during years of collaborative research. Most importantly, we would like to thank Professor William J. Schull for his mentorship and friendship to us and others at the Center for Demographic and Population Genetics in Houston. During dim times, he is an inspiration to us journeymen that science is an enjoyable and rewarding career of the highest calling.

REFERENCES

Berg K (1963) New serum type system in man: The Lp system. *Acta Pathol Microbiol Scand* 59:369–382.

Boerwinkle E, Menzel HJ, Kraft HG, Utermann G (1989) Genetics of the quantitative Lp(a) lipoportein trait. III. Contribution of Lp(a) glycoprotein phenotypes to normal lipid variation. *Hum Genet* 82:73–78.

Boerwinkle E, Sing CF (1987) The use of measured genotype information in the analysis of quantitative phenotypes in man. III. Simultaneous estimation of the frequencies and effects of the apolipoprotein E polymorphism and residual polygenetic effects on cholesterol, betalipoproteins and triglyceride levels. *Ann Hum Genet* 51:211–226.

Boerwinkle E, Utermann G (1988) Simultaneous effects of the apolipoprotein E polymorphism on apolipoprotein E, apolipoprotein B, and cholesterol metabolism. *Am J Hum Genet* 42:104–112.

Boerwinkle E, Visvikis S, Welsh D, Steinmetz J, Hanash SM, Sing CF (1987) The use of measured genotype information in the analysis of quantitative phenotypes in man. II. The role of the apolipoprotein E polymorphism in determining levels, variability, and covariability of cholesterol, betalipoprotein and triglycerides in a sample of unrelated individuals. *Am J Med Genet* 27:567–582.

Boerwinkle E, Xiong W, Fourest E, Chan L (1988) Rapid typing of tandemly repeated hypervariable loci by the polymerase chain reaction: application to the apolipoprotein B 3′ hypervariable region. *Proc Natl Acad Sci USA* 86:212–216.

Braunwald E, Isselbacher KJ, Petersdorf RG, Wilson JD, Martin JB, Fauci AS (1987) *Principles of Internal Medicine* (11th ed). New York, McGraw Hill Book Co.

Brown MS, Goldstein JL (1986) A receptor-mediated pathway for cholesterol homeostasis. *Science* 232:34–47.

Brown MS, Goldstein JL (1989) Familial hypercholesterolemia. In Scriver CR, Beaudet AL, Sly WS, Valle D (eds) *The Metabolic Basis of Inherited Disease.* New York, McGraw Hill, pp 1215–1250.

Brown ML, Inazu A, Hesler CB, Agellon LB, Mann C, Whitlock ME, Marcel YL, Milne RW, Koizumi J, Mabuchi H, Takeda, and Tall AR (1989) Molecular basis of lipid transfer protein deficiency in a family with increased high-density lipoproteins. *Nature* 342:448–451.

Chakraborty R, Boerwinkle E, Hanis CL (submitted) Effect of a marker locus on the variability of a quantitative trait. *Am J Hum Genet.*

Cooper DN, Clayton JF (1988) DNA polymorphisms and the study of diesease association. *Hum Genet* 78:299–312.

Cortese C, Marenah CB, Miller NE (1982) The effects of probucol on plasma lipoproteins in polygenic and familial hypercholesterolemia. *Atherosclerosis* 44:319–325.

Davignon J, Gregg R, Sing CF (1988) Apolipoprotein E polymorphism and atherosclerosis. *Arteriosclerosis* 8:1–21.

Drayna DT, Hegele RA, Hass PE, Emi M, Wu LL, Eaton DL, Lawn RM, Williams RR, White RL, Lalouel J-M (1989) Genetic linkage between lipoprotein(a) phenotype and a DNA polymorphism in the plasminogen gene. *Genomics* 3:230–236.

Eaves L (1984) The resolution of genotype × environment interaction is segregation analysis of nuclear families. *Genet Epid* 1:215–228.

Ehnholm C, Lukka T, Kuusi T, Nikkila E, Utermann G (1986) Apolipoprotein E polymorphism in Finnish population: gene frequencies and relation to lipid concentrations. *J. Lipid Res* 27:227–235.

Gueguen R, Visvikis S, Steinmetz J, Siest G, Boerwinkle E (1989) An analysis of genotype effects and their interactions by using the apolipoprotein E polymorphism and longitudinal data. *Am J Hum Genet* 45:793–802.

Hamsten A, Iselius L, Dahlen G, de Faire U (1986) Genetic and cultural inheritance of serum lipids, low and high density lipoprotein cholesterol and serum apolipoprotein A-I, A-II, and B. *Atherosclerosis.* 60:199–208.

Hasstedt SJ, Ash KO, Williams RR (1986) A reexamination of major locus hypotheses for high density lipoprotein cholesterol levels using 2170 persons screened in 55 Utah pedigrees. *Am J Med Genet* 24:57–66.

Hasstedt SJ, Williams RR (1986) Three alleles for quantitative Lp(a). *Genet Epid* 3:53–55.

Hegele RA, Huang L-S, Herbert PN, Conrad BB, Buring JE, Hennedens CH, Breslow JL (1986) Apolipoprotein B-gene DNA polymorphisms associated with myocardial infarction. *N Engl J Med* 315:1509–1515.

Hedrick PW (1987) Gametic disequilibrium measures: proceed with caution. *Genetics* 117:331–341.

Humphries SE (1988) DNA polymorphisms of the apolipoprotein genes—their use in the investigation of the genetic component of hyperlipidaemia and atherosclerosis. *Atherosclerosis* 72:89–108.

Langlios S, Deeb S, Brunzell JD, Kastelein JJ, Hayden MR (1989) A major insertion accounts for a significant proportion of mutations underlying human lipoprotein lipase deficiency. *Proc Natl Acad Sci USA* 86:948–952.

Lemna WK, Gerald MS, Feldman L, Bat-shema K, Fernbach SD, Zevkovich EP, O'Brien WE, Riordan JR, Collins FS, Tsui L-C, Beaudet AL (1990) Mutation analysis for heterozygote detection and the prenatal diagnosis of cystic fibrosis. *New Engl J Med* 322:291–296.

Lilienfeld AM, Lilienfeld DE (1980) *Foundations of Epidemiology* (2nd ed). New York, Oxford University Press.

Luzzatto L, Mehta A (1989) Glucose-6-phosphate dehydrogenase deficiency. In Scriver CR, Beaudet AL, Sly WS, Valle D (eds) *The Metabolic Basis of Inherited Disease*. New York, McGraw Hill, pp 2237–2265.

McLean JW, Tomlinson JE, Kuang W, Eaton DL, Chen EY, Fless GM, Scanu Am, Lawn RM (1987) cDNA sequence of human apolipoprotein(a) is homologous to plasminogen. *Nature* 330:132–137.

Mausner JS, Kramer S (1985) *Epidemiology: An Introductory Text* (2nd ed). Philadelphia, WB Saunders.

Moll PP, Michels VV, Weidman WH, Kottke BA (1989) Genetic determination of plasma apolipoprotein AI in a population-based sample. *Am J Hum Genet* 44:124–139.

Murray JC, Buetow KH, Donovan M, Hornung S, Motulsky AG, Desteche C, Dyer K, Swisshelm K, Anderson J, Giblett E, Sadler E, Eddy R, Shows TB (1987) Linkage disequilibrium of plasminogen polymorphisms and assignment of the gene to human chromosome 6q26–6q27. *Am J Hum Genet* 40:338–350.

Myant NB, Gallaghar J, Barbir M, Thompson GR, Wile D, Humphries SE (1989) Restriction fragment length polymorphisms in the apo B gene in relation to coronary artery disease. *Atherosclerosis* 77:193–201.

Nestruck AC, Bouthillier D, Sing CF, Davignon J (1987) Apolipoprotein E polymorphism and plasma cholesterol response to probucol. *Metabolism* 36:743–747.

Odenheimer DJ, Whitten CF, Rucknagel DL, Sarnaik SA, Sing CF (1983) Heterogeneity of sickle-cell anemia based on a profile of hematological variables. *Am J Hum Genet* 35:1224–1240.

Orr JD, Sing CF, Moll PP (1981) Analysis of genetic and environmental sources of variation in serum cholesterol in Tecumseh, Michigan. VI. A search for genotype by environment interaction. *J Chron Dis* 34:545–559.

Rhoads GG, Dahlen G, Berg K, Morton NE, Dannenberg AL (1986) Lp(a) lipoprotein as a risk factor for myocardial infarction. *JAMA* 256:2540–2544.

Riordan JR, Rommens JM, Kerem B, Alon KN, Rozmahel R, Grzelczak Z, Zielenski J, Lok S, Plavsic N, Chou J-L, Drumm ML, Iannuzzi MC, Collins FC, Tsui L-C (1989) Identification of the cystic fibrosis gene: cloning and characterization of complementary DNA. *Science* 245:1066–1073.

Seed M, Hoppichler F, Reaveley D, McCarthy S, Thompson GR, Boerwinkle E, Utermann G (1990) Relation of serum lipoprotein (a) concentration and apolipoprotein (a) phenotype to coronary heart disease in familial hypercholesterolaemia. *New Engl J Med* 322:1494–1499.

Sing CF, Boerwinkle E, Moll PP (1985) Strategies for elucidating the phenotypic and genetic heterogeneity of a chronic disease with a complex etiology. In Chakraborty R, Szathmary EJ (eds) *Diseases of Complex Etiology in Small Populations: Ethnic Differences and Research Approaches.* New York, Alan R Liss, Inc pp 39–66.

Talmud PJ, Barni N, Kessling AM, Carlsson P, Darnfors C, Bjursell G, Galton D, Wynn V, Kirk H, Hayden M, Humphries SE (1987) Apolipoprotein B gene variants are involved in the determination of serum cholesterol levels: A study in normo- and hyperlipidemic individuals. *Atherosclerosis* 67:81–89.

Tikkanen MJ, Huttunen JK, Ehnholm C, Pietinen P (1990) Apolipoprotein E_4 homozygosity predisposes to serum cholesterol elevation during a high-fat diet. *Arteriosclerosis* 10:285–288.

Utermann G, Hees M, Steinmetz A (1977) Polymorphism of apolipoprotein E and occurrence of dysbetalipoproteinemia in man. *Nature* 269:604–607.

Utermann G, Hoppichler F, Dieplinger H, Seed M, Thompson G, Boerwinkle E (1989) Defects in the low density lipoprotein receptor gene affect lipoprotein (a) levels: Multiplicative interaction of two gene loci associated with premature atherosclerosis. *Proc Natl Acad Sci USA* 86:4171–4174.

Utermann G, Menzel HJ, Kraft HG, Duba C, Kemmler HG, Seitz C (1987) Lp(a) glycoprotein phenotypes: inheritance and relationship to Lp(a) glycoprotein concentrations in plasma. *J Clin Invest* 80:458–465.

Xu C-F, Boerwinkle E, Tikkanen MJ, Huttunen JK, Humphries S, Talmud P (in press) Genetic variation at the apolipoprotein gene loci contribute to response of plasma lipids to dietary change. *Genet Epid.*

III

FAMILY VARIABILITY

7

Problems and Pitfalls in Linkage Mapping of Human Genetic Diseases: Illustrations from Autosomal Dominant Retinitis Pigmentosa (ADRP)

STEPHEN P. DAIGER
SUSAN H. BLANTON

During the past four decades, linkage mapping in humans has become one of the most effective methods for establishing the molecular basis of genetic diseases. By linkage mapping we mean linkage testing in families with the related goals of assigning genetic loci both to linkage groups and to chromosome regions. The first autosomal linkage in humans, between the Lutheran blood group and the ABO secretor locus, was reported in 1954 by J. Mohr. Shortly thereafter, linkage between Rh and elliptocytosis was reported and the first instance of genetic heterogeneity was demonstrated (Morton, 1956). Since then the power of linkage mapping in humans has expanded profoundly and the utility of linkage as a first step toward identification of disease loci and the cloning of mutant genes has been demonstrated repeatedly.

A vast array of DNA markers are physically mapped in the human genome. These include over 2000 restriction fragment length polymorphisms (RFLPs) (Human Gene Mapping 10.5 [1990]), more than 200 variable-number-of-tandem-repeat (VNTR) polymorphisms (Nakamura et al., 1987), and numerous DNA polymorphisms detectable by the polymerase chain reaction (PCR) such as short-tandem repeat (STR) variants (Daiger et al., 1990; Saiki et al., 1988; Weber and May, 1989). Obvious successes in the application of linkage mapping to the diagnosis and molecular characterization of human genetic diseases include localization of Huntington disease to chromosome band 4p16 (Gusella et al., 1983); mapping and cloning of the genes for Duchenne muscular dystrophy (Koenig et al., 1987), cystic fibrosis (Rommens et al., 1989) and neurofibromatosis (Wallace et al., 1990); and assignment of over 200 diseases genes to specific chromosomal sites (Human Gene Mapping 10.5 [1990]).

No one would dispute the great power of linkage mapping in unravelling human disease porcesses. Further, although the human genome initiative will eventually provide a complete catalog of human genes, linkage mapping in families with genetic diseases is currently the only practical method for determining which gene is defective in a particular family.

In spite of these successes, and the bright future promise of linkage mapping, a number of problems and potential pitfalls exist for investigators using this methodology. Problems and pitfalls include (1) the complexity of human genetic diseases: incomplete penetrance, variable expressivity, and clinical/genetic heterogeneity; (2) technical considerations in marker testing: selection of informative but unambiguous markers and means for expediting laboratory procedures; and (3) computational difficulties: appropriate choice of genetic and clinical parameters, of computer programs and procedures, and of methods for enhancing computational efficiency.

For the sake of illustrating these difficulties, and possible solutions, we have chosen examples from our current research on autosomal dominant retinitis pigmentosa (ADRP). Although the focus is necessarily narrow, limited to a dominant, clearly Mendelian disease, we have grappled with several general problems and stumbled over a few pitfalls. Therefore, our hope is that these comments will find broader application than to just within the community of investigators interested in retinitis pigmentosa.

BACKGROUND ON *ADRP*

Retinitis pigmentosa (RP) is a set of genetic diseases all of which manifest as the progressive degeneration of rod cells in the retina, accompanied by pigment deposition and culminating in the loss of retinal function and blindness. (See Heckenlively, 1988 for an extensive review.) Over 100,000 Americans are affected with these diseases. Symptoms include initial night blindness, characteristic retinal and electroretinographic (ERG) changes, loss of visual acuity, and eventual blindness. Initial symptoms can manifest as early as the first decade of life or as late as the fourth decade, and profound visual impairment can occur as early as in the teens or as late as in the 60s or 70s. That is, RP is very heterogeneous clinically, and difficulties regarding penetrance and expressivity are to be expected.

RP is also very heterogeneous genetically. There are X-linked forms, autosomal recessive forms, autosomal dominant forms, and syndromic forms including Usher syndrome (US), which involves RP in conjunction with congenital deafness (Heckenlively, 1988). Within each genetic category there are also clinical subcategories. As an example, for ADRP there is a clinical type 1 with early onset and rapid progression, a type 2 with later onset and slower progression, and further types such as "sectoral RP" in which only part of the retina is affected. These clinical distinctions appear to be consistent within families. However, in type 2 ADRP some individuals may have early onset and rapid progression of disease, though most do not.

In type 1 families, though, all affected members are severely affected. Thus even before linkage testing was initiated, considerable clinical and genetic heterogeneity was anticipated.

Linkage mapping of the various forms of RP has validated all of these expectations. Although they are not distinguishable clinically, there are at least two X-linked forms of RP, one mapping to Xp11 and the other to Xp21 (Ott et al., 1990). These also appear to be two genetic forms of US, distinguished by severe deafness (type 1) or milder deafness (type 2); type 1 US maps to 1q and type 2 US can be excluded from 1q (Kimberling et al., 1990; Smith et al., 1992).

Finally, and most directly relevant to this discussion, genetic heterogeneity has also been demonstrated in autosomal dominant RP based on linkage mapping and gene cloning. The ADRP locus in some families, both types 1 and 2, is tightly linked to an anonymous marker on 3q, D3S47, which in turn is tightly linked to rhodopsin, the rod phototransduction protein (McWilliams et al., 1989). Further, some D3S47-linked families have been shown to have specific rhodopsin mutations (Dryja et al., 1990). Although different rhodopsin mutations appear to be associated with different clinical phenotypes, one specific, frequent mutation, at rhodopsin codon 23, is usually associated with late onset or, possibly, sectoral RP (Heckenlively et al., 1990). To date at least 30 distinct rhodopsin mutations have been reported (Dryja et al., 1991; Sung et al., 1991). Rhodopsin mutations are expected to be the cause of disease in approximately one-third of ADRP families.

However, in spite of the importance of rhodopsin mutations in explaining many cases of ADRP, other forms are not rhodopsin-based. In particular, at least two ADRP families have been reported with significant but more distant linkage (8% recombination) to D3S47 (Gal et al., 1990) and other families are clearly excluded from linkage to markers on 3q (Blanton et al., 1990; Inglehearn et al., 1990).

Of the ADRP families excluded from linkage to 3q, one in particular, UCLA-RP01, has been the subject of extensive investigation by our research group and will serve as the major "case study" in the following discussion. UCLA-RP01 is a Kentucky family with late onset, type 2 ADRP (Daiger et al., 1989; Heckenlively et al., 1982). The portion of the family we are investigating consists of a large, extended pedigree with over eight affected generations with 50 living, affected members and with over 120 individuals available for linkage testing (see Figure 7.1) (Daiger 1988; Daiger et al., 1989). We have DNAs from all living members of the pedigree and have tested over 45 genetic markers for linkage (Daiger et al., 1990).

In ongoing research we excluded the ADRP locus in UCLA-RP01 from more than one-half of the human genome (Daiger et al., 1990) and, recently, we demonstrated statistically significant linkage (lod score over 12) to several markers in the pericentric region of chromosome 8 (Blanton et al., 1991). Further, recent evidence has shown that mutations in two retinal-specific genes, peripherin on 6p and ROM1 on 11p, can also cause ADRP (Farrar et al., 1991; Kajiwara et al., 1991; McInnis, 1991). Thus at least five gene loci

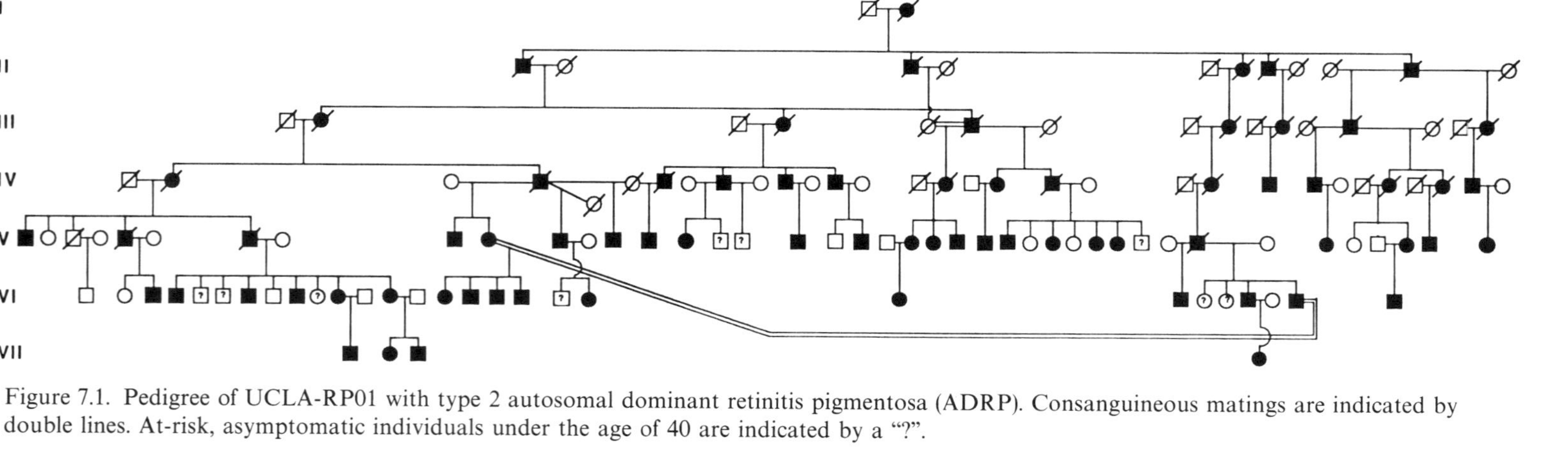

Figure 7.1. Pedigree of UCLA-RP01 with type 2 autosomal dominant retinitis pigmentosa (ADRP). Consanguineous matings are indicated by double lines. At-risk, asymptomatic individuals under the age of 40 are indicated by a "?".

can cause ADRP: rhodopsin on 3q, probably a second locus on 3q, peripherin on 6p, an as-yet-unknown gene on 8q, and ROM1 on 11p.

What follows is limited to a summary of some of the problems and pitfalls we have encountered in our studies of UCLA-RP01 (which led us to chromosome 8 linkage), with at least a few suggested solutions.

PROBLEMS IN CLINICAL DIAGNOSIS OF UCLA-RP01

It is a truism, or it should be a truism, that careful clinical characterization is the *sine qua non* of successful linkage studies. Because UCLA-RP01 is geographically centralized and has been the subject of genetic studies for over 15 years (Daiger et al., 1989; Heckenlively et al., 1982; Spence et al., 1977a,b), clinical characterization has been excellent. Specifically, many at-risk family members have been followed clinically throughout this period and many have been administered the complete clinical examination recommended for RP families (Marmor et al., 1983). An additional advantage of this family is that an extensive genealogy has been prepared by family members including, where known, details on blindness and RP (Breeding, 1982). Nonetheless, we have had difficulties with both pedigree reconstruction and clinical characterization. The pedigree we use for linkage analysis has been constructed largely from interviews of senior family members and subsequently checked in the genealogy manual. In such a geographically and ethnically circumscribed family, though, surnames recur frequently and there are several pairs of individuals with the same names. On more than one occasion this has led us to assign a deceased individual to the wrong generation. The solution to this problem, of course, is constant checking and re-checking of the pedigree. Fortunately this is largely a problem unique to large, extended families with dominant, late-onset disease. Because we have tested over 75 markers without evidence of parental exclusion, we are confident the current pedigree is correct.

The first publications on this family described it as typical type 1, i.e., early onset, ADRP (Heckenlively et al., 1982; Spence et al., 1977a,b), whereas our more recent publications have classified the family as type 2, i.e., late onset (Daiger, 1988; Daiger et al., 1989). In the initial clinical screening of the family in the 1970s, numerous affected individuals with severe visual impairment in their 40s were observed, typical of type I ADRP. However, over the subsequent decade, many at-risk individuals were re-tested and extensive clinical examinations were conducted in RP research centers. As a result, more and more family members were found with late onset, milder disease. (The terms "type 1" and "type 2" refer to families and not individuals, of course, and are based on clinical evaluation of all available at-risk individuals.) Thus the first impression, with ascertainment bias toward more severely affected individuals, was superseded with documented evidence of numerous cases of less severe disease.

At present the genetic significance of the clinical subcategories of ADRP is uncertain, except to suggest that in the presence of such obvious clinical heterogeneity the summing of linkage results between families is risky. For linkage purposes the change in clinical type of UCLA-RP01 was not substantial since the pedigree is large enough to be highly informative without combining results from other families (Daiger et al., 1990). However, this change has served as a warning to us that detailed clinical characterization is essential and that problems of late onset, variable expressivity, and incomplete penetrance can be profound.

Finally, a number of methods for dealing with delayed age-of-onset in linkage analysis have been proposed (Ott, 1985). Because patient reporting of age-of-onset is often inaccurate and because it would take decades to follow a sufficiently large cohort of at-risk individuals to determine age-of-onset empirically, we chose to base our onset corrections on informed clinical impressions. The clinicians familiar with UCLA-RP01 have concluded that individuals who are at-risk but apparently unaffected (based on clinical examination) fall into one of four categories: those under 30 are uncertain hence their genetic status is unknown, those 30 to 39 have a 20% chance of possessing an as-yet unexpressed RP gene, those 40 to 49 have a 10% chance, and those over 49 do not carry the gene. We have successfully implemented this scheme in both LIPED (Ott, 1985) and LINKAGE (Blanton et al., 1991; Daiger et al., 1990; Lathrop et al., 1984).

PROBLEMS IN SELECTION AND TESTING OF DNA MARKERS IN UCLA-RP01

Ideally, linkage markers should be optimally placed so that multipoint methods may be used, should be highly informative, and should be easy to type. To meet the goal of multipoint analysis, we often select markers that flank previously tested markers or which complete linkage groups spanning whole chromosomes or chromosome regions. An example is the exclusion of the ADRP locus in UCLA-RP01 from most of chromosome 4 (Daiger et al., 1989). The problem with this approach has been that a number of supposedly well-mapped markers have turned out to be too close together or too far apart for useful multipoint testing. (Although new linkage groups have emerged from such findings [Daiger et al., 1989].) Another problem is that to exclude the last few regions of a chromosome may take months of work. For these reasons we have concluded that it is usually more effective to select easily tested markers scattered throughout the genome than to select markers to completely exclude specific chromosome regions, though we pay careful attention to opportunities for multipoint analysis. The computer program EXCLUDE (Edwards, 1987; Sarfarazi et al., 1987) has proven particularly useful for summarizing and portraying remaining regions of the genome for further linkage testing (Blanton et al., 1991; Daiger et al., 1990).

In our experience, one of the most informative markers for linkage analysis in humans, VNTRs, have not proven useful in large, extended families such as UCLA-RP01. This is because of the difficulty in distinguishing VNTR alleles of similar molecular weights when not all parents are tested. Also, VNTR allele frequencies are uncertain in many cases.

We have found that the most useful markers on a routine basis are simple restriction-site polymorphisms where a single DNA probe, or contiguous probes, can detect more than one RFLP. Such markers are easily tested, the resulting types are unambiguous and, by constructing haplotypes, the systems can be highly informative. This approach effectively converts two-allele loci into one four-allele "haplotype" locus. (To the extent that our data or published data permit a test of linkage equilibrium, all of the systems we have used are in equilibrium.) Also, allele frequencies are usually known and haplotype frequencies are easily calculated, assuming linkage equilibrium. In using the computer program LINKAGE it is not necessary to construct haplotypes; it is only necessary to enter the multiple RFLP sites as separate markers at zero genetic distance from each other. For two-point mapping there is a potential problem of phase-unknown double heterozygotes. This difficulty can be dealt with in LIPED by defining a single phenotype combining both phases of the double heterozygotes and letting the program determine phase. That is, it is not necessary, nor desirable, in our opinion, to construct haplotypes by inspection.

Finally, the most promising development in DNA linkage testing recently has been the application of methods based on the polymerase chain reaction (PCR) to detection of polymorphic variation (Saiki et al., 1988; Weber and May, 1989). To date we have tested over 10 PCR-detectable polymorphisms in UCLA-RP01 (Daiger et al., 1990; Laidlaw et al., 1990; Rodriguez and Daiger, 1989; Rodriguez et al., 1990b). These include several loci with more than one restriction site variant and one candidate gene locus, retinol binding protein 3 (RBP3) on the short arm of chromosome 10 (Rodriguez et al., 1990a). In our experience testing "PCR polymorphisms" requires about one-third the time required by Southern gel analysis and is substantially less prone to artifacts. It is our goal to continue to develop PCR-detectable markers and to use them as the standard panel of linkage markers in future mapping projects.

PROBLEMS IN LINKAGE COMPUTATION FOR UCLA-RP01

Several computational problems arise in the analysis of families such as UCLA-RP01. When dealing with autosomal dominant diseases that have an advanced age at onset, one is often faced with a pedigree consisting of several two- to three-generation nuclear families connected by several missing generations. In the case of UCLA-RP01, the number of missing generations between any two nuclear families ranges from one to seven. These missing generations greatly influence the linkage analysis: as the marker phenotypes of these individuals are unknown, genotypes must be reconstructed during computation. All possible genotypes must be considered for each individual. The first problem is that this results in a substantial increase in computational time when dealing with even one locus with three or four alleles, and can make multipoint mapping nearly impossible. Our attempts to use a Cray X-MP supercomputer to alleviate this computational bottleneck have been only marginally successful to date (Reed-Fourquet and Daiger, 1989).

The second problem is that the frequency of each possible genotype is based on the allele frequencies provided to the computer program. When two or more distantly related families share a rare allele at a marker locus (say, with a frequency of less that 10% in a two-allele system), it is more likely that a single, shared ancestor brought this uncommon allele into the family than that it was brought in by several different individuals further down the pedigree. Thus reversing the frequencies could produce spurious evidence of linkage. The problem is exacerbated if a large, extended family is analyzed as separate nuclear families whose results are then summed. Therefore it is imperative that the correct allele frequencies be used.

Incorrect frequencies may be used mistakenly for at least two reasons. The first possible reason is simply a reversal of frequencies so that the most common allele becomes the least common and vice versa. This happened in one of the original linkage reports on UCLA-RP01, where frequencies for amylase 2 were reversed and linkage of ADRP to amylase was proposed (Spence et al., 1977a,b). More recently, this occurred with us using a marker on chromosome 4, D4S10 (K082) (unpublished). As can be seen from Table 7.1,

Table 7.1. LOD scores between D4510 and ADRP in UCLA-RP01 with varying allele frequencies.

	Recombination Frequency					
Allele Frequency	0.00	0.05	0.10	0.15	0.20	0.30
Published frequencies						
A = .094	−∞	−4.59	−2.85	−1.88	−1.26	−0.52
B = .031						
C = .656						
D = .219						
Reversed frequencies						
A = .656	−∞	−0.01	1.09	1.43	1.45	1.07
B = .219						
C = .094						
D = .031						
Spouse-derived frequencies						
A = .150	−∞	−4.49	−2.78	−1.83	−1.23	−0.52
B = .040						
C = .640						
D = .170						
Equal gene frequencies						
A = .250	−∞	−1.96	−0.67	−0.10	0.14	0.23
B = .250						
C = .250						
D = .250						

permutation of the allele frequencies alters the results dramatically, with linkage being excluded at over 10% recombination using the correct frequencies (row 1), while there is a suggestion of loose linkage (20% recombination) with reversed frequencies (row 2). The problem resulting from allele frequency reversal is easily avoided by carefully checking input data, particularly when loose linkage is suggested.

The second possible problem with allele frequencies is simply the question of exactly what the correct frequencies may be. Many frequencies are based on only 20 to 30 chromosomes (10 to 15 individuals) so that the associated standard error is high. Moreover, when dealing with a pedigree such as UCLA-RP01, where most family members have lived in the same restricted geographical area and have a common ethnic origin (United Kingdom), and where individuals who marry into the pedigree share surnames with other spouses, the question of appropriate allele frequencies becomes more critical. Published allele frequencies are often from populations of either unknown ethnic origin or dissimilar origins. A simple solution is to assume equal allele frequencies, although much information will be lost. Additionally, very rare alleles will become more frequent with resulting distortions in LOD scores. Obtaining population data from the area where the family is based would be ideal, but generally impractical.

A simpler and better solution is to use the data available from unrelated spouses, as was done in row 3 of Table 7.1. In this example, the observed frequencies estimated from 25 spouses do not differ significantly from the published frequencies, but this is not always the case, as we have found, particularly when dealing with two-allele systems. Another, more efficient approach is to estimate gene frequencies based on all pedigree members using the version of MENDEL (Lange et al., 1988) modified for this purpose by Dr Michael Boehnke (Boehnke, 1990) or the program ILINK in the LINKAGE package (Lathrop et al., 1984).

It is possible, of course, to divide the pedigree into smaller families so that fewer data are missing. Smaller families are certainly less influenced by changes in allele frequencies, but as there are almost no complete three-generation families in UCLA-RP01 (because of deaths), the amount of information lost far outweighs the problems encountered in determining correct frequencies.

SUMMARY AND CONCLUSIONS

The problems discussed here can result in two undesirable outcomes: a false positive linkage or a false negative linkage. False positive linkages, that is, proposed linkages that are subsequently withdrawn, are becoming more common in the literature. This may reflect the competitive pressure to be the first to report a new linkage, but another possible explanation is the statistical problem of multiple comparisons. It is now becoming common to screen 100 to 200 markers in a linkage study and dozens of linkage studies are under way in the scientific community at any given moment. Because the arbitrary "acceptance" criterion of a LOD score of 3.0 implies a probability of 95% that two markers are linked, the opportunity for spurious linkage is greatly enhanced. Fortunately, linkage results can be substantiated quickly, both by testing additional families and by testing flanking markers in the original families. In our case, we adopted a requirement of a LOD of 4.00 or greater and substantiation of linkage by at least one flanking marker before publicaton. This conservative policy has saved us considerable embarrassment on more than one occasion.

A false negative result, that is, failure to detect linkage with a truly linked marker, is the potential error that keeps all of us awake at night. The best defense against this error is constant vigilance: frequent rechecking of data, pedigree structure, and clinical status. The ability to build multipoint linkage systems with the many mapped markers is also a great hedge against a spurious failure to show linkage. Simply put, the best strategy, in addition to careful attention to detail, is to test as many markers as possible as quickly as possible!

ACKNOWLEDGMENTS

This research was supported by grants from the National Retinitis Pigmentosa Foundation, the George Gund Foundation, and NIH Grant EY07142.

We wish to take this opportunity to offer our sincere gratitude to Dr William J. Schull for all the generous encouragement, enthusiasm, and wisdom he has given over the years. He has made the practice of science not only intellectually satisfying but also deeply, personally rewarding. Thanks for the laughs, too, Jack!

REFERENCES

Blanton SH, Cottingham AW, Giesenschlag N, Heckenlively JR, Humphries P, Daiger SP (1990) Further evidence of exclusion of linkage between type II autosomal dominant retinitis pigmentosa (ADRP) and D3S47 on 3q. *Genomics* 11:857–869.

Blanton SH, Heckenlively JR, Cottingham AW, Friedman J, Sadler LA, Wagner M, Friedman LH, Daiger SP (1991) Linkage mapping of autosomal dominant retinitis pigmentosa (RP1) to the pericentric region of human chromosome 8. *Genomics* 11:857–869.

Boehnke M (1991) Allele frequency estimation from data on relatives. *Amer J Hum Genet* 48:22–25.

Breeding CI (1982) *Genealogy of Gideon Ison I.* Charlestown, IN, C Breeding.

Daiger SP (1988) Appendix H. The Retinitis Pigmentosa (RP) Collection. NIH Publication No. 89-2011, *1988/1989 Catalog of Cell lines, NIGMS Human Genetic Mutant Cell Repository*, 599–608.

Daiger SP, Blanton SH, Cottingham A, Laidlaw J, Rodriguez JA, Heckenlively JR (1990) Linkage mapping and molecular studies of autosomal forms of retinitis pigmentosa. In Humphries P (ed) *Degenerative Retinopathies: Advances in Clinical and Genetic Research. Proc 6th World Congress of the International Retinitis Pigmentosa Association.* CRC Press, Boca Raton, FL, pp 23–24.

Daiger SP, Humphries MM, Sharp E, McWilliams P, Farrar J, Bradley D, McConnel DC, Kenna P, Sparkes RS, Spence MA, Heckenlively JR, Humphries P (1989) Linkage analysis of human chromosome 4: exclusion of autosomal dominant retinitis pigmentosa (ADRP) and detection of new linkage groups *Cytogenet Cell Genet* 50:181–187.

Dryja TD, Hahn LB, Cowley GS, McGee TL, Berson EL (1991) Mutatin Spectrum of the rhodopsin gene among patients with autosomal dominant retinitis pigmentosa. *Proc Natl Acad Sci* 88:9370–9374.

Dryja TP, McGee TL, Reichel E, Hahn LB, Cowley GS, Yandell DW, Sandberg MA, Berson EL (1990) A point mutation on the rhodopsin gene in one form of retinitis pigmentosa. *Nature* 343:364–366.

Edwards JH (1987) Exclusion mapping. *J Med Genet* 24:539–543.

Elston RC, Stewart J (1971) A general model for the analysis of pedigree data. *Hum Hered* 21:523–542.

Farrar GJ, Kenna P, Jordan SA, Kumar-Singh R, Humphries MM, Sharp EM, Sheils DM, Humphries P (1991) A three base-pair deletion in the peripherin-RDS gene is one form of retinitis pigmentosa. *Nature* 354:478–480.

Gal A, Samanns C, Watty A, Jimenez J, Ludwig M, Chand A, Pongratz J, Colantuoni V, Gusseck H, Olsson J, Neugebauer M, Schinzel A, Denton M (1990) A gene for autosomal dominant retinitis pigmentosa is closely linked to D3S20 on 3q. *Proc 6th Congress, International Retinitis Pigmentosa Association.* Dublin, Ireland.

Gusella JF, Wexler NS, Conneally PM, Naylor SL, Anderson MA, Tanzi RE, Watkins PC, Ottina K, Wallace MR, Sakaguchi AY, Young AB, Shoulson I, Bonilla E, Martin JB (1983) A polymorphic DNA marker genetically linked to Huntington's disease. *Nature* 306:234–238.

Heckenlively JR (ed) (1988) *Retinitis Pigmentosa.* New York, JB Lippincott.

Heckenlively JR, Pearlman JT, Sparkes RS, Spence MA, Zedalis D, Field L, Sparkes MA, Crist M, Tideman S (1982) Possible assignment of a dominant retinitis pigmentosa gene to chromosome 1. *Ophthalmic Res* 14:46–52.

Heckenlively JR, Rodriguez JA, Daiger SP (1990) Autosomal dominant sector retinitis pigmentosa: two families with rhodopsin codon 23 transversion. *Arch Ophthal* 109:84–91.

Human Gene Mapping 10.5 (1990): Update to Tenth International Workshop on Human Gene Mapping. *Cytogenet Cell Genet* 55:1–785.

Inglehearn CF, Jay M, Lester DH, Bashir R, Jay B, Bird AC, Wright AF, Evans HJ, Papiha SS, Bhattacharya S (1990) No evidence for linkage between late onset autosomal dominant retinitis pigmentosa and chromosome 3 locus D3S47 (C17): evidence for genetic heterogeneity. *Genomics* 6:168–173.

Kajiwara K, Hahn LNB, Mukai S, Travis GH, Berson EL, Dryja TP (1991) Mutations in the human retinal degeneration slow gene in autosomal dominant retinitis pigmentosa. *Nature* 353:480–483.

Kimberling WJ, Weston MD, Moller C, Davenport SLH, Shugart YY, Priluck IA, Martini A, Smith RJH (1990) Localization of Usher syndrome type II to chromosome 1q. *Genomics* 7:245–249.

Koenig M, Hoffman EP, Bertelson CJ, Monaco AP, Feener C, Kunkel LM (1987) Complete cloning of the Duchenne muscular dystrophy (DMD) cDNA and preliminary genomic organization of the DMD gene in normal and affected individuals. *Cell* 50:509–517.

Laidlaw JL, Rodriguez JA, Cottingham AW, Blanton SH, Reed-Fourquet LL, Daiger SP (1990) Rapid linkage analysis of type II autosomal dominant retinitis pigmentosa (ADRP) using the polymerase chain reaction (PCR) *Amer J Hum Genet* 47:187A.

Lange K, Weeks D, Boehnke M (1988) Programs for pedigree analysis: MENDEL, FISHER, and dGENE. *Genet Epid* 5:471–472.

Lathrop GM, Lalouel JM, Julier C, Ott J (1984) Strategies for multilocus linkage analysis in humans. *Proc Natl Acad Sci* 81:3433–3436.

McInnis R (1991) Cited in *Nature* (1991) 353:798.

McWilliams P, Farrar GJ, Kenna P, Bradley DG, Humphries MM, Sharp EM, McConnell DJ, Lawler M, Shields D, Ryan C, Stephens K, Daiger SP, Humphries P (1989) Autosomal dominant retinitis pigmentosa (ADRP): localization of an ADRP gene to the long arm of chromosome 3. *Genomics* 5:619–622.

Marmor MF, et al. (1983) Retinitis pigmentosa: A symposium on terminology and methods of examination. *Ophthalmol* 90:126–131.

Mohr J (1954) A study of linkage in man. *Opera ex Domo Biologiae Hereditariae Humanae Universitatis Hafniensis* Copenhagen, Munksgaard, 33.

Morton E (1956) The detection and estimation of linkage between the genes for elliptocytosis and the Rh blood type. *Amer J Hum Genet* 8:80–96.

Nakamura Y, Leppert M, O'Connell P, Wolfe R, Holm T, Culver M, Martin C, Fujimoto E, Hoff M, Kumlin E, White R (1987) Variable number of tandem repeat (VNTR) markers for human gene mapping. *Sci* 235:1616–1621.

Ott J (1985) *Analysis of Human Genetic Linkage*. Baltimore, Johns Hopkins University Press.

Ott J, Bhattacharya S, Chen JD, Denton MJ, Donald J, et al. (1990) Localizing multiple X chromosome-linked retinitis pigmentosa loci using multilocus homogeneity tests. *Proc Natl Acad Sci* 87:701–704.

Reed-Fourquet LL, Daiger SP (1989) Implementation of LIPED on a Cray X-MP Supercomputer. *Amer J Hum Genet* 45:A158.

Rodriguez JA, Blanton SH, Daiger SP (1990a) The polymerase chain reaction and a candidate gene approach to mapping autosomal dominant retinitis pigmentosa. *Proc Ann Mtng, Austin, Texas Genetics Society* 17:21.

Rodriguez JA, Daiger SP (1989) Application of polymerase chain reaction (PCR) methodology to hypervariable loci in humans. *Proc Ann Mtng, Dallas, Texas Genetics Society* 16:17.

Rodriguez JA, Liou GI, Daiger SP (1990b) Enhanced detection of an RFLP at the interstitial retinol binding protein (RBP3) locus. Nuc Acids Res 18:5578.

Rommens JM, Iannuzzi MC, Kerem B-S, Drumm ML, Melmer G, Dean M, Rozmahel R, Cole JL, Kennedy D, Hidaka N, Zsiga M, Buchwald M, Riordan JR, Tsui L-C, Collins FS (1989) Identification of the cystic fibrosis gene: chromosome walking and jumping. *Sci* 245:1059–1065.

Saiki RK, Gelfand DH, Stoffel S, Scharf SJ, Higuchi R, Horn GT, Mullis KB, Erlich HA (1988) Primer-directed enzymatic amplification of DNA with a thermostable DNA polymerase. *Sci* 239:487–493.

Sarfarazi M, Huson SM, Edwards JH (1987) An exclusion map fpr von Recklinghausen neurofibromatosis. *J Med Genet* 24:515–520.

Smith RJH, Lee EC, Kimberling WJ, Daiger SP, Pelias MZ, Keats BJB, Jay M, Bird A, Reardon W, Guest M, Ayyagari R, Hejtmancik JF (1992) Localization of two genes for usher Syndrome Type I to chromosome 11. *Genomics*, in press.

Smith RJH, Pelias MZ, Daiger SP, Herrera CH, Kimberling WJ, Hejtmancik (1990) Clinical and genetic heterogeneity within the Acadian Usher population. *Amer J Hum Genet* X:XX–XX.

Spence MA, Sparkes RS, Heckenlively JR, Pearlman J, Zedalis D, Sparkes M, Crist M, Tideman S, (1977a) Probable genetic linkage between autosomal dominant retinitis pigmentosa (RP) and amylase (Amy_2): evidence of an RP locus on chromosome 1. *Amer J Hum Genet* 29:397–402.

Spence MA, Sparkes RS, Heckenlively JR, Pearlman J, Zedalis D, Sparkes M, Crist M, Tideman S (1977b) Erratum. *Amer J Hum Genet* 29:592.

Sung C-H, Davenport CM, Hennessy JC, Maumenee IH, Jacobson SG, Heckenlively JR, Nowakowski R, Fishman G, Goura P, Nathans J (1991) Rhodopsin mutations in autosomal dominant retinitis pigmentosa. *Proc Natl Acad Sci USA* 88:6481–6485.

Wallace MR, Marchuk DA, Andersen LB, Lectcher R, Odeh HM, Saulino AM, Fountain JM, Brereton A, Nicholson J, Mitchell AL, Brownstein BH, Collins F (1990) Type I neurofibromatosis gene: identification of a large transcript disrupted in three NF1 patients. *Sci* 249:181–186.

Weber JL, May PE (1989) Abundant class of human DNA polymorphisms which can be typed using the polymerase chain reaction. *Amer J Human Genet* 44:388–396.

8

Applications of Pedigree Analysis to Animal Models for Complex Diseases

JEAN W. MACCLUER

A major goal in genetic epidemiology is to understand the role of individual genes in disease susceptibility. As knowledge of the human gene map expands, research aimed at identifying and localizing disease genes will become increasingly successful. Single genes that affect a disease or a risk factor are often suggested first by formal genetic analysis, and later localized and characterized by molecular studies. Among the most obvious successes of this "top down" approach are single-gene disorders such as Duchenne muscular dystrophy and cystic fibrosis. Another striking example is familial hypercholesterolemia (FH), whose autosomal codominant mode of inheritance was postulated years before the molecular basis of the defect was clarified (Khachadurian, 1964).

The top down approach is increasingly being extended to risk factors for the complex diseases that are the major causes of death in Western societies, diseases such as coronary heart disease, hypertension, and diabetes. The aim in these studies is to decompose the genetic component in susceptibility into its individual elements and to determine the nature and magnitude of the contribution of each gene to the disease process. A problem is that complex diseases, by definition, are determined not only by genes, but also by environmental factors and by interactions between genes and environment. Detecting genetic effects may be difficult if the relevant environmental factors cannot be readily identified and quantified, or if the environments to which individuals are exposed change throughout their lives. The sharing of environments by family members and the differences in environments between families can further complicate attempts to identify genetic effects on diseases.

One way to address the confounding effects of environment in genetic studies of complex diseases is to use animal models for which relevant environmental variables can be controlled or at least quantified. The best choice of animal model varies with the disease being studied. Perhaps the most important criterion in selecting an animal model is that there should be convincing evidence from physiological and other studies that the disease process in the animal model is similar to that in humans. For genetic studies, access to pedigrees is critical. An advantage of animal models, particularly those with a short generation time relative to humans, is that large pedigrees can be created which have optimal structures for genetic analysis. Multiple generation pedigrees with large full and half sibships can be produced. A further advantage is that the animals are available for thorough measurement of quantitative risk factors under controlled environmental conditions, and may be experimentally subjected to environmental changes for studies of the differential action of susceptibility genes in different environments. If genetic analyses are successful in identifying genes that contribute to the disease process, experimental studies aimed at understanding gene action are much more feasible in an animal model than in humans.

The best uses of animal models in genetic studies are (1) in detecting genes that influence environmentally labile phenotypes (e.g., phenotypes that are strongly influenced by diet) and in determining which phenotypes are most subject to environmental influence; (2) in studies of traits that are difficult to investigate in humans because of inaccessibility of appropriate organs or tissues (e.g., liver-specific enzymes in studies of alcohol metabolism); (3) in studies of genes that have not shown variability in humans, but whose products are thought to be important in a disease process (e.g., LCAT, the locus determining lecithin-cholesterol acyltransferase, a candidate gene for atherosclerosis); and (4) in genetic analyses of systems that are more poorly characterized in humans than in the animal model (e.g. lipoprotein (a), a complex of low density lipoprotein cholesterol and an apolipoprotein called apo(a), for which the null phenotype is high in frequency in humans).

Several advances in understanding complex diseases have been made by studying inbred strains of mice (Ishida and Paigen, 1989). Work in non-obese diabetic (NOD) mice has led to the mapping of three genes that are necessary for insulin-dependent diabetes (Prochazka et al., 1987). Studies of aging in SAM mice (senescence accelerated mouse) have revealed that a mutation in the apolipoprotein AII gene, or in a gene that is involved in its metabolism, leads to systemic amyloidosis (Higuchi et al., 1983, 1986). Other studies have taken advantage of recombinant inbred (RI) strains. RI strains are created by crossing two different inbred strains, and continuing to inbreed until new homozygous lines are formed. If two traits that appear together in a progenitor strain also appear in all RI strains, then the traits are assumed to be determined by the same locus or by closely linked loci. Work with RI strains of mice has led to the discovery of at least three genes, *Ath-1*, *Ath-2*, and *Ath-3*, that alter susceptibility to atherosclerosis (Paigen et al., 1987, 1989; Stewart-Phillips et al., 1989).

Studies of crosses between inbred strains of rats (Hilbert et al., 1991; Jacob et al., 1991) have identified a major locus that contributes to hypertension in stroke-prone spontaneously hypertensive (SHRSP) rats. The locus has been mapped, and a likely candidate gene, coding for an enzyme important in blood pressure homeostasis, has been identified. These studies are valuable for pointing out directions of future research in humans.

Studies in swine (e.g., Rapacz, 1978; Rapacz and Hasler-Rapacz, 1989) have demonstrated that a single gene can influence resistance or susceptibility to diet-induced atherosclerosis. Swine with inherited hyper-LDL-cholesterolemia (IHLC) have proved valuable in studies of the progression of atherosclerotic lesions (Rosenfeld et al., 1988).

For many purposes, nonhuman primates provide an ideal animal model for studies of human diseases because of their close phylogenetic relationship to humans. On the other hand, nonhuman primates have a long generation time relative to mice and other small mammals, and because of their size, they are expensive to maintain. Therefore the primate colonies used in biomedical research tend to be small relative to the sample sizes that are possible with some other animal models. The breeding structure of primate colonies also differs from that of many of the smaller animals. The mice, rats, and rabbits used in biomedical research are primarily from inbred strains (Festing, 1979). These strains have the advantage that the animals within a strain are more or less genetically identical. Their disadvantages are that they are expensive to produce and maintain and many lines fail to survive. Moreover they do not provide the range of genetic variability that is needed for many genetic studies of complex diseases. For these reasons, and because the establishment of inbred lines would be prohibitively time consuming and expensive, primate breeding colonies tend not to be highly inbred. Consanguineous matings may be set up to test specific hypotheses, but the overall strategy in management of primate colonies is to minimize the loss of useful genetic variability that can result from inbreeding.

In this chapter, we summarize the biomedical uses of a nonhuman primate model, baboons of the subspecies *Papio hamadryas anubis* and *P. h. cynocephalus*, and describe statistical genetic analyses aimed at identifying genes that contribute to atherosclerosis in baboons.

BABOONS AS ANIMAL MODELS FOR HUMAN DISEASES AND METABOLIC STUDIES

Baboons are used as models in research on a variety of human diseases and disease precursors, for example, hypertension (Crawford et al., 1987), hyalin membrane disease (deLemos et al., 1987), toxic shock syndrome (Micha and Quimby, 1984), and malignant lymphoma (Gleiser et al., 1984). One of the most common uses of baboons in biomedical research is in studies of atherosclerosis and lipoprotein metabolism (Eggen, 1974; McGill et al., 1981a,b; Kammerer et al., 1984; Laber-Laird and Rudel, 1989).

The largest baboon colony in the world is maintained at the Southwest Foundation for Biomedical Research in San Antonio. The population includes more than 3300 baboons, maintained in breeding corrals and in specialized colonies for numerous biomedical research projects. A number of naturally occurring disorders have been observed in this population. Approximately 30% of baboons screened at the Southwest Foundation carry STLV-1, which causes lymphoma and leukemia (JS Allan, personal communication, 1991). Aging baboons have been shown to exhibit skeletal changes that are very similar to those observed in human osteoporosis. In addition, numerous defects in baboons have been observed to be familial, although no genetic studies have been done. These defects include adult onset diabetes, a form of congenital blindness, and focal seizures. Baboons are used in studies of immunity (e.g., Eichberg et al., 1985) and susceptibility to viral diseases such as herpes simplex (Levin et al., 1988) and endomyocarditis.

Many investigators have utilized baboons to understand the factors that govern the success of organ transplants, both within and between primate species (e.g., Kurlansky et al., 1987; Van der Riet et al., 1987). Baboons also have been used in studies of alcohol metabolism (Holmes and VandeBerg, 1987), cardiac development (Hixson et al., 1989a), obesity (Wene et al., 1981; Coelho et al., 1984), and ovariectomy as a model for human menopausal osteoporosis (Weaver et al., 1989). Studies in baboons have contributed to advances in embryology (Hendrickx, 1971), growth and development (Snow, 1967), nutrition (Lewis et al., 1983), and behavior (Bernstein et al., 1982; Goncharov, 1979).

BABOONS AS MODELS FOR RESEARCH IN ATHEROSCLEROSIS

Atherosclerosis in humans is characterized by accumulation of fatty streaks and plaques in the arteries. The development of streaks and plaques begins in young adulthood or before, and progresses throughout life. The extent and severity of arterial lesions is generally accepted to be a function of genetic factors interacting with environmental determinants such as diet, smoking, and exercise. The severity of atherosclerotic lesions is difficult to quantify in population studies; however, there are numerous easily measured quantitative correlates of atherosclerosis, including serum concentrations of lipoproteins and apolipoproteins. Many genetic and epidemiological studies of these quantitative phenotypes have been done in human families and populations. Genetic analyses of these atherosclerosis-related phenotypes are reviewed in Rao et al., (1984) and MacCluer (1989).

As with many complex diseases, reliable information on the environmental variables that influence lipoproteins and atherosclerosis is difficult to obtain in human studies, and these environmental factors seldom can be controlled by the investigator. Genetic analysis of lipoproteins in an animal model has the advantage that the relevant environmental variables tend not to vary substantially among animals, and within-family environmental correlations can be minimized by separating offspring from their dams after weaning. Even more important, a major environmental covariate, diet, can be controlled and experimentally changed.

Further advantages of baboons as an animal model for studies of atherosclerosis are that metabolic studies and experimental interventions can be carried out using animals that have particularly interesting phenotypes, or that are likely to carry specific alleles at major loci. Matings can be arranged to generate extreme phenotypes or to test genetic hypotheses, and very large pedigrees can be generated.

The major criterion in choosing an animal model for studies of lipoproteins and atherosclerosis is that the animal be similar to humans in lipoprotein metabolism and in the progression of atherosclerotic lesions. Baboons share with humans many aspects of lipoprotein metabolism (Bojanovski et al., 1978; Fox, 1933; McGill et al., 1960). Feral baboons exhibit atherosclerotic lesions even without dietary manipulation (McGill et al., 1960). When they are fed a diet rich in cholesterol and saturated fat, their serum cholesterol levels increase to the same extent as in humans, and the rate at which they develop arterial lesions is similar (McGill et al., 1981a,b). Baboons also are similar to humans in the correlations between serum cholesterol concentrations and extent of arterial lesions: animals with high serum concentrations of very low density plus low density lipoprotein cholesterol (VLDL + LDL-C) have more extensive arterial lesions, whereas those with increased high density lipoprotein cholesterol (HDL-C) have less extensive lesions (McGill et al., 1981a). As in humans, there is considerable evidence for genetic mediation of serum cholesterol concentrations in baboons (e.g., Flow et al., 1981; Kammerer et al., 1984; MacCluer et al., 1988). Metabolic studies have shown that the genetic control of total plasma cholesterol concentration in baboons is not related to cholesterol absorption, and that the biochemical mechanism varies with age (Mott et al., 1978).

THE HUMAN AND BABOON GENE MAPS

For genetic studies aimed at identifying genes that contribute to atherosclerosis, knowledge of the chromosomal locations of candidate genes is important. In comparison to the human gene map, the gene map of baboons is poorly characterized. The 10th Human Gene Mapping Workshop (Lalley et al., 1989) identifies 47 genes that have been assigned to 16 different baboon chromosomes and three unknown linkage groups. Only two studies in nonhuman primates, both of them in baboons, have demonstrated linkage by analysis of family data: VandeBerg et al. (1989) established linkage of baboon LPA (encoding apolipoprotein(a)) and PLG (plasminogen); and van Oorschot and VandeBerg (1991) showed tight linkage of baboon MPI (mannose phosphate isomerase) and NP (nucleoside phosphorylase).

In spite of the paucity of gene mapping information for baboons, genetic studies can take advantage of the considerable linkage conservation that is known to exist among human and nonhuman primates. The synteny shared by the old world primates, including baboons and humans, is striking (Lalley et al., 1987). Certain linkage groups have been strongly conserved through long periods of evolutionary time (Lalley et al., 1987; Morizot, 1984). Chromosomal mapping of genetic markers in baboons reveals extensive homology with the human gene map (Creau-Goldberg et al., 1983). The 16

Table 8.1. Comparative human and baboon gene maps.[a]

Human Chromosome	Gene Symbol	Baboon Chromosome	Human Chromosome	Gene Symbol	Baboon Chromosome
1pter-p36.13	ENO1	1	13q14.1-q14.2	ESD	17
1p36.3-p36.13	PGD	1	13q14.2	RB1	17
1p22.1	PGM1	1			
1q25 or q42	PEPC	1	14q11.2	NP	7
			14q32.3	CKBB	7
2p25	ACP1	U1[b]			
2p23	MDH1	13	15pter-q21	SORD	7
2q31-q32.3	COL3A1	12	15q21-qter	IDH2	7
2q32-qter	IDH1	12	15q22-qter	MPI	7
			15q22-qter	PKM2	7
3p21	ACY1	2pter-q1	15q23-q24	HEXA	7
4p14-q12	PGM2	5	17q21.3-q22	COL1A1	16
4p11-q12	PEPS	5	17q23	GAA	16
4q28-q31	GYPA	5	17q23.2-q25.3	TK1	16
6p21.3	MHC	4	19q12-q13.2	PEPD	20
6p21.3-p21.1	GLO1	4	19q13.1	GPI	20
6q12	ME1	4			
6q21	SOD2	4	20p	ITPA	10
			20q13.2-qter	ADA	10
7p13-q22	MDH2	3p			
7q21.2-q22	GUSB	3	21q21.2	APP	3
7q21.3-q22.1	COL1A2	3q	21q22.1	SOD1	3p
			21q22.3	ETS2	3
8p21.1	GSR	U2[b]			
			22q12.3-q13.1	PDGFB	3
10q11.1-q24	PP	9			
			Xp22.31-p22.1	AMG	X
11p15.1-p14	LDHA	14	Xq21-q22	TBG	X
12p13	TPI1	11			
12p12.2-p12.1	LDHB	11			
12q21	PEPB	U3[b]			

[a]Based on data from HGM10 (1989).
[b]U1, U2, and U3 are unknown linkage groups.

established baboon linkage groups, each containing 2 to 7 markers, correspond to 17 linkage groups in humans (Table 8.1). Comparisons of the karyotype of the baboon with those of the great apes and humans also have revealed quite similar banding patterns (Dutrillaux et al., 1978).

POLYMORPHIC GENES IN BABOONS

Polymorphisms that have been identified in baboon genes (Table 8.2) are useful in studies aimed at detecting single-gene effects on lipoproteins. Among these polymorphic loci are several candidate genes for atherosclerosis,

Table 8.2. Polymorphic loci in baboons of subspecies *P. h. anubis* and *P. h. cynocephalus*.

Locus	Abbreviation	Reference
ABO blood group	ABO	Moor-Jankowski et al., 1964
Adenosine deaminase	ADA	Shotake et al., 1977
Albumin	ALB	Kitchen et al., 1967
Aldehyde dehydrogenase 1	ALDH1	Holmes and VandeBerg, 1986
Apolipoprotein A-I	APOA1	Hixson et al., 1988a
Apolipoprotein A-IV	APOA4	Ferrell et al., 1990
Apolipoprotein B	APOB	Rogers et al., 1988
Adenine phosphoribosyl-transferase	APRT	R. Fisher[a]
Carbonic anhydrase I	CA1	Barnicot et al., 1967
Catalase	CAT	R. Fisher[a]
Complement component 3	C3	VandeBerg and Cheng, 1986
Complement component 6	C6	R. Fisher[a]
Complement component 7	C7	R. Fisher[a]
Group specific component	GC	Kitchen et al., 1967
Lecithin-cholesterol acyl-transferase	LCAT	Hixson et al., 1990
LDL receptor	LDLR	Hixson et al., 1989b
Apolipoprotein(a)	LPA	Rainwater et al., 1989
Mannose phosphate isomerase	MPI	VandeBerg and Aivaliotis, 1990
Phosphogluconate dehydrogenase	PGD	VandeBerg and Cheng, 1986
Plasminogen	PLG	VandeBerg et al., 1989
Thyroxine binding globulin	TBG	Lockwood et al., 1984
Transferrin	TF	Buettner-Janusch, 1963

[a]Personal communication to J. L. VandeBerg, 1981.

including APOA1, APOA4, APOB, LCAT, LDLR, and PLG. At least nine isoforms of apolipoprotein(a), determined by the LPA locus, have been identified in *P. h. anubis* and *P. h. cynocephalus* (Rainwater et al., 1989). No polymorphisms have been found in the baboon gene encoding apolipoprotein E (APOE). However, both APOE and APOA1 have been cloned and sequenced, and their structure and expression have been investigated (Hixson et al., 1988a,b).

Genes important in lipoprotein metabolism in humans are clustered on chromosomes 2p, 6q, 11q, 16q, and 19; the baboon chromosomal locations of these candidate loci have not yet been determined. Other markers that are polymorphic in *P. h. anubis* and *P. h. cynocephalus* (Table 8.2) are ABO, ADA, ALB, ALDH1, APRT, CA1, CAT, C3, C6, C7, GC, MPI, PGD, TBG, and TF (see Williams-Blangero et al. (1990) for allele frequencies). The major histocompatibility complex (MHC) also is polymorphic in these baboon subspecies (Templeton et al., 1981).

PEDIGREE ANALYSIS OF LIPOPROTEIN PHENOTYPES IN BABOONS

In 1975, a breeding program was established at the Southwest Foundation to produce pedigreed baboons for research on atherosclerosis. Several hundred wild-caught baboons (*P. h. anubis* and *P. h. cynocephalus*) were screened for total plasma cholesterol (TC) levels, and breeding groups were established, each consisting of a single sire and multiple dams. Six of the breeding groups had sires and dams selected for high TC concentrations, and five for low TC concentrations. Twelve control breeding groups also were established. Most offspring were nursed by their dams until weaning at approximately 16 weeks of age, separated from their dams, and placed in a common cage. Only 12% of offspring had to be nursery-reared because of inadequate maternal care or inability to nurse.

During the past 15 years, the sire families have been expanded and new breeding groups have been formed. A dietary study of all animals in the colony was begun in 1982. Serum concentrations of lipoproteins and apolipoproteins were determined for each animal on two different diets: a basal diet, and a high cholesterol, saturated fat (HCSF) diet. The basal diet has 0.03 mg cholesterol per Kcal and 10% of calories as fat; and the HCSF diet, 1.7 mg cholesterol per Kcal and 40% of calories as fat. Compared to the diet of many Western societies, the HCSF diet is similar in fat composition and at the upper limit in cholesterol content. The animals were fed the HCSF diet for 7 weeks before their blood was drawn for serum lipoprotein quantitation. Data are now available for more than 1300 pedigreed baboons.

Serum concentrations of VLDL + LDL-C, HDL-C, and apolipoproteins A-I and B were measured for each animal on each diet. Lipoprotein profiles on each diet were characterized by gradient gel electrophoresis (Cheng et al., 1987; VandeBerg et al., 1987), which permits detailed study of the distribution of lipoproteins within size classes. Animals also were typed for the candidate genes and other polymorphic markers described above. Statistical genetic analyses have been aimed at determining whether serum levels of lipoproteins and apolipoproteins, which are intervening variables for atherosclerosis, are influenced by single genes with major or minor effects. Of particular interest is whether the effects of such major and minor genes are mediated by diet. We have found evidence for major and minor genes affecting HDL-C, apoAI, and VLDL + LDL-C (hereafter referred to as LDL-C), and also for interactions between genotype and diet.

HDL-C and apolipoprotein A-I

Using univariate complex segregation analysis (PAP, Hasstedt and Cartwright, 1979) on data from 710 of the pedigreed baboons in 23 sire families, we found evidence for a codominant major gene influencing HDL-C serum concentrations in baboons fed the basal diet, and in the same animals fed the HCSF diet (MacCluer et al., 1988). We examined a model with no genetic component, a model whose only genetic component was polygenic, models with only a major gene and random environmental factors, mixed models (with major gene, polygenes, and random environment), a model with all transmission probabilities constrained to be equal, and a general model, in which the transmission probabilities were estimated. We were able to reject all models that did not include both a major gene (with Mendelian transmission frequencies) and a polygenic component.

On both diets, we detected an allele (frequency 0.20) that resulted in increased serum concentrations of HDL-C. The major gene(s) accounted for 35% of the variance in serum HDL-C concentration on the basal diet and 31% on the HCSF diet; polygenic factors accounted for an additional 12% of variance on the basal diet and 19% on the HCSF diet. On the HCSF diet, the mean HDL-C concentrations for the three major locus genotypes were 30 to 40 mg/dl higher than the corresponding means on the basal diet. There was no evidence for a major gene effect on response to diet, although polygenes contributed approximately 18% to the variance in response. Further bivariate segregation analyses, such as those described below for apoAI, are required to determine whether the same major gene is being detected on the basal and HCSF diets, and whether there is interaction between diet and genotype.

Apolipoprotein A-I (apoAI) is the major apolipoprotein in HDL-C. In complex segregation analysis of apoAI serum concentrations in the same pedigreed baboons, Blangero et al. (1987, 1990a) detected the effects of major genes and polygenic factors on apoAI levels on both the basal and HCSF diets. There also were significant effects of sex, of female age, and of nursery rearing. The models evaluated were the same as those described above for HDL-C. A further analysis of sexual dimorphism in apoAI levels in these animals (Towne et al., 1990) revealed that females show greater variation in apoAI serum concentrations than do males, and that the sex differences in apoAI levels are at least in part due to genetic factors, expecially on the basal diet.

In a subsequent bivariate analysis that incorporated apoAI concentrations on both diets into a single analysis (Blangero et al., 1990a), it was possible to demonstrate that the major genes detected on the two diets were not the same, but were two distinct loci that act additively to determine apoAI levels. In bivariate segregation analyses, the most general model assumes that there are two major loci that may interact epistatically to determine levels of two quantitative traits (apoAI concentration on basal diet and on HCSF diet). Each trait also may be influenced by polygenic and random environmental factors, and there may be residual genetic and environmental correlation between the traits. The two loci may be linked and may exhibit gametic disequilibrium. The restricted models that are evaluated are similar to those evaluated in the univariate case, but with a separate set of parameters and constraints for each of two loci, and with the additional parameters that are needed to specify interactions and linkage relationships between loci. To test whether two major loci are in fact the same locus, a restricted model is evaluated in which the allele frequencies at the two loci are constrained to be equal, the recombination fraction is constrained to zero, and gametic disequilibrium is constrained to one. To test whether the loci exhibit epistasis, digenic interaction terms are constrained to zero (Blangero et al., 1990a).

Bivariate analyses of apoAI concentrations on the two diets (basal and

HCSF) indicated that two separate major loci influence apoAI levels. The two apoAI loci account for approximately 40% of the phenotypic variance in apoAI concentration on each of the two diets, and also for 33% of the variance in response to HCSF diet. One of the loci (A) influences apoAI levels on both diets and the second (B), on the basal diet only. There was evidence for interaction between diet and genotype for both major loci: the additive and dominance effects of both the A and B loci were different on the basal diet than on the HCSF diet. The response to HCSF diet of individuals who are homozygous recessive at the B locus is a function of their A locus genotype: apoAI levels of bb individuals increase slightly in response to diet when the individual is of genotype AA or Aa, or decrease substantially when the individual is of genotype aa.

In a multivariate quantitative genetic analysis of HDL-C and apoAI on the basal and HCSF diets (Blangero et al., 1988), genetic correlations between the traits were high (0.75–1.00). Significant genotype–diet interaction was detected, due primarily to the effect of diet on HDL-C genotypes. Further bivariate segregation analyses will be required to determine whether the major genes affecting HDL-C and apoAI are identical, and if not, whether the HDL-C major gene contributes to variation in apoAI levels.

Segregation analysis provides evidence for the existence of major genes and estimates of the magnitudes of gene effects, but it given no information on the identities or chromosomal locations of these genes. Additional analyses are required to learn whether the major genes are linked to (or are identical to) specific candidate loci. Among the candidate loci of interest are the APOA1 structural locus, as well as LCAT and CETP (cholesteryl ester transfer protein), two loci that code for enzymes involved in reverse cholesterol transport. A sib pair test (Haseman and Elston, 1972) indicated possible linkage of one of the baboon major apoAI genes to APRT, the locus determining adenine phosphoribosyl-transferase (Kammerer et al., 1987). In humans, APRT is on the short arm of chromosome 16 and is linked to LCAT and CETP. Results of subsequent quantitative trait linkage analyses suggested that this major apoAI gene may be linked to LCAT, but not to an APOA1 structural locus RFLP (Kammerer et al., 1988). On the HCSF diet, there was significant gametic disequilibrium between the apoAI major locus genotypes and the APOA1 structural locus RFLP.

A subsequent analysis (Kammerer et al., 1988) revealed evidence for a small "measured" genotype effect of LCAT on serum apoAI concentrations, but only on the HCSF diet. Because both of the major apoAI genes detected by Blangero et al. (1990a) account for a considerably larger portion of the variance in apoAI, it is unlikely that LCAT is one of these major genes. Further work is underway to determine whether this measured genotype effect on apoAI is due to LCAT or to another, linked locus; a prime candidate is CETP.

Hixson et al. (1988a) detected associations of the APOA1 RFLP with altered distribution of apoAI in HDL size classes. Current efforts are focused on measured genotype analyses of the APOA1 structural locus RFLP and its effects on apoAI levels on the two diets. Results of these analyses will permit the dissection of the variance in HDL-C and apoAI serum concentrations into components attributable to specific candidate genes and to other, as yet unidentified loci.

LDL-C

Hixson et al. (1989b), in studies of these same pedigreed animals, found evidence for a measured genotype effect of an LDL receptor (LDLR) RFLP on serum levels of both LDL-C and apoB, on both the basal and the HCSF diet. Homozygotes for the rarer allele had lower concentrations of LDL-C and apoB than homozygotes for the common allele; and with the exception of apoB levels on the basal diet, heterozygotes were intermediate. The LDLR RFLP accounted for approximately 3% to 7% of the variation in LDL-C and apoB levels. Subsequent multivariate analyses revealed that individuals with different LDLR RFLP genotypes responded differently to the high-cholesterol, saturated fat diet (Blangero et al., 1989), thus providing evidence for a genotype by diet interaction. This interaction was apparent not only at the level of the measured genotype (LDLR), but also in the polygenic background.

Quantitative genetic analysis of LDL-C levels in these baboons (Konigsberg et al., 1989) indicated a relatively low genetic correlation between LDL-C on the basal and HCSF diets. This polygenotype by diet interaction is most easily detected when information is available for each animal on each of two diets. However, an analysis that utilized information on diet in related individuals (a single randomly selected diet for each animal) also revealed significant interaction between genotype and diet (Blangero et al., 1990b). Thus, detection of genotype by environment interaction may be feasible even if data are not available for each individual in two different environments, although within-family environmental correlations could tend to confound results in analyses of human family data.

Differences in LDL-C between subspecies also have been found (Williams-Blangero et al., 1990). *P. h. cynocephalus* had significantly lower LDL-C than *P. h. anubis* on the basal diet. The difference in LDL-C levels between these two baboon subspecies is consistent with the genetic distance estimated from polymorphic markers, which is in the range reported for differentiation among the major human races. Quantitative genetic analyses also revealed interaction between subspecies and diet (Konigsberg et al., 1989): Animals with greater *P. h. cynocephalus* ancestry had a significantly greater LDL-C response to HCSF diet than did those with greater *P. h. anubis* ancestry.

Evidence for major gene effects has been found for LDL-C levels on the basal diet (Konigsberg et al., 1991). The models evaluated were similar to those described above for HDL-C, but with several additional parameters. In particular, subspecies was included as a covariate, and the major genotype-specific mean LDL-C level was made a function of age. This analysis is instructive because the major gene effect on LDL-C was manifested primarily as a genotype-specific difference in the regression of LDL-C level on age, rather than as an overall difference in LDL-C levels between different major locus genotypes. It is not yet known to what extent the LDLR RFLP or other candidate loci contribute to this major gene effect. Likewise, we have not yet done bivariate segregation analysis of LDL-C and other lipoproteins or apolipoproteins, to determine whether the LDL-C major locus has pleiotropic effects.

CONCLUSION

All other things being equal, humans are the organism of choice for studies of a human disease process. Whenever an animal model is used, the population of inference must be carefully specified and caution must be used in extrapolating results to humans. However, as described here, studies of animal models can complement human studies and lead to greater understanding of complex phenotypes that are common across species. If genetic analyses of complex phenotypes in human families yield confusing or ambiguous results, parallel studies in an animal model, with careful control of relevant environmental variables, can provide insights that will guide subsequent studies.

These analyses illustrate some of the advantages of animal models for investigating genetic effects on lipoproteins, and the value of accurate dietary information for detecting and characterizing major genes. When the most important environmental determinant of a complex phenotype is accurately quantified and individual responses to changes in environment are included in analyses, finer resolution of the genetic determinants of the phenotype becomes possible. The discovery of two separate apoAI major loci would have been unlikely if apoAI had not been measured for each individual in two different, controlled environments. Similarly, the heritable sexual dimorphism in apoAI levels and the different age-specific changes in LDL-C levels associated with different major locus genotypes may not have been detected in the presence of dietary heterogeneity. The infant diet effects that have been demonstrated on adult apoAI concentrations in baboons emphasize the importance of early environment in determining complex phenotypes and underscore the value of incorporating dietary history in analyses of human data. These results in nonhuman primates also offer the troubling prospect that important major gene effects on lipoproteins, and interactions between major genotypes and diet, will be missed in studies of human families that cannot be subjected to carefully controlled diets.

The results of segregation analysis of LDL-C, HDL-C, and apoAI concentrations in baboons are consistent with results of human studies in which single major loci have been detected (e.g., Hasstedt et al., 1986; Moll et al., 1989; Morton et al., 1978). Segregation analysis of human apoAI (Moll et al., 1989) indicated heterogeneity among families. It is tempting to speculate that this heterogeneity is attributable to a second major locus. Following up on the results of segregation analysis of apoAI levels in baboons, two locus segregation analyses of these human families could clarify the basis of the apparent heterogeneity.

The chromosomal localization of one of the apoAI major genes in baboons is being aided by the existence of a polymorphism in the LCAT gene, a locus for which no variability has yet been detected in humans. These results can help to guide attempts to localize the apoAI major gene or genes in humans. Measured genotype analyses and quantitative trait linkage analyses of CETP and other linked markers in humans and baboons may indicate whether the same major genes influence apoAI serum concentrations in the two species.

It would be surprising if genetically mediated responses to increased dietary fat and cholesterol, as found in baboons, do not also exist in humans. It remains to be seen whether the heterogeneity in response to diet that exists between baboon subspecies also is found between human populations. Such effects are difficult to demonstrate unless each individual is exposed to two dietary environments. Nevertheless, analyses of baboon data show that genotype–environment interaction is detectable in analyses using related individuals exposed to different diets. Verification of the utility of this method may be important if interactions of major locus genotypes with dietary and other environmental effects are to be studied in humans.

ACKNOWLEDGMENTS

This research was supported by NIH grants HL28972, HL06700, and HL35444, and contract HV53030. Many people contributed to the work described here and/or made useful comments on the manuscript, including Drs Henry McGill, John Blangero, Bennett Dyke, James Hixson, Candace Kammerer, Lyle Konigsberg, Glen Mott, David Rainwater, Bradford Towne, John VandeBerg, and Sarah Williams-Blangero.

REFERENCES

Barnicot NA, Jolly CJ, Huehns ER, Dance N (1965) Red cell and serum protein variants in baboons. In Vagtborg H (ed) *International Symposium on the Baboon and Its Use as an Experimental Animal. Proceedings. The Baboon in Medical Research* (vol 2). Austin, TX, University of Texas Press, pp 323–338.

Bernstein IS, Gordon TP, Rose RM (1982) The interaction of hormones, behavior and social context in nonhuman primates. In Svare BB (ed) *Hormones and Aggressive Behavior*. New York, Plenum Press, pp 535–561.

Blangero J, Kammerer C, Konigsberg L, Hixson J, MacCluer J (1988) Statistical detection of genotype-environment interaction: a multivariate measured genotype approach. *Am J Hum Genet* 43:A234.

Blangero J, Kammerer C, MacCluer J (1989) Multivariate quantitative genetic analysis of genotype–environment interaction. *Am J Hum Genet* 45:A211.

Blangero J, MacCluer JW, Kammerer CM, Mott GE, Dyer TD, McGill HC Jr (1990a) Genetic analysis of apolipoprotein A-I in two dietary environments. *Am J Hum Genet* 47:414–428.

Blangero J, MacCluer JW, Mott GE (1987) Genetic analysis of apolipoprotein A-I in two environments. *Am J Hum Genet* 41:A250.

Blangero J, Williams-Blangero S, Konigsberg LW (1990b) Analysis of genotype-environment interaction using related individuals in different environments. *J Am Phys Anthropol* 81:195.

Bojanovski D, Alaupovic P, Kelley JL, Stout C (1978) Isolation and characterization of the major lipoprotein density classes of normal and diabetic baboon (*Papio anubis*) plasma. *Atherosclerosis* 31:481–487.

Buettner-Janusch J (1963) Hemoglobins and transferrins of baboons. *Folia Primat* 1:73–87.

Cheng M-L, Kammerer CM, Lowe WF, Dyke B, VandeBerg JL (1988) Method for quantitating cholesterol in subfractions of serum lipoproteins separated by gradient gel electrophoresis. *Biochem Genet* 26:657–681.

Coelho AM, Glassman DM, Bramblett CA (1984) The relationship of adiposity and body size to chronological age in olive baboons. *Growth* 48:445–454.

Crawford MH, Walsh RA, Cragg D, Freeman GL, Miller J (1987) Echocardiographic left ventricular mass and function in the hypertensive baboon. *Hypertension* 10:339–345.

Creau-Goldberg N, Turleau C, Cochet C, deGrouchy J (1983) New gene assignments in the baboon and new chromosome homologies with man. *Ann Genet* 26:78–88.

deLemos RA, Coalson JJ, Gerstmann DR, Kuehl TJ, Null DM (1987) Oxygen toxicity in the premature baboon with hyalin membrane disease. *Ann Rev Respiratory Disease* 136:677.

Dutrillaux B, Viegas-Perquignot E, Dubos C, Masse R (1978) Complete or almost complete analogy of chromosome banding between the baboon (*Papio papio*) and man. *Hum Genet* 43:37–46.

Eggen DE (1974) Cholesterol metabolism in rhesus monkey, squirrel monkey, and baboon. *J Lipid Res* 15:139–145.

Eichberg JW, Heberling RL, Kalter SS, Morrison JD, Lawlor DA (1985) The influence of age on immune responses of baboons to mitogens and the baboon endogenous virus. *Devel Compar Immunol* 4:135–144.

Ferrell RE, Sepehrnia B, Kamboh MI, VandeBerg JL (1990) Highly polymorphic apolipoprotein A-IV locus in the baboon. *J Lipid Res* 31:131–135.

Festing MFW (1979) *Inbred Strains in Biomedical Research.* London, Macmillan.

Flow BL, Cartwright TC, Kuehl TJ, Mott GE, Kraemer DC, Kruski AW, Williams JD, McGill HC Jr (1981) Genetic effects on serum cholesterol concentrations in baboons. *J Hered* 72:97–103.

Fox H (1933) Arteriosclerosis in lower mammals and birds: its relation to the disease in man. In Cowdry EV (ed) *Arteriosclerosis: A survey of the Problem.* New York, Macmillan Co, pp 153–193.

Gleiser CA, Carey KD, Heberling RL (1984) Malignant lymphoma and Hodgkin's disease in baboons (*Papio* sp.). *Laboratory Animal Science* 34:286–289.

Goncharov N, Taranov A, Antonichev A, Gorlushkin V, Aso T, Cekan S, Diezfalusy E (1979) Effect of stress on the profile of plasma steroids in the baboon (*Papio hamadryas*). *Acta Endocrinol* 90:372–384.

Haseman JK, Elston RC (1972) The investigation of linkage between a quantitative trait and a marker locus. *Behav Genet* 2:3–19.

Hasstedt SJ, Ash KO, Williams RR (1986) A reexamination of major locus hypothesis for high density lipoprotein cholesterol level using 2,170 persons screened in 55 Utah pedigrees. *Am J Med Genet* 24:57–67.

Hasstedt SJ, Cartwright PE (1979) *PAP—Pedigree Analysis Package*. Technical Report No. 13, Dept Medical Biophysics and Computing, University of Utah, Salt Lake City, UT.

Hendrickx A (1971) *Embryology of the Baboon*. Chicago, University of Chicago Press.

Higuchi K, Matsumura A, Hashimoto K, Honma A, Takeshita S, Higuchi K, Kohno A, Matshuchita M, Hosokawa M, Takeda T (1983) Isolation and characterization of senile amyloid-related antigenic substance (SAS_{SAM}) from mouse serum. *J Exp Med* 158:1600–1614.

Higuchi K, Yonezu T, Kogishi K, Matsumura A, Takeshita S, Higuchi K, Kohno A, Matschuchita M, Hosokawa M, Takeda T (1986) Purification and characterization of a senile amyloid-related antigenic substance (apo SAS_{SAM}) from mouse serum. *J Biol Chem* 25:12834–12840.

Hilbert P, Lindpaintner K, Beckmann JS, Serikawa T, Soubrier F, Dubay C, Cartwright P, De Gouyon B, Julier C, Takahashi S, Vincent M, Ganten D, Georges M, Lathrop GM (1991) Chromosomal mapping of two genetic loci associated with blood-pressure regulation in hereditary hypertensive rats. *Nature* 353:521–529.

Hixson JE, Borenstein S, Cox LA (1990) PvuII RFLP for the lecithin: cholesterol acyltransferase gene (LCAT) in baboons. *Nuc Acids Res* 18:384.

Hixson JE, Borenstein S, Cox LA, Rainwater DL, VandeBerg JL (1988a) The baboon gene for apoA-I: characterization of a cDNA clone and identification of DNA polymorphisms for genetic studies of cholesterol metabolism. *Gene* 74:483–490.

Hixson JE, Cox LA, Borenstein S (1988b) The baboon apo E gene: structure, expression, and linkage with the gene for apo C-I. *Genomics* 2:315–323.

Hixson JE, Henkel RD, Britten ML, Vernier DT, deLemos RA, VandeBerg JL, Walsh RA (1989a) Alpha-myosin heavy chain cDNA structure and gene expression in adult, fetal, and premature baboon myocardium. *J Mol Cell Cardiol* 21:1073–1086.

Hixson JE, Kammerer CM, Cox LA, Mott GE (1989b) Identification of LDL receptor gene marker associated with altered levels of LDL cholesterol and apolipoprotein B in baboons. *Arteriosclerosis* 9:829–835.

Holmes RS, VandeBerg JL (1986) Aldehyde dehydrogenase, aldehyde oxidase and xanthine oxidase from baboon tissues: subcellular distribution in liver and brain, tissue distribution and phenotypic variability. *Alcohol* 3:205–214.

Holmes RS, VandeBerg JL (1987) Baboon alcohol dehydrogenase isozymes: phenotypic changes in liver following chronic consumption of alcohol. *Isozymes: Current Topics in Biological and Medical Research* 16:1–20.

Human Gene Mapping 10 (1989) Tenth International Workshop on Human Gene Mapping. *Cytogenet Cell Genet* 51, Nos. 1–4.

Ishida BY, Paigen B (1989) Atherosclerosis in the mouse. In Lusis AJ, Sparkes SR (eds) *Genetics Factors in Atherosclerosis: Approaches and Model Systems, Monogr Hum Genet* (vol 12). Basel, Karger, pp 189–222.

Jacob HJ, Lindpaintner K, Lincoln SE, Kusumi K, Bunker RK, Mao Y-P, Ganten D, Dzau VJ, Lander ES (1991) Genetic mapping of a gene causing hypertension in the stroke-prone spontaneously hypertensive rat. *Cell* 67:213–224.

Kammerer CM, Hixson JE, Blangero J, MacCluer JW (1988) Relationship between RFLP's for apo A-I and LCAT and probable major gene genotypes for high density lipoprotein cholesterol (HDL-C) and apo A-I serum concentrations in baboons. *Am J Hum Genet* 43:A216.

Kammerer CM, MacCluer JW, VandeBerg JL, Mott GE (1987) Possible linkage between APRT and a major gene(s) for serum concentrations of lipoproteins and apolipoproteins in baboons. *Am J Hum Genet* 41:A257.

Kammerer CM, Mott GE, Carey KD, McGill HC Jr (1984) Effects of selection for serum cholesterol concentrations on serum lipid concentrations and body weight in baboons. *Am J Med Genet* 19:333–345.

Khachadurian AK (1964) The inheritance of essential familial hypercholesterolemia. *Am J Med* 73:402–407.

Kitchen FD, Barnicot NA, Jolly CJ (1967) Variations in the group-specific (Gc) component and other blood proteins of baboons. In Vagtborg H (ed) *International Symposium on the Baboon and Its Use as an Experimental Animal. Proceedings: The Baboon in Medical Research* (vol 2). Austin, TX, University of Texas Press, pp 637–657.

Konigsberg L, Kammerer C, Blangero J, Williams-Blangero S, MacCluer J (1989) Quantitative genetic analysis of low density lipoprotein serum cholesterol levels in baboons fed two different diets. *Am J Hum Genet* 45:A242.

Konigsberg LW, Blangero J, Kammerer CM, Mott GE (1991) Mixed model segregation analysis of LDL-C concentration with genotype-covariate interaction. *Genet Epidemiol* 8:69–80.

Kurlansky PA, Sadeghi AM, Michler RE, Smith CR, Marboe CC, Thomas WA, Coppey L, Rose EA (1987) Comparable survival of intra-species and cross-species primate cardiac transplants. *Transplant Proc* 19:1067–1071.

Laber-Laird K, Rudel LL (1989) Genetic aspects of plasma lipoprotein and cholesterol metabolism in nonhuman primate models of atherosclerosis. In Lusis AJ, Sparkes SR (eds) *Genetic Factors in Atherosclerosis: Approaches and Model Systems, Monogr Hum Genet* vol 12. Basel, Karger, pp 170–188.

Lalley PA, Davisson MT, Graves JAM, O'Brien SJ, Womack JE, Roderick TH, Creau-Goldberg N, Hillyard AL, Doolittle DP, Rogers JA (1989) Report of the committee on comparative mapping. *Cytogenet Cell Genet* 51:503–532.

Lalley PA, O'Brien SJ, Creau-Goldberg N, Davisson MT, Roderick TH, Echard G, Womack JE, Graves JM, Doolittle DP, Guidi JN (1987) Report of the committee on comparative mapping. In Human Gene Mapping 9 (1987): Ninth International Workshop on Human Gene Mapping. *Cytogenet Cell Genet* 46:367–389.

Levin JL, Hilliard JK, Lipper SL, Butler TM, Goodwin WJ (1988) A naturally occurring epizootic of Simian Agent 8 in the baboon. *Lab Animal Sci* 38:394–397.

Lewis DS, Bertrand HA, Masoro EJ, McGill HC Jr, Carey KD, McMahon CA (1983) Preweaning nutrition and fat development in baboons. *J Nutrition* 113:2253–2259.

Lockwood DH, Coppenhaver DH, Ferrell RE, Daiger SP (1984) X-linked, polymorphic genetic variation of thyroxin-binding globulin (TBG) in baboons and screening of additional primates. *Biochem Genet* 22:81–88.

MacCluer JW, Kammerer CM, Blangero J, Dyke B, Mott GE, VandeBerg JL, McGill HC Jr (1988) Pedigree analysis of HDL cholesterol concentration in baboons on two diets. *Am J Hum Genet* 42:401–413.

MacCluer JW (1989) Statistical approaches to identifying major locus effects on disease susceptibility. In Lusis AJ, Sparkes SR (eds) *Genetic Factors in Atherosclerosis: Approaches and Model Systems. Monographs in Human Genetics.* Basel, Karger, pp 50–78.

McGill HC Jr, McMahan CA, Kruski AW, Mott GE (1981a) Relationship of lipoprotein cholesterol concentrations to experimental atherosclerosis in baboons. *Arteriosclerosis* 1:3–12.

McGill HC Jr, McMahan CA, Wene JD (1981b) Unresolved problems in the diet–heart issue. *Arteriosclerosis* 1:164–176.

McGill HC Jr, Strong JP, Holman RL, Werthessen NT (1960) Arterial lesions in the Kenya baboon. *Circulation Res* 8:670–679.

Micha JP, Quimby F (1984) Baboon cervical colposcopy, histology, and cytology. *Gynecol Oncol* 17:308–313.

Moll PP, Michels VV, Weidman WH, Kottke BA (1989) Genetic determination of plasma apolipoprotein AI in a population-based sample. *Am J Hum Genet* 44:124–139.

Moor-Jankowski J, Wiener AS, Gordon EB (1964) Blood groups of apes and monkeys. The ABO blood groups in baboons. *Transfusion* 4:92–100.

Morizot DC (1984) Tracing linkage groups from fishes to mammal. *Cytogenet Cell Genet* 37:543.

Morton NE, Gulbrandsen CL, Rhoads GG, Kagan A, Lew R (1978) Major loci for lipoprotein concentrations. *Am J Hum Genet* 30:583–589.

Mott GE, McMahan CA, McGill HC Jr (1978) Diet and sire effects on serum cholesterol and cholesterol absorption in infant baboons (*Papio cynocephalus*). *Circ Res* 43:364–371.

Paigen B, Albee D, Holmes P, Mitchell D (1987) Genetic analysis of murine strains C57BL/6J and C3H/Hej to confirm the map position of *Ath-1*, a gene determining atherosclerosis susceptibility. *Biochem Genet* 23:501–511.

Paigen B, Nesbitt M, Albee D, Mitchell D, LeBoeuf R (1989) *Ath-2*, a second gene affecting atherosclerosis susceptibility and high density lipoprotein levels in mice. *Genetics* 122:163–168.

Prochazka M, Leiter E, Serreze D, Coleman D (1987) Three recessive loci required for insulin-dependent diabetes in nonobese diabetic mice. *Science* 237:286–289.

Rainwater DL, Manis GS, VandeBerg JL (1989) Hereditary and dietary effects on apolipoprotein(a) isoforms and Lp(a) in baboons. *J Lipid Res* 30:549–558.

Rao DC, Elston RC, Kuller LH, Feinleib M, Carter C, Havlik R (1984) *Genetic Epidemiology of Coronary Heart Disease, Past, Present and Future. Progress in Clinical and Biological Research* (vol 147). New York, Alan R Liss.

Rapacz J (1978) Lipoprotein immunogenetics and atherosclerosis. *Am J Med Genet* 1:377–405.

Rapacz J, Hasler-Rapacz J (1989) Animal models: the pig. In Lusis AJ, Sparkes SR (eds) *Genetic Factors in Atherosclerosis: Approaches and Model Systems, Monogr Hum Genet* (Vol 12). Basel, Karger, pp 139–169.

Rogers J, Kidd JR, Murphy PD, Castiglione CM, Hixson JE, Phillips-Conroy J, Kidd KK (1988) RFLPs for APOB in two subspecies of savanna baboons (*Papio hamadryas cynocephalus* and *Papio hamadryas anubis*). *Cytogenet Cell Genet* 46:683.

Rosenfeld ME, Prescott MF, McBride C, Rapacz J, Hasler-Rapacz J, Attie AD (1988) The composition and distribution of atherosclerotic lesions in the Lpb-5 spontaneously hypercholesterolemic pig closely resemble human atherosclerotic lesions (Abstract). Federation of American Societies for Experimental Biology—Pathology.

Shotake T, Nozawa K, Tanabe Y (1977) Blood protein variations in baboons. I. Gene exchange and genetic distance between *Papio anubis, Papio hamadryas* and their hybrid. *Japan J Genet* 52:223–238.

Snow CC (1967) Some observations on the growth and the development of the baboon. In Vagtborg H (ed) *The Baboon in Medical Research* (volume 2). Austin, TX, University of Texas Press.

Stewart-Phillips J, Lough J, Skomene E (1989) Ath-3, a new gene for atherosclerosis in mice. *Clin Invest Med* 12:121–126.

Templeton JW, Kuehl TJ, Sharp RM (1981) Population characterization of postpartum serums detecting baboon lymphocyte antigens BabLA. *Am J Primatol* 1:320.

Towne B, Blangero J, Mott GE (1990) *A Genetic Analysis of Sexual Dimorphism in Apolipoprotein AI Serum Levels in Baboons.* Austin, TX, Texas Genetics Society.

VandeBerg JL, Aivaliotis MJ (1990) Mannose-6-phosphate isomerase polymorphism in baboon erythrocytes. *Biochem Genet* 28:495–501.

VandeBerg JL, Cheng ML (1986) Genetics of baboons in biomedical research. In Else JG, Lee PC (eds) *Congress of the International Primatological Society, 10th.* Cambridge, Cambridge University Press, pp 317–327.

VandeBerg JL, Cheng M-L, Kammerer CM, MacCluer JW, McGill HC Jr (1987) Genetic and dietary mediation of lipoprotein phenotypes in baboons. In Hauss WH et al. (eds) *Proceedings, Fourth Münster Arteriosclerosis Symposium.* Dusseldorf, West Deutcher Verlag, pp 71–86.

VandeBerg JL, Weitkamp LR, Kammerer CM, Weill P, Rainwater DL (1989) Baboon gene mapping: linkage of plasminogen and apo(a). *Cytogenet Cell Genet* 51:1098.

Van der Riet FS, Human PA, Cooper DK, Reichart B, Fincham JE, Kalter SS, Kanki RJ, Essex M, Madden DL, Lai-Tung MT, et al. (1987) Virological implications of the use of primates in xenotransplantation. *Transplant Proc* 19:4068–4069.

van Oorschot RAH, VandeBerg JL (1991) Tight linkage between MPI and NP in baboons. *Genomics* 9:783–785.

Weaver DS, Jayo MJ, Jerome CP (1989) Bone structure and serum chemistry eighteen months after ovariectomy in baboons (*Papio anubis*). *Am J Phys Anthropol* 78:321.

Wene JD, Mitchell DS, Barnwell GM, Coelho AM (1981) Flavor preferences and weight changes in baboons. *Am J Primatol* 1:313.

Williams-Blangero S, VandeBerg JL, Blangero J, Konigsberg L, Dyke B (1990) Genetic differentiation between baboon subspecies: relevance for biomedical research. *Am J Primatol* 20:67–81.

9

Genetics of Common Diseases that Aggregate, but Do Not Segregate, in Families

CHARLES F. SING
SHARON L. REILLY

The common diseases of humans that have a multifactorial etiology make a substantial contribution to morbidity (Baird et al., 1988; Trimble and Doughty, 1974; Trimble and Smith 1977) and mortality in the population at large. Because they are more common, and morbidity for each may extend over long periods of time, diseases like coronary artery disease, arthritis, diabetes, hypertension, and the psychiatric disorders require a greater demand on health services than the simply inherited inborn errors of metabolism and the chromosomal abnormalities combined (Baird et al., 1988). These common diseases also provide the greatest research challenge because etiological links between multiple genetic and environmental causes and health status are complex.

Appreciation for the role that principles of genetics might play in unraveling the complexity of the common chronic diseases of humans began to take form in 1954 with the publication of a classic volume, *Human Heredity*, co-authored by Jim Neel and Jack Schull (1954). These pioneers in human genetics were among the first to recognize the utility of genetic analysis in explaining the epidemiology of diseases that aggregate in families but do not segregate as if determined by a single gene. The proceedings of a conference organized by Jim Neel, Margery Shaw, and Jack Schull (1965) nurtured the early dialogue between geneticists and epidemiologists. The exchange between these two groups of scientists has grown over the years to include the involvement of clinicians, biochemists, and molecular biologists (Berg et al., 1990; Bock and Collins, 1987; Morton and Chung, 1978; Rao et al., 1984; Sing and Skolnick, 1979). A journal, *Genetic Epidemiology*, has emerged to report the progress of these interactions. Recent organization of an international society documents the scope of interest in the discipline.

For the past 30 years, Jack Schull has been a catalyst and a cofactor in the scientific interactions that have nurtured the emergence of genetic epidemiology of common diseases as an integrative field of study. We have all benefited from the insights and guidance he has generously imparted. His commitment to understanding the biology of the whole organism, rather than focusing entirely on defining the parts, has provided encouragement and inspiration for the work discussed in this chapter. We begin by reviewing how complexity enters into studies of the genetic architecture of the common multifactorial diseases (Sing and Moll, 1990a). Then we present how family studies can be the basis for a syncretism (Schull and Weiss, 1980) between genetics and epidemiology to understand this complexity.

COMPLEXITY IN THE STUDY OF COMMON CHRONIC DISEASES

Despite 30 years of effort to combine the power of genetics and epidemiology, promises of an integrated concept of how the effects of genetic and environmental variation combine to determine the distribution of predisposition to a common chronic multifactorial disease among individuals of the population at large have not been fulfilled. Why has progress been so slow? Reviews of where we have been and where we are going (Murphy, 1979; Schull, 1979; Schull and Weiss, 1980) have served to underscore the difficulty of the challenge. Research over the past 10 years on the biological effects of the molecules associated with these diseases has convinced most that the paradigm established for the study of single-gene diseases is not appropriate for the multifactorial diseases (Davignon et al., 1983; Sing et al., 1988; Sing and Moll, 1990a, 1990b; Sing et al., 1992). Such diseases are characterized by extreme complexity in (1) the biological processes involved in the development of disease; (2) the genetic architecture of inter-individual variability in initiation, progression, and severity of disease; and (3) the research strategies to unravel the relationships between genetic variability and variability in disease outcomes. In this section we first summarize the nature of the biological complexity using coronary artery disease (CAD) as an example. Then, we present alternative models for the complex relationships between genetic variation and variation in risk of a common chronic disease like CAD. We believe these considerations are a necessary preamble for discussions of the role that family studies might play in understanding the genetic architecture of CAD.

Biological complexity of CAD

Figure 9.1 presents a reminder of the hierarchical nature of the organization of units that are typically studied by geneticists. In this chapter, we consider the biological complexity of the relationships between characteristics of the genotype and the intermediate quantitative measures of the biochemistry and physiology of cells and tissues that link genetic and environmental variation with inter-individual variation in risk of CAD. The role of pedigree studies to understand these relationships cannot be evaluated without some

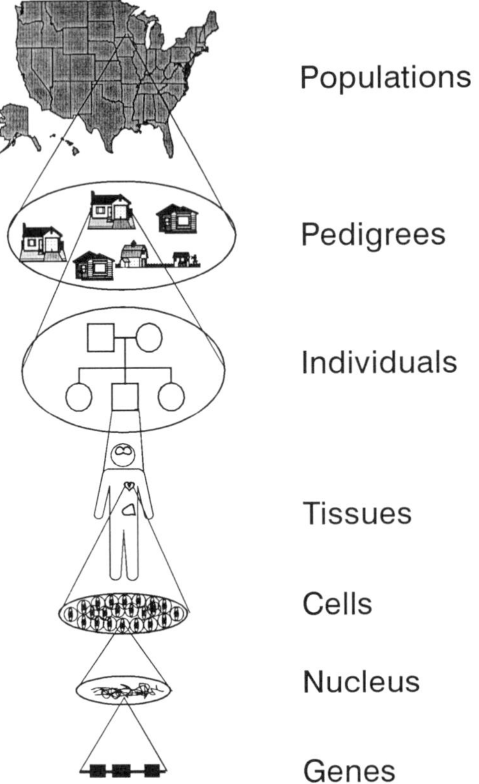

Figure 9.1. The hierarchy studied by geneticists.

a priori notion about the nature of the complexity upon which hypotheses to be tested are built.

At the individual level and below, diseases like CAD share four basic attributes that characterize all complex systems (Salthe, 1985; Simon, 1962). First, initiation and development of disease is determined by the same *coherent network* of interrelated biological traits that define normal biochemical and physiological processes. Structural characteristics of tissues, measures of intracellular metabolism, plasma measures of lipid metabolism, carbohydrate metabolism, thrombosis and thrombolysis, and blood pressure are traits that have been implicated as having separate and joint effects on the development of CAD (Anderson et al., 1991; Davignon et al., 1983; Flack and Sowers, 1991). Such traits are "agents" in a complex network of interdependent cellular and molecular processes that determine the behavior of tissues that make up the individual. A schematic diagram that portrays the nature of this complexity is presented in Figure 9.2. Coherence of the network implies that variation in each trait is bounded and covariation between traits is constrained in order to maintain homeostatic responses to external perturbations.

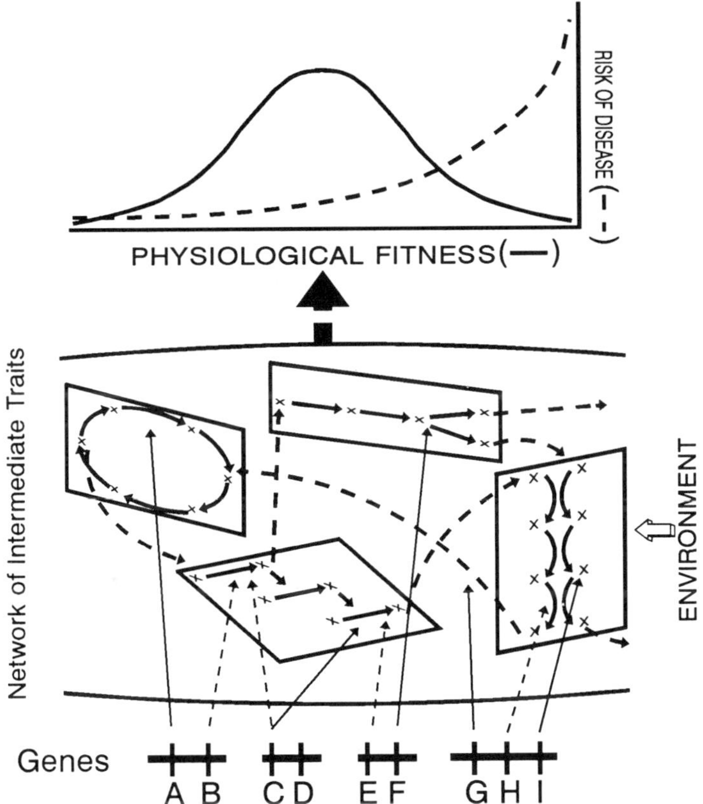

Figure 9.2. Condensed hierarchy consisting of a level of causes (genes), a coherent network of intermediate traits, and the emergent levels of physiological fitness and risk of disease.

Coherence in the network is manifest as strong forces (depicted by thick solid lines) between traits within a subsystem (tissue, cell, organelle, or metabolic pathway) and weaker forces (depicted by thick dashed lines) between traits of the different subsystems. The biological relationships between traits in the same or different subsystems are usually nonlinear (Savageau, 1976). It should be obvious that no single trait can be considered an independent predictor or risk of disease because of the inherent biological interactions with other traits in the network. Zannis et al. (1992) give an excellent review of the complexity of the relationships among traits that are measures of lipid metabolism. Guyton (1991) reviews the nature of the coherence between biochemical and physiological determinants of blood pressure regulation. Ferrannini (1991) describes the behavior of the network at the macro level for hypertension, "in a network, stretching any knot (for instance, blood pressure) out of the tolerance boundaries (resulting in hypertension) will drag neighboring knots [i.e., plasma cholesterol, triglycerides, glucose, and insulin are examples] along, with different solidarity, eventually producing a mixture of clinical and subclinical abnormalities." Establishing which knot is dragging which neighboring knots will depend on where in the network the responsible genetic or environmental perturbation is acting.

Second, CAD is an *emergent property* of the network of interrelated biological traits (Figure 9.2). All complex systems have emergent properties, i.e., qualities or attributes that cannot be predicted by the "sum of the parts." CAD is an emergent property because there is no value of a particular trait, or values for a combination of traits, that is always associated with the presence or absence of disease. Relationships between levels of traits in the network, as well as the actual levels of particular traits, may be involved in determining risk of disease (Austin et al., 1990; Krauss 1991). Consequently, risk of disease is expected to be a complex function of the N-dimensional network of traits that is continuously distributed among individuals in the population at large. The number of traits in the network that determines the initiation, progression, and severity of CAD is large. For a two-dimensional example, risk of CAD increases as the plasma levels of low density lipoprotein cholesterol increases and high density lipoprotein cholesterol decreases. The impact of this relationship on risk may be different in smokers or individuals on a very high fat diet (see Grundy, 1991). Disease may also emerge as a consequence of loss of coherence among traits in the network. For example, genetic or environmental factors may push the network of traits into a pattern of non-homeostatic relationships that increases the risk of initiation or influences the rate of progression and the degree of severity of disease. Our knowledge of the exact mechanisms responsible for the emergence of disease is far from complete. There are some who argue that until a general theory for relating the organization of leptons and quarks etc. at the micro level of the atomic nucleus to the organization of celestial bodies at the macro level of the universe is developed, we cannot hope to understand the mechanisms responsible for emergent biological properties (Hawking, 1988; Renteln, 1991). In the meantime, probability models play a central role in describing the relationships between the properties of the network and the onset, progression, and severity of disease.

Third, the genetic and environmental causes of disease act in a *hierarchical* fashion through the network of intermediate biological traits that are associated with the development of disease. Cells make up tissues, tissues make up organs, and organs make up individuals. Morphological and physiological variations of cells and tissues are emergent properties determined by interactions of variations in the micro properties of DNA with variations in external (to the cell, tissue, or organism) environments. Ferrell (chapter 3, this volume) discusses the nature of intraorganismal interactions that lead to cancer and possibly CAD. The hierarchical nature of the biology of common chronic dieseases involves three key features shared by all complex systems (Simon, 1962) that make crossing between levels a special challenge.

(1) Each level in the hierarchy (Figure 9.1) is characterized by fewer agents than exist at the level below. For example, plasma level of total cholesterol is one of several well-established biological risk factors for CAD. It is determined by the cholesterol present in at least four lipoprotein particles. The metabolism of the lipoprotein particles is under the control of many genes (perhaps as many as 100) that code for enzymes, ligands, and receptors. Each gene is encoded by thousands of nucleic acids. For example, there are over 150 mutations in the gene coding for the low density lipoprotein receptor protein that are associated with the extreme values of the low density lipoprotein cholesterol characteristic of the familial hypercholesterolemia phenotype (Hobbs et al., 1990).

(2) The hierarchical structure necessitates that there will be a slower rate of reaction at each higher level in the hierarchy. For instance, cellular processes occur on the order of micro-seconds, whereas the reaction of the kidneys to regulate blood pressure is on the order of hours or days (Guyton, 1991).

(3) Effects at each level in the hierarchy exist in different scales and are measured with very different yardsticks (Albrecht-Buehler, 1990). For example, square millimeters of abnormal coronary artery surface, mg/dl plasma cholesterol, and number of kringles in the apo (a) molecule that is associated with the Lp(a) lipoprotein particle are currently common metrics for studying the genetic architecture of CAD.

For research purposes, the hierarchy can be summarized by a level of causes (DNA and environments), a coherent network of traits at an intermediate level, and the emergent level of disease manifestation (Figure 9.2). The key features of hierarchical structures listed above suggest several heuristics about how genetic variation may be associated with variation in risk of disease. For instance, some of the relationships between genetic variation and variations in the network of intermediate traits are expected to be strong (thin solid lines in Figure 9.2), while others are expected to be weak (thin dashed lines in Figure 9.2). However, on average, traits at each level in the hierarchy are expected to be more strongly associated than are traits between levels. Hence, we expect many of the associations between variations at the genotype level and variations in the network of intermediate traits (thin dashed lines in Figure 9.2) to be on the order of the weak relationships between traits of different subsystems of the network. These weak relationships between levels in the hierarchy contribute to the inability to precisely know when a macro event (coronary occlusion) will occur using only information about a particular micro property (amino acid sequence of a protein or DNA sequences of a gene) at a lower level in the hierarchy. Whether genetic variation has a greater influence on the strong or the weak relationships between traits in the network is an important issue for understanding the genetics of diseases like CAD. The hierarchical nature of the linkages between genetic and environmental causes and disease endpoints is not reflected in the models that are currently being used to study the genetic epidemiology of CAD. There is some solace in the fact that the problem of crossing hierarchical levels is shared by all research disciplines (Bak and Chen, 1991; Kauffman, 1991; Renteln, 1991).

Fourth, like all common chronic diseases, CAD is a *dynamic process.* Age of initiation, rate of progression, and degree of clinical severity are major features that may vary from one individual to another (McGill, 1979). Inter-individual variation in these features is a consequence of the interactions of genetic variation with environmental variations that are distributed in time and space. The intermediate traits that determine age of initiation, rate of progression, and degree of clinical severity are often more easily measured and are also dynamic. Levels of these intermediate traits are a consequence of physiological adaptation to environmental changes over the lifetime of the individual and to different environmental niches at a particular age. The relationships between intermediate traits may also be a major feature of adaptation that varies within and among individuals as a consequence of the interactions of genetic variation with environmental variations in time and space. Physiological adaptation to environmental variation in time and space may be occurring at different rates in different individuals, and this property may be under genetic control. Additionally, genetic analysis is complicated because many of the genes that are involved in the etiology of the common chronic diseases are programmed to turn off (or on) during a lifetime and the genotype of a tissue may change over time as a consequence of somatic mutation (see Ferrell, Chapter 3, this volume).

A summary of the four major features of a complex disease like CAD is given in Figure 9.3. Age of onset, progression, and severity are emergent properties of a complex network of intermediate traits under the influence of genetic and environmental factors. Change in the pattern of relationships among traits in the network of intermediate traits over a lifetime may be a primary force determining the development of disease. This pattern is expected to be influenced by an individual's genotype and exposures to atherogenic (or therapeutic) environments. Because disease is a dynamic process, genetic variations that influence inter-individual variation in the responses of intermediate traits to changes in particular environments may be better predictors of risk of disease than genetic variation that influences the level of the trait averaged over all environmental circumstances (Berg, 1988, 1990; Nestruck, et al., 1987; Reilly et al., 1991; Tikkanen et al., 1990). Genetic analysis will be complicated by the possibility that different genes and different combinations of intermediate traits are expected to be relevant at different times in the life cycle.

Examples of the nature of the complexity of the genotype–phenotype relationships abound from experimental work with plants and animals. Waelsch (1989) gives a particularly sobering account of 50 years of effort to understand the complexity of the relationships between mutations at the T locus and phenotypic variations of the mouse. Tauber and Sarkar (1992) raise serious doubts about the scientific merit of the Human Genome Project because its reductionist strategy ignores the complexity of the genotype-to-

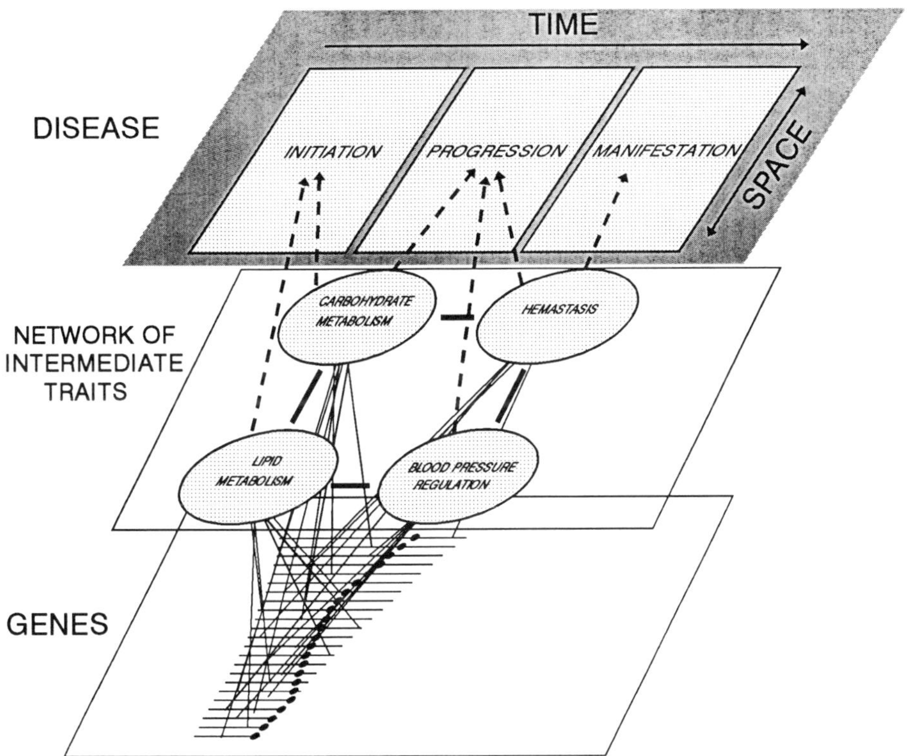

Figure 9.3. A summary of the four major features of the complex disease CAD—coherent network, emergent properties, hierarchy, and dynamic.

phenotype relationship. Such accounts should forewarn geneticists and epidemiologists of the difficulty of the problem of crossing the levels between genetic variation and phenotypes of traits that define the multifactorial diseases.

Modeling the complex biological relationships between the genotype and risk of disease through the network of intermediate traits

A mathematical model that includes biological and environmental parameters that reflect all characteristics of CAD discussed above does not exist. Much effort has been directed toward developing a class of linear models for prediction of inter-individual variation in intermediate traits or risk of disease. The traditional biometrical models used by geneticists in path analysis, complex segregation analysis, linkage analysis, and logistic regression analyses do not fully parameterize the voluminous information about the relationships between genes (and environments), intermediate traits, and endpoints that has come from research over the past decade on the molecular biology, biochemistry, and pathobiology of the disease process. The absence of relevant parameters in these models deters one from investigating some of the most interesting hypotheses about etiological relationships within and between levels of the hierarchy. Research questions are often chosen because there is an available statistical model rather than vice versa. Only crude insights into causes and their effects can be expected from the interpretation of estimates of correlation and regression coefficients, variance components, and probability parameters (e.g., allele frequencies, Mendelian ratios, recombination fraction, and relative risk) that characterize these models.

It may not be necessary for a useful prediction model to include all etiological relationships between causes and risk of disease. A model that provides a strategy for finding a subset of traits that optimizes the sensitivity and specificity of assignment of individuals and families to lower, or higher, risk groups would have great utility in developing public health programs. Most candidate traits will be suggested by laboratory studies. Such a prediction model should primarily facilitate a test of the null hypothesis that it is not necessary to include measures of the genotype and the environment when the phenotypes of intermediate traits that link these causes with disease risk are available. An alternative hypothesis states that genes and environments have pleiotropic effects on intermediate traits that are difficult, or impossible, to measure. Prediction is improved by measuring variation in the more accessible measures of genes and exposures to causal environments. The role of families in sorting between these two alternatives is discussed below.

At present there are only a few models that may be used to describe relationships known to exist between traits at a particular level or between traits from different levels in the hierarchy. These are relatively simple models that do not include much of the biological information that has accumulated in the past quarter century. For example, Weir and Cockerham (1989) present a complete model for the organization of genetic variation in pairs of polymorphic loci. Hartl (1989) reviews work in progress to model the evolution of genetic control of complex metabolic systems in *E. coli*. The work of Kacser and Burns (1979) and Savageau (1979) with complex metabolic pathways offers a step in the direction of defining models that may be useful for relating genetically determined parameters to quantitative variation of the intermediate traits that are involved in determining disease. These are examples of efforts to incorporate knowledge about the etiological complexity of biological relationships at one level of the hierarchy from DNA to disease manifestation.

Studies across two levels in the hierarchy have employed traditional regression models. For example, Kaprio et al. (1991) studied the relationship between genetic variation and inter-individual variation in intermediate traits that are involved in determining risk of CAD. Turner et al. (1992) use logistic regression to study the relationship between levels of physiological and biochemical intermediates and the occurrence of hypertension. No general model for prediction of a complex disease has been devised that includes all of the known biological interconnections between traits measured at each of the three major levels in the hierarchy and the relationships that are known to exist between levels. It is appropriate to question whether a set of mathematical equations can be developed that incorporates the behavior of all traits, from DNA to coronary pathology, that are involved. And, if they can be written, will it be possible to estimate the parameters that define the interesting questions? Could some simpler set of expressions, like those offered by a weatherman (Gleick, 1987; Stewart, 1989) or suggested by the non-traditional work of a few geneticists (Murphy and Trojak, 1986) provide utility, at least for prediction?

A useful model for prediction of CAD must provide the most accurate statement of probability (risk) of disease that is possible given the intermediate traits and genotypic and environmental stratifications that are hypothesized to determine risk. Some genotypes and environments will reflect information about risk of disease that cannot be measured at the intermediate trait level. Some intermediate traits will reflect emergent properties of genotype-by-genotype and genotype-by-environmental interaction effects that cannot be measured at either the genotype or environmental levels. What follows explores the first steps in developing a prediction model that considers the biological complexity of diseases like CAD that have been presented above.

We employ a nonparametric graphical visualization to convey the properties of the levels and relationships in the N-dimensional network of intermediate traits that connect variation in causes with variation in risk of disease. Figures 9.4–9.7 illustrate these properties for the distribution of five intermediate traits in the population at large. Each horizontal line is a profile of trait levels for an individual. The points on the vertical lines that intersect this horizontal line denote the values of the traits for the individual. Most

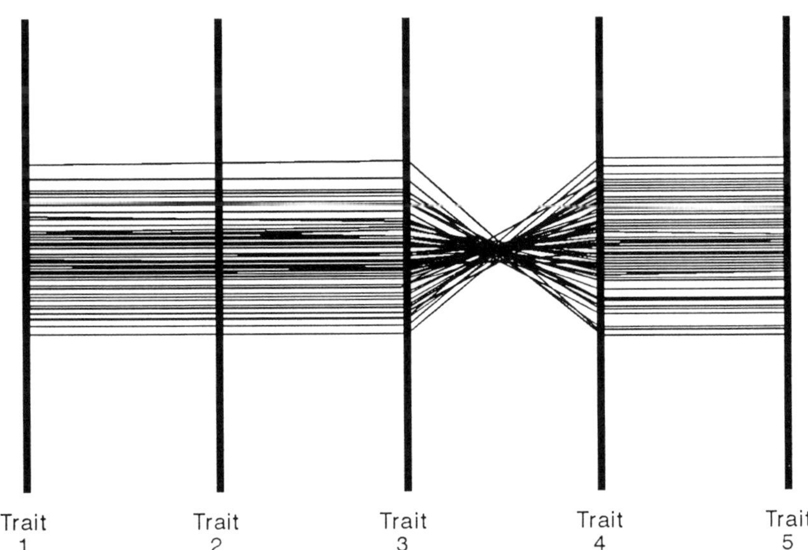

Figure 9.4. Theoretical representation of the healthy low-risk individuals from the population at large.

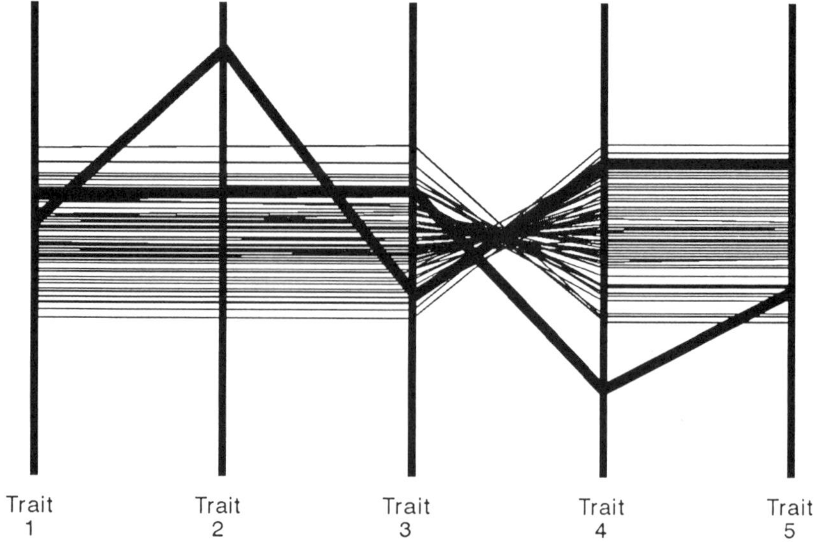

Figure 9.5. Theoretical representation of risk as a consequence of an extreme value in only one trait (bold lines = high risk individuals).

individuals cluster about the centroid of the five distributions. Figure 9.4 gives a theoretical representation of the healthy low-risk individuals from the population at large. Traits 1, 2, and 3 are positively correlated and each is negatively correlated with traits 4 and 5. These relationships illustrate only one of many possible correlation structures that might be characteristic of

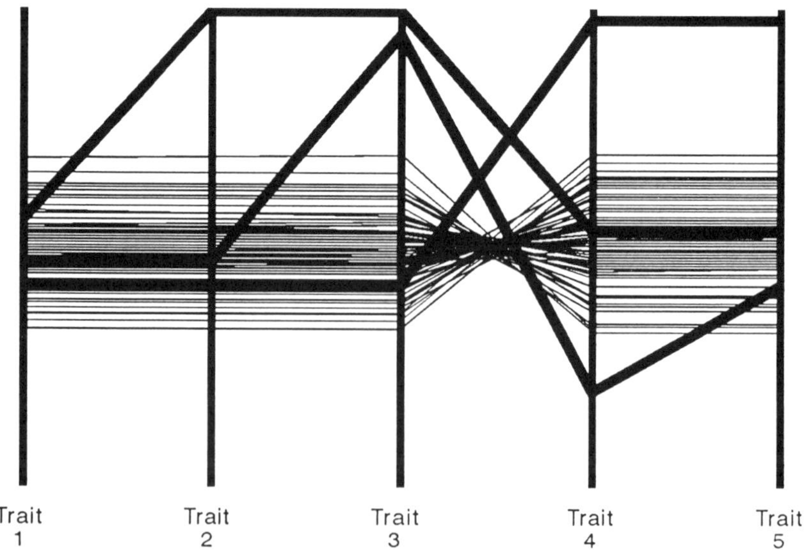

Figure 9.6. Theoretical representation of risk as a consequence of extreme values in more than one trait (bold lines = high-risk individuals).

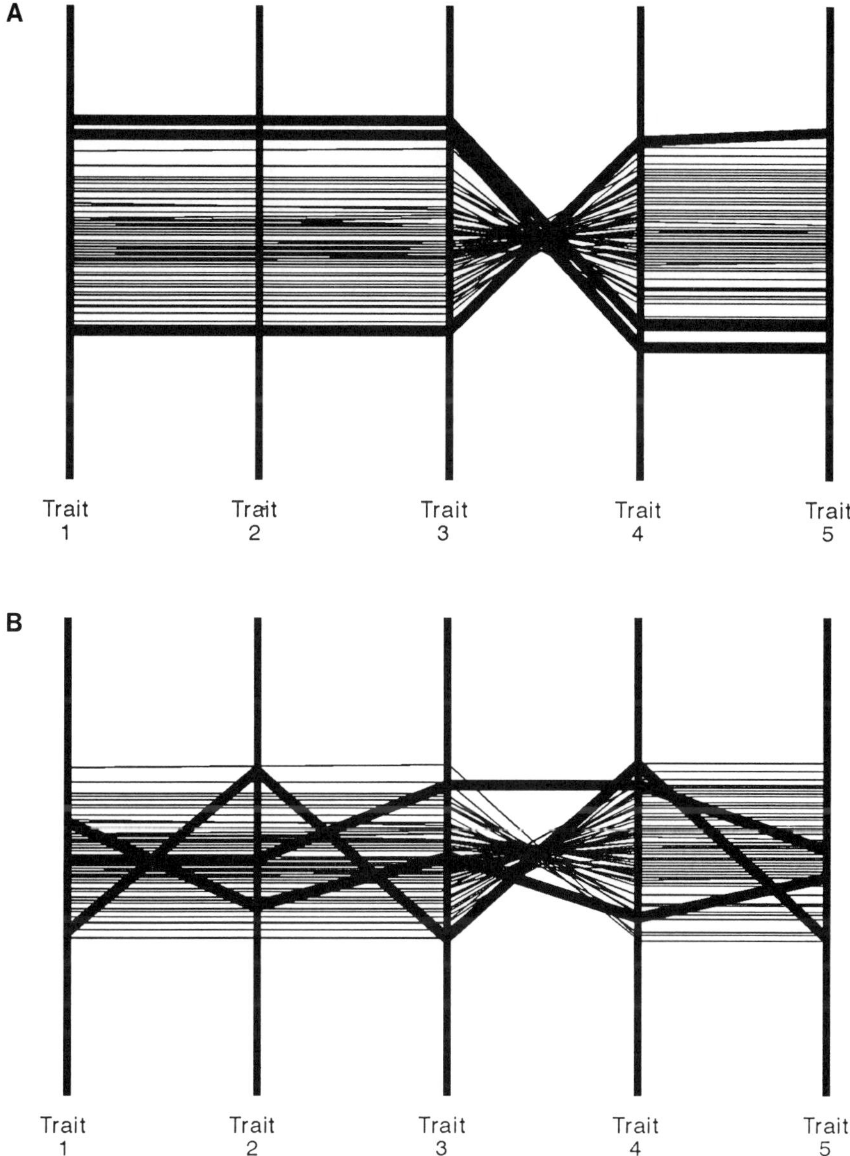

Figure 9.7. Theoretical representation of risk as a consequence of (A) values in the upper or lower tail of five traits; and (B) altered relationships between trait levels. Bold lines = high-risk individuals.

low-risk individuals. In reality, the correlation structure may be more complex and vary from one network of intermediate traits to another (Reilly and Sing, 1991; Roy et al., 1991).

Figures 9.5–9.7 present possible scenarios for relationships of measures of inter-individual variations in intermediate traits that are associated with increased risk of disease. In each case the heavy lines connect trait levels of individuals who are at higher risk to disease superimposed upon the distribution of profiles of the lower risk members of the population presented in Figure 9.4. Figure 9.5 presents two examples of the *univariate* model for causation of risk. Being extreme for one intermediate risk factor trait is sufficient for increasing risk of developing disease. Different individuals are at higher risk because they have extreme values for different intermediate traits. This multiple univariate explanation for a fraction of the cases of disease has been the etiological paradigm for many working on the genetics of complex diseases such as CAD (Mahley et al., 1991). A rare allelic variation in a particular gene is hypothesized to explain the large phenotypic deviation in a particular intermediate trait that is associated with risk of disease. Lifton et al. (1992) give an example of a rare mutation in a single gene that explains the segregation of extreme levels of blood pressure among members of a large pedigree. The fraction of CAD cases described by this model must be small as nearly every intermediate risk factor trait that has been described is approximately normally distributed in the population at large.

Figures 9.6 and 9.7 present *multivariate* models for the causation of risk of disease. Figure 9.6 portrays risk as a consequence of extreme values of more than one trait. In this example, different individuals are at risk because they have extreme values for different combinations of two intermediate traits. Because intermediate traits are usually involved in a metabolic pathway, combinations of alleles for genes that each influence a step in the pathway or an allelic variation in a single gene that has pleiotropic effects on more than one trait in the pathway may be responsible. Individuals with elevated cholesterol and triglycerides who are diagnosed as having combined familial hyperlipidemia serve as examples of this model (Brunzell et al., 1987). As with the first model presented in Figure 9.5, the fraction of cases of disease explained by extreme values of two traits is expected to be small.

Figure 9.7 presents two *multivariate* models for increased risk where all combinations of intermediate traits fall within the so-called normal range of variation defined by the profile of levels presented in Figure 9.4. The model in Figure 9.7A portrays increased risk as being a consequence of exceeding the upper (or lower) 95th percentile for each trait while the relationship between the traits is the same as found in healthy individuals (Figure 9.4). More than one genetic etiology could explain such a vector of high-risk values. Allelic variation at a single gene could have pleiotropic effects on all five traits or a multilocus genotype may be responsible. Figure 9.7B models increased risk as a consequence of altered relationships between levels in the profile of risk factor traits that again fall in the normal range of variation established by healthy individuals. These altered relationships indicate that these individuals have metabolically over- or under-compensated in response to specific genetic or environmental perturbations. Most cases of common chronic diseases like CAD will be associated with these two types of models.

A realistic model for describing the distribution of disease in the population at large must consider all four classes of models described above. Because most of the intermediate risk factor traits are approximately normally distributed in the population at large, the models portrayed in Figures 9.5 and 9.6 can explain only a fraction of the total prevalence of disease. For instance, if we assume (1) 10 statistically independent risk factor traits, (2) an individual has increased risk of disease if any one trait is above the 99th percentile (or below the 1st percentile), (3) a small fraction of these individuals will not develop disease because of compensating factors that lower risk, and (4) the prevalence of disease in the population is 20%, then we expect less than 50% of all cases of disease will be explained by these two models. The models portrayed in Figure 9.7 are necessary to explain the remainder of cases. Sing and Moll (1989) suggest that the genetic architecture of each of the intermediate traits that link genetic and environmental causes of risk of CAD involves a few genes with large effects like those modeled in Figures 9.5 and 9.6 and many genes, each having allelic variations with small effects, that combine to determine the normal range of phenotypic variation that is modeled in Figures 9.4 and 9.7.

THE ROLE OF MULTIGENERATION PEDIGREES IN THE STUDY OF COMMON CHRONIC DISEASES HAVING A COMPLEX GENETIC ARCHITECTURE

There are more questions about complex diseases than there are research laboratories that are currently actively seeking answers. Our ability to answer most of these questions depends critically on the ability to make appropriate measurements and to sample individuals in a way that allows one to estimate relevant parameters and test meaningful hypotheses. The range of sampling designs for non-experimental, observational studies of the causes of variation in risk in the population at large is limited by the social organization of the human species. We briefly review here unique strengths of multigeneration pedigrees for unraveling the complex genetic architecture of diseases like CAD.

Schull and Weiss (1980) have pointed out that the family is a unit of study that can serve as a basis for syncretism between geneticists and epidemiologists. The nature of the interface was stated well by Francis (1965), "so when the human geneticist turns to disease and disorder in the population as his basis of genetic analysis, he is promptly in epidemiology.... Conversely, where the epidemiologist seeks explanations for familial or other group aggregations of health or disease, he is immediately involved in genetic problems." Neither perspective is adequate for investigating all questions that might be asked about the relationships between causes, intermediate traits, and risk of disease that are presented above. Epidemiological studies typically ignore the clustering of genes and environments in families. On the other hand, a single pedigree can represent only a small window into the causes of the distribution of disease in the population at large.

We consider here three major questions that benefit from the use of extended families that are sampled to be representative of the population at large. First, what is the relative contribution of a particular gene to explaining the total phenotypic variance, and the genetic variance, of an intermediate trait that is continuously distributed in the population at large? Second, what fraction of diseases is explained by each of the many possible different etiologies that are illustrated by the four models presented in Figures 9.5–9.7? And, third, is information about genetic variation necessary for prediction of a late-onset, common, disease like CAD when measures of intermediate traits are available?

Randomly ascertained multigeneration pedigress provide a strategy for partitioning the phenotypic variance of a quantitative intermediate trait

Traditional biometrical models can impart only crude insights into the nature of the genetic and environmental causes of phenotypic variation. Until more realistic models are developed, these biometrical models provide the only method for partitioning the causes of inter-individual variation in quantitative traits. The pedigree unit plays a key role in application of these models to human data.

The distribution of phenotype values among related and unrelated individuals in randomly ascertained pedigree units provides appropriate statistical contrasts for simultaneously estimating the relative contribution of four causes of inter-individual variance (and familial aggregation) in the population at large. They are (1) allelic variations in a single gene having large phenotypic effects, (2) variation in polygenes, (3) variation in environmental factors shared by family members, and (4) variation in environmental factors (including measurement error) specific to each member of the pedigree. Although, the mathematical models (introduced by Elston and Stewart, 1971) and the computational software for accomplishing this goal exists (Moll, personal communication, 1992), no study of randomly ascertained pedigrees has been carried out to simultaneously estimate all four partitions. Annest et al. (1979) estimated the contribution polygenes, shared environments, and individual specific components of blood pressure variation using randomly ascertained nuclear families. Moll et al. (1979) estimated the same components of plasma cholesterol variation using pedigrees sampled to represent the population at large. Perusse et al. (1991) estimated the relative contribution of a single major gene, polygenes, and individual specific components of blood pressure variation using multigeneration pedigrees sampled to represent the population at large. Moll et al. (1991) give the only example of estimation of all four components of variation, but they used pedigrees that were ascertained through probands who were not representative of the population at large. This top-down, unmeasured genotype approach is a logical first step in a search for genes responsible for large phenotypic effects. Linkage to random markers, or measures of genes that are candidates because they are involved in the metabolism of the trait, would complete the phenotype-to-genotype link. However, only a very small fraction of the total number of genes involved in determining the genetic architecture of a quantitative trait will be revealed by this method because genotypes associated with such large effects will be rare in the population at large.

Boerwinkle et al. (1986) presented an extension of the Elston and Stewart (1971) model for simultaneous estimation of the frequencies and effects of measured genotypes at a single locus and the contribution of residual unmeasured polygenes to quantitative variation of a trait among individuals in the population at large. In an application of this strategy to randomly ascertained nuclear families, Boerwinkle and Sing (1987) estimated that 11% of the genetic variance of total serum cholesterol in the Nancy, France population may be assigned to genotypic variation determined by polymorphic alleles of the gene that codes for apolipoprotein E. This bottom-up, measured genotype approach makes possible the estimation of phenotypic effects determined by common alleles that are much smaller than might be detected by the top-down approach.

Randomly ascertained multigeneration pedigrees provide a natural sampling design for sorting out mapping relationships between genetic variation, variation in the network of intermediate traits, and risk of disease that exist in the population at large

Each pedigree with disease cannot be segregating for every one of the many genes involved in determining the genetic component of inter-individual variation in the network of intermediate biological traits that determine risk of a common disease like CAD. Different combinations of genetic and environmental factors responsible for the distribution of risk of disease in the population at large are expected to aggregate in different subsets of the pedigrees having diseased individuals. This truth presents the opportunity to allocate similar etiological relationships and similar subsets of the genetic and environmental causes of variation in intermediate traits to a particular group of pedigrees. The possible allocation strategies include sorting on the basis of similarity of disease endpoints, levels of intermediate traits, segregation of the same subset of genes, or sharing the same environmental factors. Major features of complex diseases (discussed above), such as the mapping between elements at different levels in the hierarchy is many-to-many and each level in the hierarchy is characterized by fewer elements than exist below it, dictate that pedigrees be sorted according to the segregating genes they share. In addition, pedigrees whose members share the same exposure to environmental factors that influence risk of disease would be desirable but is likely unattainable in an observational study.

For a disease like CAD, only a small fraction of pedigrees with disease are expected to be segregating for a single gene with large allelic effects, or a combination of alleles for a few genes, that cause extreme values of one or two of the intermediate traits (Figures 9.5 and 9.6). Because extreme values of the intemediate traits are much less frequent than the prevalence of a common disease like CAD, most cases of disease will occur in pedigrees whose members exhibit more common alterations of the relationships between intermediate trait values (Figures 9.7a and b) that fall within the normal range of variation (Figure 9.4) associated with healthy individuals. For example, epidemiological studies have convincingly established that most cases of CAD are associated with elevated levels of plasma cholesterol that fall within two standard deviations of the population mean. The pedigrees of these CAD patients are expected to be segregating for only a subset (say 3 to 5 of the possible 20 to 100) of the polymorphic genes that each have relatively small influences on alterations of the network of intermediate risk factor traits.

A paradox in this strategy exists that threatens our complete understanding using the pedigree approach: the greater the number of genes involved in determining risk, the fewer the number of pedigrees that will be segregating for the same subset of genes. The difficulty that we face cannot be fully realized until a complete list of CAD genes has been assembled and their genetic structure fully defined for the population of interest. The assignment of candidate genes to this list is progressing rapidly (Kessling et al., 1992).

Multigeneration pedigrees offer a cross sectional sampling strategy for estimating the relative contribution of genetic variation and variation in intermediate risk factor traits to the prediction of late onset endpoints

Common complex chronic diseases like CAD are characterized by clinical manifestations that appear later in life. Although clinically defined diseases will aggregate in pedigrees, most cases will occur in the older generations. The bulk of information about genetic variation and intermediate trait variation will be obtained from asymptomatic individuals in the parental and offspring generations. Furthermore, information about disease is often only available on deceased individuals who cannot provide tissues for genotyping and measurement of intermediate traits. Cross sectional studies of pedigrees offer the opportunity to investigate the association between genetic, biochemical, and physiological information collected on children and their parents with disease endpoint information collected on the grand-parental generation. Such a study takes advantage of the fact that allelic variations, intermediate trait values, and disease endpoints aggregate in families. The pedigree rather than the individual becomes the unit of study. The objective of a statistical analysis of such data is to determine if variation among pedigrees in frequency of disease in the older generations is predicted by the constellation of genotypes and intermediate trait values obtained from the asymptomatic parental and offspring generations. At a considerable cost savings, compared to a longitudinal study of unrelated individuals, such a cross sectional study offers a test of the null hypothesis that information about genetic variation does not improve our ability to predict risk of disease beyond that provided by information about intermediate traits that link genetic variation with variation in risk.

Recent developments in non-invasive methods to measure coronary artery disease in clinically asymptomatic individuals (Breen et al., in press) promises to shift the focus back to the individual as the unit of study. However, because different genes are likely to be involved at different stages in the development of disease and the establishment of the endpoint, information about genes, intermediate traits, and endpoints collected in pedigrees should provide complementary insights into the genetic architecture of a complex disease like CAD.

SUMMARY

The technological explosion of the 20th century has provided an incredible amount of information about every level of the biological hierarchy. One result of this technological explosion is an increase in our awareness of the difficulty in understanding the genetic architecture of common diseases such as CAD. The four main features of all complex systems that may have the greatest impact on our research strategies for understanding genetics of.CAD are coherence, emergence, hierarchy, and dynamics. Consequently, research strategies must consider the possibility that the endpoint is an *emergent property* of a *coherent network* of intermediate traits that tie together genetic variation and variation in risk to form a *hierarchical* system of partially decomposable subsystems that respond *dynamically* to exposures to environmental factors that vary in time and space. A new research paradigm is needed that will recognize the complexity of the genetic architecture that determines the initiation, progression, and severity of disease. The pedigree unit offers a data collection strategy for partitioning measured and unmeasured causes of phenotypic variation in intermediate traits, identifying subsets of individuals that share the same genetic etiology and evaluating the relative contribution of information about genetic variation and variation in intermediate biochemical and physiological traits to the prediction of common chronic diseases like CAD that have late age-of-onset. The relevant measures of genetic variation and intermediate traits that link genetic variation with variation in risk that have been provided by the technological explosion are crucial to the success of using pedigrees to answer these key questions.

ACKNOWLEDGMENTS

The central theme of this chapter developed as a consequence of our association with the Santa Fe Institute. We wish to thank Brian Athey, John Holland, Martha Haviland, Ken Weiss, and Kim Zerba for stimulating discussions about the issues that are presented in this manuscript. This work was supported in part by National Institutes of Health grants HL-30428 and HL-39107. We also wish to thank Dave Sing for the graphic representations of the ideas presented here.

REFERENCES

Albrecht-Buehler G (1990) In defense of "nonmolecular" cell biology. *Inter Rev Cytol* 120:191–241.

Anderson KM, Wilson PWF, Odell PM, Kannel WB (1991) An updated coronary risk profile: a statement for health professionals. *Circulation* 83:356–362.

Annest JL, Sing CF, Biron P, Mongeau JG (1979) Familial aggregation of blood pressure and weight in adoptive families. II. Estimation of relative contributions of genetic and common environmental factors to the phenotypic correlations between family members. *Am J Epidemiol* 110:492–503.

Austin MA, Brunzell JD, Fitch WL, Krauss RM (1990) Inheritance of low density lipoprotein subclass patterns in familial combined hyperlipidemia. *Arteriosclerosis* 10:520–530.

Baird PA, Anderson TW, Newcombe HB, Lowry RB (1988) Genetic disorders in children and young adults: a population study. *Am J Hum Genet* 42:677–693.

Bak P, Chen K (1991) Self-organized criticality. *Sci Amer* 264:46–53.

Berg K (1988) Variability gene effect on cholesterol at the Kidd blood group locus. *Clin Genet* 33:102–107.

Berg K (1990) Level genes and variability genes in the etiology of hyperlipidemia and atherosclerosis. In Berg K, Retterstol N, Refsum S (eds) *From Phenotype to Gene in Common Disorders.* Copenhagen, Denmark, Munksgaard A/S International Publishers, pp. 77–91.

Berg K, Retterstol N, Refsum S (eds) (1990) *From Phenotype to Gene in Common Disorders.* Copenhagen, Denmark, Munksgaard A/S International Publishers.

Bock G, Collins GM (eds) (1987) Molecular Approaches to Human Polygenic Disease. Wiley, Chichester, Ciba Foundation Symposium 130.

Boerwinkle E, Chakraborty R, Sing CF (1986) The use of measured genotype information in the analysis of quantitative phenotypes in man. I. Models and analytical methods. *Ann Hum Genet* 50:181–194.

Boerwinkle E, Sing CF (1987) The use of measured genotype information in the analysis of quantitative phenotypes in man. III. Simultaneous estimation of the frequencies and effects of the apolipoprotein E polymorphism and residual polygenetic effects on cholesterol, betalipoprotein and triglyceride levels. *Ann Hum Genet* 51:211–226.

Breen JF, Sheedy PF, Schwartz RS, Stanson AW, Kaufmann RB, Moll PP, Rumberger JA (in press) Coronary calcification detected with fast CT as a marker of coronary artery disease: works in progress. Radiology.

Brunzell JD, Chait A, Albers JJ, Foster DM, Failor RA, Bierman EL (1987) Metabolic consequences of genetic heterogeneity of lipoprotein composition. *Am Heart J* 113(2 Pt 2):583–588.

Davignon J, DuFour R, Cantin M (1983) Atherosclerosis and hypertension. In Genest J, Kuchel O, Hamet P, Cantin M (eds) *Hypertension, physiopathology and Treatment* (2nd ed). New York, McGraw-Hill, pp 810–852.

Elston RC, Stewart J (1971) A general model for the genetic analysis of pedigree data. *Hum Hered* 21:523–542.

Ferrannini E (1991) Metabolic abnormalities of hypertension. *Hypertension* 18:636–639.

Flack JM, Sowers JR (1991) Epidemiologic and clinical aspects of insulin resistance and hyperinsulinemia. *Am J Med* 91 (suppl 1A):11S–21S.

Francis T Jr (1965) Genetics and epidemiology opening comments. In Neel JV, Shaw MW, Schull WJ (eds) *Genetics and the Epidemiology of Chronic Disease.* U.S. Department of Health, Education, and Welfare (Public Health Service Publication No. 1163). Washington, DC, GPO, pp 1–6.

Gleick J (1987) Chaos, Making a New Science. New York, Viking.

Grundy SM (1991) Multifactorial etiology of hypercholesterolemia. *Arteriosclerosis and Thrombosis* 11:1619–1635.

Guyton AC (1991) Blood pressure control—special role of the kidneys and body fluids. *Science* 252:1813–1816.

Hartl DL (1989) Evolving theories of enzyme evolution. *Genetics* 122:1–6.

Hawking SW (1988) *A Brief History of Time.* New York, Bantam Books.

Hobbs HH, Russell DW, Brown MS, Goldstein JL (1990) The LDL receptor locus in familial hypercholesterolemia: mutational analysis of a membrane protein. *Annu Rev Genet* 24:133–170.

Kacser H, Burns JA (1979) Molecular democracy: who shares the controls? *Trans Biochem Soc* 7:1149–1160.

Kaprio J, Ferrell RE, Kottke BA, Kamboh MI, Sing CF (1991) Effects of polymorphism in apolipoproteins E, A-IV, and H on quantitative traits related to risk for cardiovascular disease. *Arteriosclerosis and Thrombosis* 11:1330–1348.

Kauffman SA (1991) Antichaos and adaptation. *Sci Amer* 264:78–84.

Kessling A, Ouellette S, Bouffard O, Chamberland A, Betard C, Selinger E, Xhignesse M, Lussier-Cacan S, Davignon J (1992) Patterns of association between genetic variability in apolipoprotein (apo) B, apo AI-CIII-AIV, and cholesterol ester transfer protein gene regions and quantitative variation in lipid and lipoprotein traits: influence of gender and exogenous hormones. *Am J Hum Genet* 50:92–106.

Krauss RM (1991) Low-density lipoprotein subclasses and risk of coronary artery disease. *Curr Opin Lipid* 2:248–252.

Lifton RP, Dluhy RG, Powers M, Rich GM, Cook S, Ulick S, Lalouel J-M (1992) A chimaeric 11β-hydroxylase/aldosterone synthase gene causes glucocorticoid-remediable aldosteronism and human hypertension. *Nature* 355:262–265.

McGill HC Jr (1979) Atherosclerosis: problems in endpoints for genetic analysis. In Sing C, Skolnick M (eds) *Genetic Analysis of Common Diseases: Applications to Predictive Factors in Coronary Disease.* New York, Alan Liss, pp. 27–49.

Mahley RW, Weisgraber KH, Innerarity TL, Rall SC (1991) Genetic defects in lipoprotein metabolism. *Science* 265:78–83.

Moll PP, Burns TL, Lauer RM (1991) The genetic and environmental sources of body mass index variability: The Muscatine ponderosity family study. *Am J Hum Genet* 49:1243–1255.

Moll PP, Powsner R, Sing CF (1979) Analysis of genetic and environmental sources of variation in serum cholesterol in Tecumseh, Michigan. V. Maximum likelihood estimates of genetic variance components. *Ann Hum Genet* 42:343–354.

Morton NE, Chung CS (eds) (1978) *Genetic Epidemiology.* New York, Academic Press.

Murphy EA (1979) Where are we going? In Sing C, Skolnick M (eds) *Genetic Analysis of Common Diseases: Applications to Predictive Factors in Coronary Disease.* New York, Alan Liss, pp 3–23.

Murphy EA, Trojak JL (1986) The genetics of quantifiable homeostasis. I. The general issues. *Am J Med Gen* 24:159–169.

Neel JV, Schull WJ (1954) *Human Heredity*. Chicago, University of Chicago Press.

Neel JV, Shaw MW, Schull WJ (eds) (1965) *Genetics and the Epidemiology of Chronic Disease*. Washington, DC, GPO, pp VI and 395.

Nestruck AC, Bouthillier D, Sing CF, Davignon J (1987) Apolipoprotein E polymorphism and plasma cholesterol response to probucol. *Metabolism* 36:743–747.

Perusse L, Moll PP, Sing CF (1991) Evidence that a single gene with gender- and age-dependent effects influences systolic blood pressure determination in a population-based sample. *Am J Hum Genet* 49:94–105.

Rao DC, Elston RC, Kuller LH, Feinleib M, Carter C, Havlik R (eds) (1984) *Genetic Epidemiology of Coronary Heart Disease: Past, Present, and Future*. New York, Alan Liss.

Reilly SL, Ferrell RE, Kottke BA, Kamboh MI, Sing CF (1991) The gender-specific apolipoprotein E genotype influence on the distribution of lipids and apolipoproteins in the population of Rochester, MN. I. Pleiotropic effects on means and variances. *Am J Hum Genet* 49:1155–1166.

Reilly S, Sing C (1991) The impact of genotype specific variances and covariances on the evaluation of risk to a common disease having complex etiology. *Am J Hum Genet* 49 (suppl):480.

Renteln P (1991) Quantum gravity. *Amer Sci* 79:508–527.

Roy M, Sing CF, Betard C, Davignon J (1991) Impact of a > 10 kb deletion in the low density lipoprotein receptor (LDLR) gene on means, variances and correlations of measures of lipid metabolism in French-Canadians. *Am J Hum Genet* 49 (suppl):106.

Salthe SN (1985) *Evolving Hierarchical Systems*. New York, Columbia University Press.

Savageau MA (1976) *Biochemical Systems Analysis: A Study of Function and Design in Molecular Biology*. Reading, MA, Addison-Wesley.

Savageau MA (1979) Growth of complex systems can be related to the properties of their underlying determinants. *PNAS* 76:5413–5417.

Schull WJ (1979) And now what? In Sing C, Skolnick M (eds) *Genetic Analysis of Common Diseases: Applications to Predictive Factors in Coronary Disease*. New York, Alan Liss, pp 733–743.

Schull WJ, Weiss KM (1980) Genetic epidemiology: four strategies. *Epidemiol Rev* 2:1–18.

Simon HA (1962) The architecture of complexity. *Proc Amer Phil Soc* 106:467–482.

Sing CF, Boerwinkle E, Moll PP, Templeton AR (1988) Characterization of genes affecting quantitative traits in humans. In Weir BS, Eisen EJ, Goodman MM, Namkoong G (eds) *Proceedings of the 2nd International Conference on Quantitative Genetics*. Sunderland, MA, Sinauer, pp 250–269.

Sing CF, Moll PP (1989) Genetics of variability of CHD risk. *Int J Epidemiol* 18 (suppl 1):S183–S185.

Sing CF, Moll PP (1990a) Genetics of atherosclerosis. *Ann Rev Genet* 24:171–187.

Sing CF, Moll PP (1990b) Strategies for unravelling the genetic basis of coronary artery disease. In Berg K, Retterstol N, Refsum S (eds) *From Phenotype to Gene in Common Disorders*. Copenhagen, Denmark, Munksgaard A/S International Publishers, pp 37–59.

Sing C, Skolnick M (eds) (1979) *Genetic Analysis of Common Diseases: Applications to Predictive Factors in Coronary Disease*. New York, Alan Liss.

Sing CF, Zerba KE, Haviland MB (1992) Genetic architecture of interindividual variation in plasma cholesterol. In Bearn AG (ed) *Genetics of Coronary Heart Disease*. Oslo, Norway, Institute of Medical Genetics, University of Oslo, pp 143–162.

Stewart I (1989) *Does God Play Dice? The Mathematics of Chaos*. New York, Basil Blackwell.

Tauber AI, Sarkar S (1992) The human genome project: has blind reductionism gone too far? *Perspectives in Biology and Medicine* 35:220–235.

Tikkanen MJ, Huttunen JK, Ehnholm C, Peitinen P (1990) Apolipoprotein ε_4 homozygosity predisposes to serum cholesterol elevation during high fat diet. *Arteriosclerosis* 10:285–288.

Trimble BK, Doughty JH (1974) The amount of hereditary disease in human populations. *Ann Hum Genet, Lond* 38:199–223.

Trimble BK, Smith ME (1977) The incidence of genetic disease and the impact on man of an altered mutation rate. *Can J Genet Cytol* 19:375–385.

Turner ST, Rebbeck TR, Sing CF (1992) Sodium-lithium countertransport and probability of hypertension in Caucasians 47 to 89 years of age. *Hypertension* 20:841–850.

Waelsch SG (1989) In praise of complexity. *Genetics* 122:721–725.

Weir BS, Cockerham CC (1989) Complete characterization of disequilibrium at two loci. In Feldman MW (ed) *Mathematical Evolutionary Theory*. Princeton, NJ, Princeton University Press, pp 86–110.

Zannis VI, Kardassis D, Cardot P, Hadzopoulou-Cladaras M, Zanni EE, Cladaras C (1992) Molecular biology of the human apolipoprotein genes: gene regulation and structure/function relationship. *Curr Opin Lipid* 3:96–113.

IV

POPULATION VARIABILITY

10

Pathodemes: Heredity, Environment, and Populations of Disease Susceptibility

ALFRED G. KNUDSON, JR.

The burden of disease falls unequally upon the populations of the world, and unequally upon the individuals in those populations. Much of this variation is a function of time and place, as in the example of the toll of tetanus of the newborn in Haiti. Some of it relates to the genetic composition of a population, as with the incidence of skin cancer and melanoma among fair-skinned persons, especially in the southern United States and in Australia. Among the populations of first-world countries, most mortality is a function of age; the heavy burdens of cardiovascular disease and cancer are cumulative. Yet here too there is considerable geographic and familial variation. How much disease is determined by environment, how much by genetic variation, how much by chance? This question has been the target of much of the thought and effort of Jack Schull, and I suspect that more of my own thoughts on the subject are derived from him than I realize.

ONCODEMES: THE EXAMPLE OF CANCER

My own perspective on the problem has been largely shaped by the set of diseases known collectively as cancer. For many cancers, the incidence increases with a power of time, i.e., $I = kt^r$, where r is 5 to 6 for several common cancers of adults. For some cancers, for example, those of stomach and colon, this relationship is true for countries with strikingly different incidences; the value of k may vary considerably, but that of r generally does not. Most of the international differences have been ascribed to environmental rather than ethnic factors, particularly because the incidence of a common cancer typically changes among migrants and their offspring. In some instances, as with the hepatitis B virus and liver cancer, and smoking and lung cancer, a specific environmental agent accounts for most of the world's cases, but even here variations in host susceptibility may play an important role. Both heredity and environment can modify the age-specific incidence curves.

In lung cancer an environmental factor, cigarette smoking, clearly contributes most of the world's cases. Determination of the agent(s) responsible has been an unsuccessful task so far, but there are many carcinogenic chemicals in tobacco smoke. Many of these chemicals are metabolized by detoxication mechanisms that involve their oxidation by heme proteins of the P-450 class, followed by conjugation and excretion. The intermediate metabolite may interact with DNA, producing mutations. When these mutations are in certain critical genes, cancer may result. Attention has been paid to the fact that these enzymatic activities are polymorphic in human populations. In some cases, as with the metabolism of the anti-hypertensive agent debrisoquine, severe symptoms may occur in persons failing to oxidize a compound. On the other hand, such individuals do not produce usual levels of intermediate metabolites via this particular P-450 enzyme and seem to be at considerably lower risk of lung cancer, if they smoke, than are actively metabolizing homozygotes (Ayesh et al., 1984). It is now thought that a majority of lung cancer cases may occur among actively metabolizing smokers; i.e., lung cancer may in most cases result from an interaction of heredity and environment. The hereditary component appears to be highly polymorphic, so such predisposition is common, a matter discussed by Weiss in this volume (chapter 1). Most of the rest of lung cancer probably occurs in smokers who are not genetically predisposed.

Some persons with genetic susceptibility to environmentally induced cancer are rare in the population because of strong selection. The classical example is xeroderma pigmentosum, a rare recessively inherited disorder.

Also rare are individuals who have a genetic predisposition to cancer that is virtually independent of their environmental exposures. These persons often experience early death, and as a group have reduced fitness. Inheritance of this susceptibility is typically dominant and specific for one or a few types of cancer. Such an inherited group exists for most cancers, including many adult carcinomas (lung, colon, stomach, pancreas, breast, ovary, endometrium, kidney, and bladder).

Retinoblastoma has been used as a prototypic hereditary cancer (Knudson, 1971, 1978, 1985). The hypothesis was developed that the inherited mutation is one of two necessary events for production of the tumor, the second being a somatic mutation that mutates or eliminates the second copy of the gene in the homologous chromosome. The inherited mutation is a step on the path to cancer, in contrast to cases of genetic predisposition to environmental cancer, in which the genetic factor leads to an increased probability of mutation in some critical gene. The implication is that the gene in question in a dominantly inherited cancer is recessive in oncogenesis, at the cellular level. Indeed this has been demonstrated for retinoblastoma. The gene has been cloned, and defects or losses of both copies of the gene have been found in tumors (Friend et al., 1986; Fung et al., 1987; Lee et al., 1987). Furthermore,

Table 10.1. Four "oncodemes" or categories of persons with respect to roles for genetics and environmental agents in cancer causation.

Onceodeme	Heredity	Environment	Comment
1	−	−	"Spontaneous," background cancer
2	+	−	Strong genetic predisposition
3	−	+	"Purely" environmental origin
4	+	+	Interactive group

introduction of a normal cDNA into tumor cells in vitro has been associated with loss of tumorigenicity in immunodeficient mice (Huang et al., 1988). The recessive mechanism operates in non-hereditary cases as well, although here both events are somatic.

In non-hereditary retinoblastoma the mutations are most often "spontaneous," since the worldwide incidences of this tumor are rather similar. It is presumed that all cancers can be caused in this way. Such cancers represent a "background" incidence that cannot be reduced. On the other hand, there are surely some cancers that are induced by environmental agents in persons without genetic predisposition.

There are, then, four categories of persons with respect to roles for genetics and environmental agents in cancer causation. I have coined the term "oncodemes" for those groups (Knudson, 1985), which are defined in Table 10.1. The distribution of cases among these groups varies according to the cancer. For example, in retinoblastoma there is little evidence for a role of environment, and the four classes may be, respectively, 60%, 40%, 0%, and 0%. We know very little about most cancers in adults, but for lung cancer the distribution in the United States could be as extreme as 15%, 0%, 35%, and 50%.

The utility of assigning cases of various cancers to these oncodemes is twofold: Assignment is part of etiology and it provides a basis for a program in prevention. For example, environmental exposures should be very different for oncodemes 3 and 4. Screening for oncodeme 2 and intensive surveillance might be in order. A theoretical goal would be to achieve a distribution of 100%-0%-0%-% for the four oncodemes. The realization of the existence of the first oncodeme underscores the futility of any effort to "eliminate" cancer, because spontaneous mutations promise to produce the first oncodeme in the best of all possible worlds. There will always be a need for the diagnosis and treatment of cancer.

PATHODEMES: A GENERALIZATION FOR DISEASE

Cancer is certainly not the only disease for which more than one category of susceptibility exists. I wish to propose that the concept of oncodeme can be extended to diseases of all kinds, under the rubric *pathodeme*. We can begin by supposing that for most diseases the relative importance of the different groups is even less well known than for cancer. Yet there are numerous examples. For coronary artery disease there is considerable epidemiologic evidence for a causative role of dietary fat. And some kinds of hereditary hypercholesterolemia place the host at risk no matter what diet is selected. It may also be that many people would develop the disease regardless of heredity or diet if they live long enough.

In the case of infectious diseases the first two pathodemes do not exist; cases are probably distributed among the third and fourth pathodemes, according to the infection. Thus, meningococcal infections are much more likely among persons with certain abnormalities in complement, which goes far to explain why only a minority of the organism's carriers are clinically affected, and the latter are severely so (Fijen et al., 1989). Specific immunization of such persons is warranted. Hypersensitivity is another example of the interaction of heredity and environment in the origin of disease; most people affected by common allergies apparently belong to the fourth pathodeme, although it seems to be true that even "normal" persons can be sensitized to an environmental agent under certain conditions. Here again, all cases would seem to belong to the third or fourth pathodeme.

Dietary deficiency diseases similarly involve environment in all cases, but interaction with environment in some cases. Thus, rickets can occur in any person deprived of sunlight and dietary vitamin D, but also occurs in persons who inherit a gene for resistant rickets. An early study of vitamin A deficiency noted that many children dying of the condition were found to have lesions characteristic of the disease we now call cystic fibrosis, because untreated pancreatic insufficiency caused loss of ingested fat-soluble vitamin A in feces (Blackfan and Wolbach, 1933).

Heredity seems to have a strong involvement in some diseases. Examples are gout and diabetes mellitus. It may even be that all cases involve genetic predisposition, this being a virtually complete determinant in a minority, as with the Lesch-Nyhan syndrome and juvenile diabetes mellitus, but an interactive determinant in a majority, with diet interacting with a weaker genetic predisposition. In some patients with so-called "hereditary" disease, environmental factors can play a decisive role, as with dietary milk in galactosemic individuals.

CONCLUSIONS

It is no longer useful to classify most diseases as genetically or environmentally determined. The examples of cancer and of some other major diseases illustrate how the concept of pathodemes can be used in the analysis of disease causation. Many other examples could be given, but it is not my purpose here to present a catalog. Rather, my intention is to suggest that most disease falls non-randomly upon a population, according to genetic susceptibility and environmental exposure. It may be that most premature deaths occur in individuals who are genetically predisposed to environmental agents of disease, and that future programs of disease prevention will rely heavily upon the identification of those at genetic risk. Those with extremely high relative risk (the second pathodeme), as with "purely" hereditary disease, can never contribute much to the attributable risk. In some diseases, "background" cases (first pathodeme) may constitute a small fraction, or even a zero fraction, of all cases. The fourth pathodeme can include a large fraction of cases when a common polymorphism imparts a relatively low relative risk but high attributable risk. The ability to characterize genetic differences with the use of molecular genetic techniques could revolutionize preventive medicine by identifying persons in the second and fourth pathodemes.

ACKNOWLEDGMENTS

This work was supported in part by an appropriation from the Commonwealth of Pennsylvania and by Public Health Service Grants CA-06927 and CA-43211.

REFERENCES

Ayesh R, Idle JR, Ritchie JC, Crothers MJ, Hetzel MR (1984) Metabolic oxidation phenotypes as markers for susceptibility to lung cancer. *Nature* 312:169–170.

Blackfan KD, Wolbach SB (1933) Vitamin A deficiency in infants; clinical and pathological study. *J Pediat* 3:679–706.

Fijen CAP, Kuijper EJ, Hannema AJ, Sjöholm AG, vanPutten JPM (1989) Complement deficiencies in patients over ten years old with meningococcal disease due to uncommon serogroups. *Lancet* 2:585–588.

Friend SH, Bernards R, Rogelj S, Weinberg RA, Rapaport JM, Albert DM, Dryja TP (1986) A human DNA segment with properties of the gene that predisposes to retinoblastoma and osteosarcoma. *Nature* 323:643–646.

Fung Y-KT, Murphree AL, T'Ang A, Qian J, Hinrichs SH, Benedict WF (1987) Structural evidence for the authenticity of the human retinoblastoma gene. *Science* 236:1657–1661.

Huang H-JS, Yee J-K, Shew J-Y, Chen P-L, Bookstein R, Freidman T, Lee EY-HP, Lee W-H (1988) Suppression of the neoplastic phenotype by replacement of the RB gene in human cancer cells. *Science* 242:1563–1566.

Knudson AG (1971) Mutation and cancer: statistical study of retinoblastoma. *Proc Natl Acad Sci USA* 68:820–823.

Knudson AG (1978) Retinoblastoma: a prototypic hereditary neoplasm. *Semin Oncol* 5:57–60.

Knudson AG (1985) Hereditary cancer, oncogenes, and antioncogenes. *Cancer Res* 45:1437–1443.

Lee W-H, Bookstein R, Hong F, Young L-J, Shew J-Y, Lee EY-HP (1987) Human retinoblastoma susceptibility gene: cloning, identification, and sequence. *Science* 235:1394–1399.

11

Genetic Inferences from Epidemiologic Investigations

CRAIG L. HANIS

Genetic epidemiology is a rapidly burgeoning field focusing on the distribution of disease in families and the impact of inherited disease in populations (Morton, 1982). While there is fruitful ground for collaboration between epidemiologists and geneticists (Neel et al., 1965), there seems to be an unbalanced contribution to the field from genetics (Khoury et al., 1986). Furthermore, while genetic epidemiology encompasses inherited disease in populations, emphasis has been on disease in collections of relatives. This emphasis has led to the development of increasingly complex, sometimes elegant, analytical methods for pedigree data that are often perceived as genetic epidemiology rather than its tools. Partially explaining this trend is the desire to find well-behaved biological and statistical models that explain the observed patterns of disease. Unfortunately, biological and statistical order come at the expense of the level of inference. For example, there is immense interest and discussion of the genetic epidemiology of coronary heart disease in the population, but nearly all analyses relate to risk factors such as the lipids, lipoproteins, and apolipoproteins (see Bock and Collins, 1987; Rao et al., 1984; Sing and Skolnick, 1979). It is also axiomatic that the clearer the statistical story, the more restricted is the population of inference.

In this chapter, non-familial based sampling designs and the types of genetic inferences they permit are examined in the context of the common chronic diseases. In this way, focus is on more traditional epidemiologic designs and the impact of inherited variation on disease in populations. Here too, it will quickly become evident that understanding risk factors is a less formidable task than understanding disease endpoints. Designs examined include population surveys, investigations of migrant populations, association and case-control studies, and longitudinal follow-up of specific cohorts. Longitudinal designs should be particularly informative in the examination of genetic and environmental interactions. The intent is not to provide a comprehensive review, but rather to indicate the range of inferences that are practicable.

Where possible, issues will be illustrated using data on chronic disease among the Mexican-American population of Starr County, Texas.

POPULATION SURVEYS

Within descriptive epidemiology, a first step is often establishing the frequency of disease. With a cross-sectional design, the appropriate measure is prevalence. By itself, prevalence provides little information regarding etiology, but this changes with additional detail regarding the population (e.g., age- and sex-specific frequencies) and comparisons to other populations. Comparisons of males and females and examination of age trends document the impact of the disease and partially delimit the scope of possible etiologic mechanisms.

Non-insulin-dependent diabetes mellitus (NIDDM) is a chronic disease that affects nearly all populations. Among Mexican-Americans in Starr County, Texas, NIDDM is rare in those under age 35 years and rises to a peak frequency of approximately 17% in those 55 to 64 years old. Above age 65, the frequency levels off or even declines and this is most likely a manifestation of differential mortality. Frequencies in males and females are similar (Hanis et al., 1983, 1993). Whatever the causal mechanisms responsible for diabetes, they must account for this age distribution that is consistent with a "developmental" type mechanism. Beyond this, little can be said regarding possible genetic involvement from single population prevalence data, though it is important to note that the prevalence pattern has profound implications for determining familial-based sampling designs.

Comparing the frequencies of diabetes in Starr County to other populations begins to provide added perspective on the disease process. It is now clear that Hispanics, in general, and Mexican-Americans, in particular, have markedly elevated frequencies of diabetes. Age- and sex-specific frequencies of NIDDM are three- to five-fold higher than that of the general population (Hanis et al., 1983a, 1993; Stern et al., 1984) and this appears to hold true for Mexican-Americans in New Mexico (Samet et al., 1988) and Colorado (Hamman et al., 1989). The homogeneity of disease frequencies in space indicates similar disease mechanisms in the different groups and clustering (at the population level) of etiologic agents. Genes are obvious candidates, but then, so too are cultural and environmental factors.

Among populations that have been examined for the occurrence of NIDDM, the Pima Indians are perhaps the most comprehensively documented and have among the highest prevalences reported (Knowler et al., 1978). The Pima Indian age- and sex-specific NIDDM frequencies parallel those of the Mexican-American population, but are several-fold higher than the already high Mexican-American prevalences. This places the Mexican-American frequencies intermediate between the Pima Indian and general population frequencies. Because of the known historical admixture between Spanish and Native American groups giving rise to the contemporary Mexican-American population, a compelling argument has been made that genetic factors are largely responsible for the distribution and occurrence of diabetes (Gardner et al., 1984; Weiss et al., 1984a). To examine this pattern further, genetic markers have been determined and the tools of population genetics used to compute the proportion of alleles in the Mexican-American population that are Native American- or Anglo-derived (Chakraborty et al., 1986; Hanis et al., 1986). It has been shown that diabetes frequencies in the Mexican-American population can be predicted based on admixture and disease frequencies in the putative ancestral populations (Hanis et al., 1991). It is ironic that these observations are population-based and remain among the best evidence for the role of genes in the distribution of diabetes. While family-based studies certainly document the familial aggregation of diabetes in all populations, genetic models that explain the distribution of NIDDM in families and populations remain to be resolved.

Diabetes is not the only disease for which population data and admixture arguments have been used to implicate genetic factors. Gallbladder disease and obesity demonstrate the same distribution pattern as diabetes in Anglos, Mexican-Americans, and Native Americans although the frequencies in each population are higher than diabetes. Focusing on gallbladder disease, approximately 45% of Mexican-American women above age 45 have either had their gallbladder removed or have gallstones visualized on ultrasound (Hanis et al., 1993). Nearly 90% of female Pima Indians in this age group are so affected (Sampliner et al., 1970). These population observations currently are the only evidence for a role of genetic factors in the etiology of gallbladder disease (see also Weiss et al., 1984b). They also make clear the need to account for population frequencies in the selection of families for more formal genetic analysis because simple sampling rules applied among the Pima would result in reduced power due to the scarcity of segregation events.

Diabetes, obesity, and gallbladder disease have been shown to increase risk for cardiovascular disease, but, paradoxically, cardiovascular disease does not appear to be elevated in Mexican-Americans or to parallel admixture. In fact, the frequency of cardiovascular disease among the Pima Indians is less than that found in the general population (Ingelfinger et al., 1976) and that among Mexican-Americans appears to be similar to the general population (Weiss et al., 1984a). This is not evidence for the lack of a genetic role in the etiology of cardiovascular disease; rather, it strengthens the argument for an important role of genes in cardiovascular disease, diabetes, gallbladder disease, and obesity. This results from the notion that if cultural and environmental factors were simply explaining the increased frequencies, then one would expect a concomitant increase in the occurrence of cardiovascular disease (Weiss et al., 1984a). There may also be insufficient power to detect the admixture pattern for cardiovascular disease because it occurs with frequency high enough to preclude observing increases across these populations of sufficient magnitude to be detected.

The strategy of examining population disease frequencies and ancestral affinities is not limited to those common conditions that have been resistant to implicating simple genetic machanisms. Insulin-dependent diabetes (IDDM) is an order of magnitude less frequent than NIDDM and has been shown to have more clear direct evidence of genetic involvement; though it is by no means "simple" (Foster, 1989). Among populations, a 60-fold difference in risk has been reported (DERI, 1989). Besides investigations of other populations, Spanish and Portuguese admixed groups appear to be particularly useful groups to follow (DERI, 1989). This stems from the known distributions of these population groups throughout the world as a consequence of extensive exploration and colonization. As a result, a possibly homogeneous genetic background has been distributed across widely varying environmental strata. Examination of genetic structure, particularly as revealed at the HLA locus, may provide important insight into the genetics and epidemiology of IDDM (DERI, 1989).

MIGRANT STUDIES

In this section, two classes of studies are discussed; first, the repeated sampling of the same population across time, and second, the examination of more traditional migrant studies wherein space rather than or in addition to time is the classification factor. A common feature to both types of investigations is that they begin to impose a pseudo-experimental design to studies of natural populations that permit contrasting groups. Obviously, there are many pitfalls to investigations of populations over time due to the nature of disease definition, sampling protocols, and exposures; but for diabetes, obesity, and gallbladder disease among Native Americans, the data are persuasive in showing a marked change in disease frequencies in the last 50 years (Weiss et al., 1984a). Prior to this time, evidence indicates that these diseases were infrequent, wherein now they are epidemic. This change points to the need to investigate genetic and environmental interactions because in this period of rapidly changing disease frequencies there could not have been changes of the same magnitude in the population genetic structure. That genetic susceptibility must be involved can be concluded from studies of other groups that have undergone "Westerization." A general pattern observed is an increase in the frequencies of chronic conditions with Westernization, but none come close to the magnitude of changes seen in the Native American and their admixed groups (Weiss et al., 1984a).

More traditional migrant studies involve following a cohort of individuals that move from one area to another or conducting contemporaneous investigations of relatively similar populations that have moved in the past. It is assumed that the migrant group is representative of the parental population. Migrant studies have been used to examine changes in frequency of diabetes (Taylor and Zimmett, 1983), cardiovascular disease (Robertson et al., 1977b), or changes in levels of quantitative risk factors such as blood pressure or lipids (Robertson et al., 1977a; Salmond et al., 1985). These investigations are extremely informative in that they begin to uncover the array of changes in disease or risk factors that is possible within the context of normal genetic variability. The implication is that environmental modifications will be able to alter risk or disease profiles at least to the degree of change seen with migration. Even though this discussion is limited to non-familial based sampling, it is worth noting that incorporation of a familial sampling scheme in the context of migration studies provides one approach for examining genotype and environmental interactions (Ward et al., 1980).

ASSOCIATION STUDIES

A prime objective of epidemiology is the identification of factors that alter disease risk (Kleinbaum et al., 1982). Such an objective can be achieved from a variety of sampling designs. For simplicity, we refer to these as association studies and note that they include case-control studies, cohort studies, retrospective studies, and the like. Where "exposures" represent genotypes, there is potential for identifying genetic factors that may be involved in the etiology of disease or, in the case of quantitative risk factors, determining the impact of genetic differences on levels of the risk factors. Numerous recent reviews of such genetic association studies have amply illustrated and discussed the various issues involved (see e.g., Breslow 1988; Gibbs and Caskey, 1989; Humphries 1988). Among the best described has been the determination of the role of the apolipoprotein E polymorphism in determining variation in lipid levels and associations with disease (Davignon et al., 1988). This class of study is probably the most common example of that aspect of genetic epidemiology dealing with impact of genes in populations. The attraction to association studies in populations is the ability to examine genetic variability directly at genes known to be important in the metabolism of disease risk factors (e.g., the genes involved in lipid, carbohydrate, and bile acid metabolism). Having the direct measures simplifies the application of statistical procedures and permits testing of hypotheses relating to gene action and interaction using stratification strategies. For example, we have recently demonstrated markedly different effects of amino acid substitutions at position 158 of apolipoprotein E in determining apolipoprotein E levels in pre-menopausal females compared to males or post-menopausal females (Hanis et al., 1991b). This is consistent with a physiologic interaction that is likely mediated by hormonal action. Such studies, while providing important population considerations, also point to specific mechanisms that can be studied in the laboratory to determine the basis of the observation.

LONGITUDINAL FOLLOW-UP

As intimated in the foregoing, a primary goal of genetic epidemiology is the understanding of genetic and environmental interactions. Heuristically, genetic and environmental interaction can be thought of in rather broad terms as the differential action of genes or environments in differing strata.

Being able to examine this phenomena statistically requires either the direct determination of genes or environments (preferably both). Most studies have relied on the experimental design to set up contrasts in which genes or environments can be assumed to be different across strata. Now, however, the ability to characterize genes more fully (compared to characterization of environments) permits more formal study of interaction. An example of the investigation and documentation of interaction is the finding of differential effects of probucol in altering lipids according to apolipoprotein E genotype (Nestruck et al., 1987).

Longitudinal designs seem to offer even greater potential in the investigation of interaction. Their appeal is that genotype is held constant (Hanis et al., 1983b). Therefore, differential changes in disease state or risk factor levels across time according to genotype would be a direct measure of interaction. Longitudinal studies also more closely replicate real life in that genes are not static entities. An understanding of the dynamic response of genes across time will be required if the findings of genetic epidemiology relating to populations are to be useful in the design of public health intervention strategies.

CONCLUSIONS

To date, great strides have been made in identifying genetic variation at loci known to be important in the metabolism of chronic disease risk factors, but at this juncture the impact on public health policy and practice has been minimal (Schull and Hanis, 1990). This largely reflects the fact that the variation so far identified has been at loci coding for structural proteins rather than at the loci involved in regulation. Progress on these latter will obviously be a much more extensive task, but ultimately these genes and their respective roles will be described. As they are, non-familial based sampling will play an important role in examining the impact of variation and genotype by environmental interaction in the population. This information will indicate the range of probable effects that may be achieved within the array of genetic variability, permit identification of populations, families, and individuals at whom intervention may be targeted, and allow design and assessment of public health strategies aimed at reducing the burden of chronic diseases.

ACKNOWLEDGMENTS

This work was supported in part by Grant AM34666 and Research Career Development Award DK01748 from the National Institute of Diabetes and Digestive and Kidney Diseases, Grant HL34823 from the National Heart, Lung, and Blood Institute and Grant EY06231 from the National Eye Institute.

REFERENCES

Bock G, Collins GM (1987) *Molecular Approaches to Human Polygenic Disease.* New York, John Wiley & Sons.

Breslow JL (1988) Apolipoprotein genetic variation and human disease. *Physiological Reviews* 68:85–132.

Chakraborty R, Ferrell RE, Stern MP, Haffner SM, Hazuda HP, Rosenthal M (1986) Relationship of prevalence of non-insulin-dependent diabetes mellitus with Amerindian admixture in the Mexican Americans of San Antonio, Texas. *Genet Epidemiol* 3:435–54.

Davignon J, Gregg RE, Sing CF (1988) Apolipoprotein E polymorphism and atherosclerosis. *Arteriosclerosis* 8:1–21.

Diabetes Epidemiology Research International Group (DERI) (1989) Evaluation of epidemiology and immunogenetics of IDDM in Spanish- and Portuguese-Heritage registries. *Diabetes Care* 12:487–493.

Foster DW (1989) Diabetes mellitus. In Scriver CR, Beaudet AL, Sly WS, Valle D (eds) *The Metabolic Basis of Inherited Disease.* New York, McGraw-Hill Information Services Company, pp 375–397.

Gardner LI, Stern MP, Haffner SM, Gaskill SP, Hazuda HP, Relethford JH, Eifler CW (1984) Prevalence of diabetes in Mexican Americans: relationship to percent of gene pool derived from Native American sources. *Diabetes* 33:86–92.

Gibbs RA, Caskey CT (1989) The application of recombinant DNA technology for genetic probing in epidemiology. *Ann Rev Public Health* 10:27–48.

Humman RF, Marshall JA, Baxter J, Kahn LB, Mayer EJ, Orleans M, Murphy JR, Lezotte DC (1989) Methods and prevalence of non-insulin-dependent diabetes mellitus in a biethnic Colorado population. *Am J Epidemiol* 129:295–311.

Hanis CL, Chakraborty R, Ferrell RE, Schull WJ (1986) Individual admixture estimates: disease associations and individual risk of diabetes and gallbladder disease among Mexican-Americans in Starr County, Texas. *Am J Phys Anthropol* 70:433–441.

Hanis CL, Ferrell RE, Barton SA, Aguilar L, Garza-Ibarra A, Tulloch BR, Garcia CA, Schull WJ (1983a) Diabetes among Mexican-Americans in Texas. *Am J Epidemiol* 118:659–672.

Hanis CL, Hewett-Emmett D, Bertin TK, Schull WS (1991a) The origins of U.S. Hispanics: implications for diabetes. *Diab Care* 14:618–627.

Hanis CL, Hewett-Emmett D, Douglas TC, Bertin TK, Schull WJ (1991b) Effects of the apolipoprotein E polymorphism on levels of lipids, lipoproteins and apolipoproteins among Mexican-Americans in Starr County, Texas. *Arteriosclerosis and Thrombosis* 11:362–370.

Hanis CL, Hewett-Emmett D, Kubrusly LF, Maklad MN, Douglas TC, Mueller WH, Barton SA, Yoshimaru, H, Kubrusly, DB, Gonzalez R, Schull WJ (1993) An ultrasound survey of gallbladder disease among Mexican-Americans in Starr County, Texas: Frequencies and risk factors. Ethnicity and Disease (to appear).

Hanis CL, Sing CF, Clarke WR, and Schrott HG (1983b) Multivariate models for human genetic analysis: aggregation, coaggregation and tracking for systolic blood pressure and weight. *Am J Hum Genet* 35:1196–1210.

Humphries SE (1988) DNA polymorphisms of the apolipoprotein genes—their use in the investigation of the genetic component of hyperlipidemia and atherosclerosis. *Atherosclerosis* 72:89–108.

Ingelfinger JA, Bennett PH, Liebow IM, Miller M (1976) Coronary heart disease in the Pima Indians: electrocardiographic findings and postmortem evidence of myocardial infarction in a population with a high prevalence of diabetes mellitus. *Diabetes* 25:561–565.

Khoury MJ, Beaty TH, Cohen BH (1986) Interaction of genetics and epidemiology in the literature. *Genet Epidemiol* 3:269–277.

Kleinbaum DG, Kupper LL, Morgenstern H (1982) *Epidemiologic Research: Principles and Quantitative Methods.* New York, Van Nostrand Reinhold Company.

Knowler WC, Bennett PH, Hamman RF, Miller M (1978) Diabetes incidence and prevalence in Pima Indians: a 19-fold greater incidence than in Rochester, Minnesota. *Am J Epidemiol* 108:497–505.

Morton NE (1982) *Outline of Genetic Epidemiology.* New Yrok, Karger.

Neel JV, Shaw MW, Schull WJ (1965) *Genetics and the Epidemiology of Chronic Disease.* Washington, DC, US Government Printing Office.

Nestruck AC, Bouthillier D, Sing CF, Davignon J (1987) Apolipoprotein E polymorphism and plasma cholesterol response to probucol. *Metabolism* 8:743–747.

Rao DC, Elston RC, Kuller LH, Feinleib M, Carter C, Havlik R (1984) *Genetic Epidemiology of Coronary Heart Disease: Past, Present, and Future.* New York, Alan R. Liss, Inc.

Robertson TL, Kato H, Gordon T, Kagan A, Rhoads GG, Land CE, Worth RM, Belsky JL, Dock DS, Miyanishi M, Kawamoto S (1977a) Epidemiologic studies of coronary heart disease and stroke in Japanese men living in Japan, Hawaii and California: coronary heart disease risk factors in Japan and Hawaii. *Am J Cardiol* 39:239–243.

Robertson TL, Kato H, Rhoads GG, Kagan A, Marmot M, Syme SL, Gordon T, Worth RM, Belsky JL, Dock DS, Miyanishi M, Kawamoto S (1977b) Epidemiologic studies of coronary heart disease and stroke in Japanese men living in Japan, Hawaii and California: incidence of myocardial infarction and death from coronary heart disease. *Am J Cardiol* 39:239–243.

Salmond CE, Joseph JG, Prior IAM, Stanley DG, Wessen AF (1985) Longitudinal analysis of the relationship between blood pressure and migration: The Tokelau Island migrant study. *Am J Epidemiol* 122:291–301.

Samet JM, Coultas DB, Howard CA, Skipper BJ, Hanis CL (1988) Diabetes, gallbladder disease, obesity and hypertension among Hispanics in New Mexico. *Am J Epidemiol* 128:1302–11.

Sampliner R, Bennett PH, Comess L, Rose F, Burch T (1970) Gallbladder disease in Pima Indians: demonstration of high prevalence and early onset by cholecystography. *N Engl J Med* 283:1358–1364.

Schull WJ, Hanis CL (1990) Genetics and public health in the 1990s. *Ann Rev Public Health* 11:105–125.

Sing CF, Skolnick M (1979) *Genetic Analysis of Common Diseases: Applications to Predictive Factors in Coronary Heart Disease.* New York, Alan R. Liss, Inc.

Stern MP, Rosenthal M, Haffner SM, Hazuda HP, Franco LJ (1984) Sex differences in the effects of sociocultural status on diabetes and cardiovascular risk factors in Mexican Americans: the San Antonio Heart Study. *Am J Epidemiol* 120:834–851.

Taylor R, Zimmet P (1983) Migrant studies in diabetes epidemiology. In Mann JI, Pyrl K, Teuscher A (eds) *Diabetes in Epidemiological Perspective.* New York, Churchill Livingstone, pp 58–77.

Ward RH, Chin PG, Prior IAM (1980) Tokelau Island migrant study: effect of migration of the familial aggregation of blood pressure. *Hypertension* 2 (suppl I):143–154.

Weiss KM, Ferrell RE, Hanis CL (1984a) A New World syndrome of metabolic diseases with a genetic and evolutionary basis. *Yearbook Phys Anthropol* 27:153–178.

Weiss KM, Ferrell RE, Hanis CL, Styne PN (1984b) Genetics and epidemiology of gallbladder disease in New World native peoples. *Am J Hum Genet* 36:1259–1278.

12

Generalized Occupancy Problem and Its Applications in Population Genetics

RANAJIT CHAKRABORTY

Analysis of categorical observations constitutes one of the principal ways of handling genetic data, evidenced in the contributions of the honoree of this symposium in relation to his studies on inbreeding effects (Schull and Neel, 1965), genetic effects of radiation exposure (Schull et al., 1981) and adaptation effects of high altitude hypoxia (Schull and Rothhammer, 1990). In the context of hypothesis testing based on specific genetic models in such data analysis, often we need to contrast frequencies of different endpoints in samples of unequal sizes or predict expected frequencies in populations from observations in samples of small sampling proportions. For example, Neel (Chapter 4, this volume) illustrates one such problem where the frequencies of different genetic variants are contrasted in two Japanese cities, Hiroshima and Nagasaki, where the sample sizes are widely different. In a previous study (Chakraborty et al., 1988) the issue of sample size adjustment was addressed by deriving the expectation and variance of the observed number of alleles in samples of fixed sizes.

In this chapter, a more complete solution will be given from a combinatorial approach, whereby the sampling distribution can be completely specified for an arbitrary sample size. Although the problem is formulated and solved in general terms of a multinomial distribution with known number of classes, it will be illustrated that the general theory can be used in solving several types of genetic problems, such as, non-randomness of mutagen-induced mutations across loci; tests of Hardy-Weinberg equilibrium in the presence of a large number of alleles in samples of comparatively moderate sizes; and global tests of gametic phase disequilibria within a defined DNA segment. Because the combinatorial approach used in this context emerges from the combinatorics of the classical occupancy problem, I call this class of problems *Generalized Occupancy Problems* in population genetics.

FORMULATION OF THE GENERALIZED OCCUPANCY PROBLEM

Consider a multinomial distribution with K classes (K can be arbitrarily large), with class probabilities represented by the vector $\pi' = (\pi_1, \pi_2, \ldots, \pi_K)$.

Obviously, the elements of π form a simplex on a K-dimensional space satisfying

$$0 < \pi_i < 1, \quad \text{and} \sum_{i=1}^{K} \pi_i = 1. \qquad [12.1]$$

When a random sample (with replacement) of size n is drawn from such a distribution, several characteristics of the sample are of interest; e.g., the number of classes represented in the sample, or the number of classes are represented with a specified number of sampling units in each of these classes; and the sampling distributions of such variables. For brevity, I shall only discuss the sampling distribution of the observed number of classes, although the combinatorial solution can be easily extended to many other variables. Denote the observed number of classes that are represented in the sample by X. Obviously, X is a random variable that can take values between 1 and min (K, n), since in the event $K > n$, we cannot observe more than n classes in any sample of size n. Before attempting to solve the problem, note several interesting properties of the random variable X.

First, when all π_i's are equal ($\pi_i = 1/K$ for all i), a situation more commonly known as the classical occupancy problem, Arnold and Beaver (1988) showed that X is a sufficient statistic for the parameter K, and its sampling distribution is given by

$$P_{[m]} = \text{Prob.}\,(X = m) = \binom{K}{m} \frac{m!\, S_n^{(m)}}{K^n}, \qquad [12.2]$$

for $m = 1, 2, \ldots, \min(K, n)$, where $S_n^{(m)}$ is a Stirling number of the second kind (Abramowitz and Stegun, 1965; p 824). Also note that in the case of the classical occupancy problem, a value of X together with the sample size n, and number of possible classes K, uniquely determines the number (R) of sampling units that are "repeated" within the observed classes in a sample. In this context, a sampling unit is called a "repeat" if the class in which this sampling unit belongs already contains another sampling unit in a given sample, Therefore, the sampling distribution of R can be uniquely specified from that of X, and going through the computation of equation [12.2] is preferable, since X is a sufficient statistic for the only parameter (K) of the multinomial distribution with equi-probable classes.

These abstract definitions may be better understood through the following example. Consider an experiment where $20\,(= n)$ mutants are observed which could belong to $30\,(= K)$ loci numbered from 1 through 30. Suppose that there are 3 loci each containing 4 mutants (i.e., $n_1 = n_2 = n_3 = 4$), two loci containing 2 mutants each ($n_4 = n_5 = 2$), and four loci containing one mutant each ($n_6 = n_7 = n_8 = n_9 = 1$). The remaining loci (10 through 30) do not contain any mutants (i.e., $n_{10} = \cdots = n_{30} = 0$). In this example, the observed number of loci containing at least one mutant (X) is 9, and the number of "repeat" mutants is 11. Noting that $11 = 20 - 9$, (i.e., $R = n - X$, more generally), it is clear that the probability distribution of R is completely specified by that of X for any fixed sample size (n).

Second, in the general case of unequal π_i's, the expectation and variance of X have been derived using the indicator variable approach (Chakraborty et al., 1988), which are simple functions of n, K, and the π_i's. I shall show in the following that both of these properties are compatible with the general probability distribution function derived by the combinatorial approach used here.

COMBINATORIAL SOLUTION OF THE PROBABILITY FUNCTION OF X

Instead of working with the variable X, the observed number of non-empty classes, it is easier to work with its complement, $Y = K - X$, which is equivalent to the number of classes not represented in a sample of size n. Define events $A_1, A_2, \ldots, A_K$, such that A_i is the collection of all partitions of n into K segments, such that $n_i = 0$ (but $\sum n_i = n$). In other words, A_i represents the collection of sample points where the ith class ($i = 1, 2, \ldots, K$) remains empty in a specific sample. Note that A_i's are not exclusive of each other; i.e., there can be sample configurations where more than one n_i can be simultaneously zero.

For sampling with replacement, the following equations hold:

$$P_i = \text{Prob}(A_i) = (1 - \pi_i)^n, \qquad [12.3a]$$

$$P_{ij} = \text{Prob}(A_i A_j) = (1 - \pi_i - \pi_j)^n, \qquad [12.3b]$$

$$P_{ijk} = \text{Prob}(A_i A_j A_k) = (1 - \pi_i - \pi_j - \pi_k)^n, \qquad [12.3c]$$

etc., for all $i \neq j \neq k = 1, 2, \ldots, K$.

Following Feller (1968, p 99), define a sequence of summations $\{T_1, T_2, \ldots, T_K\}$ where

$$T_1 = \sum_i p_i,\ T_2 = \sum_i \sum_j p_{ij},\ T_3 = \sum_i \sum_j \sum_k p_{ijk}, \text{ etc.}$$

where the summations are taken such that $i < j < k < \cdots \leqslant K$, so that each combination appears once and only once; hence, the summation $T_r (1 \leqslant r \leqslant K)$ contains ${}^K C_r$ terms. The last term T_K reduces to only one term,

$$T_K = \text{Prob}(A_1 A_2 \cdots A_K) = P_{123\cdots K},$$

which is the probability of simultaneous occurrences of all K events A_1 through A_K. Invoking condition [12.1] on equations [12.3a–c], we note that $T_K = 0$, and furthermore,

$$T_{K-1} = \sum_{i=1}^{K} \pi_i^n, \qquad [12.4a]$$

$$T_{K-2} = \sum_{i>j=1}^{K} \sum^{K} (\pi_i + \pi_j)^n, \qquad [12.4b]$$

$$T_{K-3} = \sum_{i>j>k=1}^{K} \sum^{K} \sum^{K} (\pi_i + \pi_j + \pi_k)^n, \qquad [12.4c]$$

etc.

Applying Feller's theorem (Feller, 1968, p 106), we obtain

$$P_{[K-m]} = \text{Prob}(X = K - m) = \text{Prob}(Y = m)$$
$$= \sum_{i=m}^{K} (-1)^{i-m} \binom{i}{m} T_i, \qquad [12.5]$$

for $K - \min(K, n) \leqslant m \leqslant K - 1$, giving the sampling distribution of X, the number of non-empty classes in a sample of size n. Note that in [12.5], T_0 is conventionally defined as unity (see also Feller, 1968).

EXPECTATION AND VARIANCE OF THE NUMBER OF NON-EMPTY CLASSES IN A SAMPLE

As mentioned before, the expectation and variance of X can be derived by an indicator variable approach. For this we define K indicator variables, $Y_1, Y_2, \ldots, Y_K$, so that

$$Y_i = \begin{cases} 1, \text{if the } i\text{th class is unobserved in the sample,} \\ 0, \text{otherwise.} \end{cases}$$

Note that $X = K - Y$, and $Y = \sum_{i=1}^{K} Y_i$.

Therefore, the expectation of X is given by

$$E(X) = K - E(Y) = K - \sum_{i=1}^{K} E(Y_i)$$
$$= K - \sum_{i=1}^{K} \text{Prob}(Y_i = 1)$$
$$= K - \sum_{i=1}^{K} (1 - \pi_i)^n = K - T_1, \qquad [12.6]$$

derived by Emigh (1983) and Chakraborty et al. (1988).

Furthermore, the variance of X is given by

$$V(X) = V(Y) = \sum_{i=1}^{K} V(Y_i) + \sum_{i \neq j = i}^{K} \sum^{K} \text{Cov}(Y_i, Y_j). \qquad [12.7]$$

Since Y_i's are Bernoulli variables,

$$V(Y_i) = E(Y_i^2) - [E(Y_i)]^2 = (1 - \pi_i)^n [1 - (1 - \pi_i)^n], \qquad [12.8]$$

and

$$\text{Cov}(Y_i, Y_j) = E(Y_i Y_j) - E(Y_i) \cdot E(Y_j)$$
$$= (1 - \pi_i - \pi_j)^n - (1 - \pi_i)^n (1 - \pi_j)^n. \qquad [12.9]$$

Substituting equations [12.8] and [12.9] into equation [12.7], we obtain

$$V(X) = \sum_{i=1}^{K} (1-\pi_i)^n [1-(1-\pi_i)^n] + \sum_{i \neq j=1}^{K}\sum^{K} [(1-\pi_i-\pi_j)^n - (1-\pi_i)^n(1-\pi_j)^n]$$

$$= \sum_{i=1}^{K} (1-\pi_i)^n \cdot [1-\sum(1-\pi_i)^n] + \sum_{i \neq j=1}^{K}\sum^{K} [(1-\pi_i-\pi_j)^n$$

$$= T_1(1-T_1) + T_2. \qquad [12.10]$$

Note that when n is large, and each π_i small, we may approximate each term of the summations T_r's [12.3a–c] by $(1-\pi_i)^n \approx e^{-n\pi_i}$, $(1-\pi_i-\pi_j)^n \approx e^{-n(\pi_i+\pi_j)}$, etc., so that the variance of X can be approximated by

$$V(X) \approx \sum_{i=1}^{K} e^{-n\pi_i}(1-e^{-n\pi_i}), \qquad [12.11]$$

as shown in Chakraborty et al. (1988). Equation [12.10] is, however, exact, and not difficult to compute numerically even if the number of classes is large.

Let us now establish that equations [12.6] and [12.10] are compatible with the probability function [12.5]. To do this, first note that invoking equation [12.5] we have

$$\sum_m \text{Prob}(Y=m) = \sum_m \sum_{i=m}^{K} (-1)^{i-m}\binom{i}{m} T_i$$

$$= \sum_{i=0}^{K}\left[\sum_{m=0}^{i} (-1)^{i-m}\binom{i}{m}\right] T_i$$

$$= 1 + \sum_{i=1}^{K}\left[\sum_{m=0}^{i} (-1)^{i-m}\binom{i}{m}\right] T_i,$$

since $T_0 = 1$. Furthermore, the summation within the parenthesis is the binomial expansion of $(1-1)^i$, for $i = 1, 2, \ldots, K$. Therefore, we establish that [12.5] is a proper probability function since the entire probability mass equals unity.

The expectation of Y, from [12.5], then becomes

$$E(Y) = \sum_m m \cdot \text{Prob}(Y=m)$$

$$= \sum_m \sum_{i=m}^{K} (-1)^{i-m} m \binom{i}{m} T_i$$

$$= \sum_m \left[\sum_{i=m}^{K} (-1)^{i-m}\binom{i-1}{m-1} iT_i\right]$$

$$= T_1 + \sum_{i=2}^{K}\left[\sum_{m-1=0}^{i-1} (-1)^{i-m}\binom{i-1}{m-1}\right] iT_i$$

$$= T_1, \qquad [12.12]$$

since the summation within the parenthesis vanishes for all $i = 2, 3, \ldots, K$. Similar algebraic manipulations show that the second moment of Y can be written as

$$E(Y^2) = E(Y) + \sum_m m(m-1)\cdot \text{Prob}(Y = m)$$

$$= T_1 + 2T_2. \qquad [12.13]$$

Since $X = K - Y$, I therefore complete the proof of equations [12.6] and [12.10] starting from [12.5]. Furthermore, this computational logic also yields the rth factorial moment of Y, $\mu_{[r]}(Y)$, given by

$$\mu_{[r]}(Y) = E[Y(Y-1)\cdots(Y-r+1)] = r!\cdot T_r, \qquad [12.14]$$

for any $r \geqslant 1$, giving the complete characterization of the probability function [12.5] through its moments.

When all π_i's are equal (i.e., $\pi_i = 1/K$ for all i), note that

$$T_1 = K[(K-1)/K]^n, \text{ and } T_2 = K(K-1)\cdot[(K-2)/K]^n/2,$$

and hence,

$$E(X) = K[1 - \{(K-1)/K\}^n], \qquad [12.15]$$

and

$$V(X) = K\cdot\{(K-1)/K\}^n\cdot[1 - K\cdot\{(K-1)/K\}^n] + K(K-1)\cdot\{(K-2)/K\}^n, \qquad [12.16]$$

which are derived in Arnold and Beaver (1988) while studying the sampling properties of the observed number of classes in the context of the classical occupancy problem. When $\pi_i = 1/K$, note also that the summations $\{T_i\}$ take the form

$$T_i = \binom{K}{i}\{(K-i)/K\}^n,$$

so that the probability function [12.5] reduces to

$$P_{[K-m]} = \sum_{i=m}^{K} (-1)^{i-m}\binom{i}{m}\binom{K}{i}(K-i)^n/K^n$$

$$= \sum_{i=m}^{K} (-1)^{i-m}\binom{K}{K-m}\binom{K-m}{i-m}(K-i)^n/K^n$$

$$= \binom{K}{K-m}\cdot\sum_{i=m}^{K} (-1)^{i-m}\binom{K-m}{i-m}(K-i)^n/K^n$$

$$= \binom{K}{K-m}\cdot\frac{m!\,S_n^{(K-m)}}{K^n},$$

invoking the definition of a Stirling number of the second kind (Abramowitz and Stegun, 1965; p 824).

The above derivations, therefore, show that the sampling distribution of the number of observed classes in a finite sample can be analytically specified for any arbitrary multinomial distribution. This generalizes the special case solution of the problem discussed in Arnold and Beaver (1988) in the context of the classical occupancy problem. The algebraic solutions of other relevant random variables (e.g., the number of classes containing a specified number of sampling units within each of them) are also similar, although more cumbersome to compute numerically.

APPLICATIONS

I mention here three applications of this generalized occupancy problem, each of which has considerable genetic implications.

Are mutagen-induced mutations equally likely to occur at all loci?

Hanash et al. (1988) recently demonstrated that somatic cell gene mutations altering protein structure do not occur with equal probability at all loci when cultured human lymphoblastoid cell lines are treated with mutagens like ethylnitrosourea. To show this, they used the technique of two-dimensional polyacrylamide gel electrophoresis, and found 65 mutants occurring at 49 of the 263 loci scored in their experiments. The locus-specific distributions of the mutation frequencies in their work were: three mutants observed at each of five loci ($n_1 = n_2 = n_3 = n_4 = n_5 = 3$), two mutants at each of six loci ($n_6 = \cdots = n_{11} = 2$), and one mutant at each of 38 loci ($n_{12} = \cdots = n_{49} = 1$). No mutation was detected at each of the remaining 214 loci ($n_{50} = \cdots = n_{263} = 0$). The total number of mutations ($n = 65$) was, thus, distributed in $K = 263$ classes. The null hypothesis to be tested in H_0: $\{\pi_i = \text{the probability of mutation occurring at the } i\text{th locus} = 1/K = 1/263, \text{ for all } i\}$. In their work, the authors defined the concept of "repeat" mutations (R), noting that 16 mutations occurred at loci each of which contained already one mutation (i.e., $R = n - X$, where X is the number of loci containing at least one mutant). Under the null hypothesis of equiprobable mutation frequencies across loci, the number of "repeat" mutations should be small, since $K = 263$ is much larger than the sample size $n = 65$. Through a simulation experiment of the occupancy problem, they determined that the probability of 16 or more "repeat" mutants is below 0.0005, and hence, they conclude that mutagen-induced mutations are not equally likely to occur at all loci.

The theory described above provides a complete analytical solution, avoiding any simulation. Note that the observed value of X in the above experiment is $m = 49$, and hence the observed number of empty classes (number of loci having no occurrences of mutations, Y) is 214 ($= 263 - 49$). With $n = 65$, and $K = 263$, the range of possible X values in this experiment is 1 through 65, and consequently the possible values of Y (number of empty classes) are from 198 ($= 263 - 65$) through 262. Figure 12.1 shows the exact probability distribution (shown by the histogram) of non-empty classes (X, the number of loci at which one or more mutants can occur) under the null hypothesis of equiprobable mutations across the 263 loci. The observed value of X ($m = 49$) is marked with an arrow, the area below which is the total probability ($P = 0.0005$) of all other sample configurations which represent deviations from the hypothesis more extreme under the null hypothesis. Note that Hanash et al.'s simulation also resulted in a P-value consistent with the present result, suggesting that the qualitative conclusion of their analysis is the same as the one obtained by the present analytical solution.

Since under the null hypothesis ($\pi_i = 1/K = 1/263$ for all i) the observed number of non-empty classes (X) is the sufficient statistic for K, I further checked to see if the exact distribution of X can be approximated by any standard distribution. Employing equations [12.15] and [12.16], the mean and variance of X for this experiment are given by $E(X) = 57.687$ and $V(X) = 5.289$. The smooth curve of Figure 12.1 represents the density function

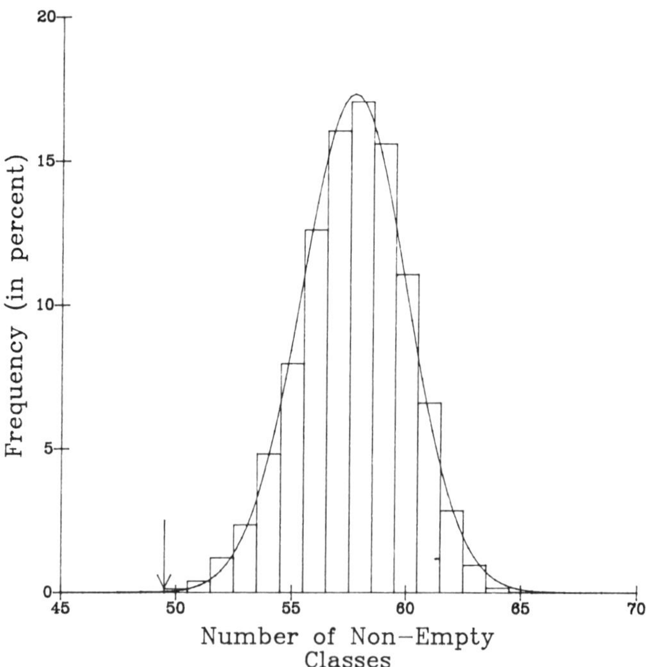

Figure 12.1. The sampling distribution of the number of non-empty classes in a sample of 65 observations drawn from a multinomial distribution with 263 equi-frequent classes. The histogram shows the exact analytical distribution evaluated from equation [12.5] and the smooth curve is the normal approximation using the expectation and variance given in equations [12.6] and [12.10]. The arrow indicates the observed value $m = 49$ (See text for details).

of a normal distribution with these mean and variance values. Note that the normal deviate for $m = 49$, then becomes $z = -3.78$, giving a corresponding P value of 0.00002, which is somewhat smaller than the p value obtained from the exact distribution. Nevertheless, the normal approximation is quite satisfactory, when compared with the histogram shown in Figure 12.1.

Test of Hardy-Weinberg expectation based on the observed numbers of distinct genotyeps in a finite sample

The generalized occupancy problem can also be used to examine whether or not the genotype distribution of a given number of alleles follows the Hardy-Weinberg expectation (HWE). Generally, this is done by either a likelihood ratio test or a goodness of fit chi-square test, contrasting the observed and expected frequencies of all possible genotypes. However, there are occasions when the number of alleles are so large that many of the genotypes are either not observed in a sample, or the observed frequencies of several genotypes are so small that the large sample approximation of these test statistics in unwarranted. The recently discovered VNTR polymorphisms provide examples of this nature, where the number of possible alleles is often so large that no reasonably sized survey can encompass all possible genotypes in any given sample. Assuming that there are K segregating alleles at a locus, there are K possible homozygote genotypes and $K(K-1)/2$ possible distinct heterozygote genotypes that can be encountered. One might ask, what would be the distribution of the numbers of distinct genotypes (of homozygote and heterozygote types, separately) observed in a sample of n individuals. Under the Hardy-Weinberg expectation of genotypic probabilities given by p_i^2 for homozygotes and $2p_ip_j$ for heterozygotes, where p_i represents the allele frequencies in the population, we can use the above analytical formulation to compute the exact distributions of the distinct numbers of homozygote and heterozygote genotypes seen in a sample.

Figure 12.2 shows a numerical example of such computations. Deka et al. (1991) recently surveyed the New Guinea population for VNTR polymorphisms at six loci. At the D1S76 locus, they discovered 6 alleles in a sample of 35 individuals. Gene counting showed that in the sample of 70 genes at this locus, the allele counts of these 6 alleles are 1, 3, 7, 9, 25, and 25. In total they observed 20 heterozygous individuals (consisting of 7 distinct genotypes). However, under the HWE assumption, the expected frequency of heterozygotes from the above allele counts is 25.4, showing a significant deficiency of heterozygotes ($P < 0.05$). Since the observed numbers of distinct homozygote and heterozygote genotypes in their sample were 4 and 7, respectively, we can ask if these observations deviate from their respective expectations under the HWE assumption. Figure 12.2a shows the exact distribution of the observed number of distinct homozygote genotypes (drawn as histogram) and Figure 12.2b gives the same for the observed number of distinct heterozygote genotypes, under the HWE assumption for the given allele frequencies. The arrows represent the observed statistics. Clearly, the observed number of distinct homozygote genotypes ($m = 4$) is not at variance with the HWE, since the probability of observing four or more distinct homozygote genotypes is 0.411. Under HWE the probability of observing seven or less distinct heterozygote genotypes is 0.017, suggesting that a significant deficiency is observed in the total number of heterozygotes as well as in the number of distinct heterozygote genotypes. Of course, as in the case of traditional likelihood ratio or chi-square tests, this test cannot ascertain the real cause of such heterozygote deficiency.

From equations [12.6] and [12.10], the mean and variance of the number of distinct genotypes were computed as 3.329 and 0.578 for the homozygotes, and 10.106 and 1.652 for the heterozygote genotypes, respectively. The expected distributions under the normality approximation are also shown by the smooth curves in both panels of Figure 12.2. As in the earlier case, it shows that the normal approximation is fairly adequate for the distribution of distinct heterozygote genotypes. This is not so for the homozygotes because of the narrow range of variation in the number of distinct homozygote genotypes. Under the normality approximation, the normal deviate corresponding to observing seven or less distinct heterozygote genotypes is $z = -2.41$, with a P-value of 0.008, which is again smaller than the exact P-value shown above.

Global test of disequilibrium based on multiple-locus haplotype data

As a third application, consider the haplotype frequency data surveyed by Wainscoat et al. (1986) at the β-globin gene cluster detected by five polymorphic restriction sites, at each of which there are two segregating alleles. This results in $2^5 = 32$ possible haplotypes at this gene region, but in a sample of 55 chromosomes sampled from a Polynesian population, these authors found only 5 observed haplotypes (see Table 1 of Wainscoat et al., 1986). One might ask, what is the expected distribution of the number of haplotypes given that these five sites are independently segregating. Figure 12.3 shows the exact distribution (represented by histogram), following the general analytical formulation (equation 12.5), where the expected haplotype frequencies are assumed to follow the independent segregation rule. Clearly, almost the entire distribution is to the right of the observed number ($m = 5$) of haplotypes, giving a rare probability of observing five or less haplotypes ($P < 10^{-5}$), suggesting that the observed number of haplotypes is incompatible with the assumption of independent segregation. Under independent

Figure 12.2. The sampling distributions of the number of distinct homozygote (A) and heterozygote (B) genotypes at the D1S76 VNTR locus in a sample of 35 individuals from Papua New Guinea (Deka et al., 1990) under the assumption of Hardy-Weinberg equilibrium frequencies of genotypic proportions. The histograms are exact computations (equation 12.5) and the smooth curves are the normal approximations based on mean and variance, given in equations [12.6] and [12.10]. The arrows indicate the observed numbers of distinct homozygote (4) and heterozygote genotypes (7) found in the sample.

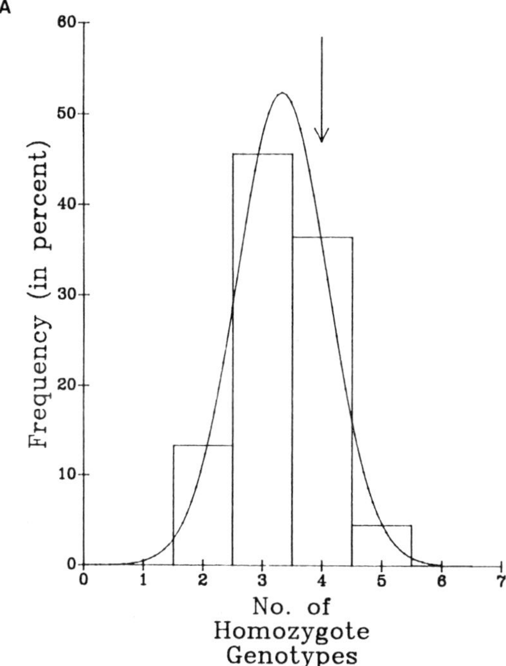
A
Frequency (in percent)
0
10
20
30
40
50
60
0 1 2 3 4 5 6 7
No. of
Homozygote
Genotypes

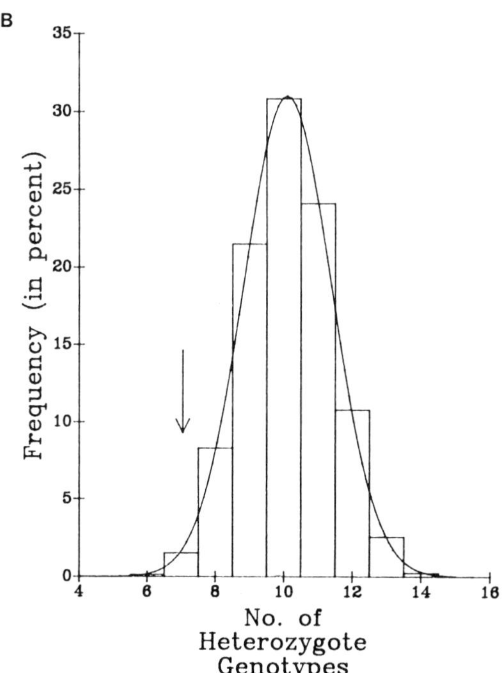
B
Frequency (in percent)
0
5
10
15
20
25
30
35
4 6 8 10 12 14 16
No. of
Heterozygote
Genotypes

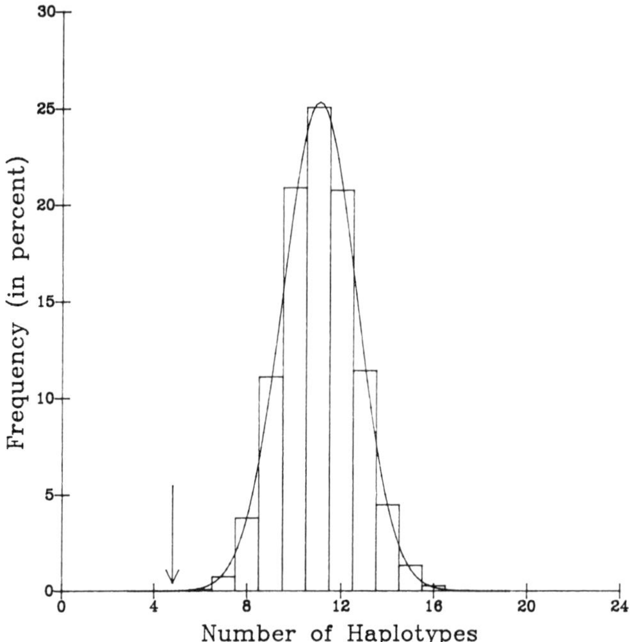

Figure 12.3. The sampling distribution of the number of DNA haplotypes at the β-globin gene cluster, defined by 5 restriction site polymorphisms (Wainscoat et al., 1986), in a sample of 55 chromosomes from a Polynesian population under the assumption of complete linkage equilibrium. The histogram is the exact distribution based on equation [12.5] and the smooth curve is its normal approximation based on mean and variance given by equations [12.6] and [12.10]. The arrow indicates the observed number of 5 different haplotypes found in the sample.

segregation, the expected mean and variance of the observed number of haplotypes in a sample of 55 chromosomes are 11.059 and 2.477, respectively. The normal approximation of the sampling distribution is again shown by the smooth curve of Figure 12.3. While the normal approximation appears satisfactory, the normal deviate corresponding to the observed number 5 is $z = -3.85$, giving a P-value of 0.00006, which is larger than the exact P-value. Note that Blanton and Chakravarti (1987) suggested this global test for disequilibrium, although unlike here their sampling distribution was obtained by simulation.

DISCUSSION

The analytical theory presented here along with the specific applications indicate that the generalized occupancy problem has a number of interesting applications in population genetics. This is particularly true in the context of sparse data, where by the very nature of the problem, the exact sampling distribution must be evaluated and no adequate large sample approximation is available. This theory enables comparison of occurrences of several biological endpoints in cross-survey comparisons, adjusting for sample size differences, as shown in Chakraborty et al. (1988), and in this chapter three other applications are mentioned. In the first application, no loss of information is attendent to the consideration of the sample statistic X, the observed number of non-empty classes, since under the equiprobable mutation rate (across loci), X is a sufficient statistic of the only parameter of the underlying distribution. In the other two cases, the consideration of observed number of classes raise the possibility of some loss of information, since the frequencies of the different observed categories do not enter into the present analysis. However, when the number of categories is large compared to the sample size, most of the observed categories are likely to have one or a few sample points in them and such loss of information is not critical. As shown through the applications here, the exact distribution evaluation does indeed detect deviations from the null hypothesis even when the sample size is larger than the total number of possible classes. A comprehensive power analysis of this approach to deal with such specific genetic applications will be attempted in the future.

In closing I should note the close resemblance of the methodology of this presentation with a percentage testing problem that Schull and I resolved several years ago (Chakraborty and Schull, 1976), where we evaluated the sampling distribution of the numbr of loci with reference to which a randomly accused man could be excluded if this man is not the father of a child born to a specific mother. Although all of these problems can be resolved by simulation, the real advantage of the present theory is that such problems can be addressed analytically, avoiding the natural bias and tediousness of computer simulations.

ACKNOWLEDGMENTS

This chapter is dedicated to Professor William J. Schull, whose encouragement and motivation primarily led to this work. This research is partially supported by Grants GM 41399 and 90-IJ-CX-0038 from the National Institutes of Health and National Institute of Justices, respectively.

REFERENCES

Abramowitz M, Stegun IA (1965) *Handbook of Mathematical Functions.* New York, Dover.

Arnold BC, Beaver RJ (1988) Estimation of the number of classes in a population. *Biometrical Journal* 30:413–424.

Blanton SH, Chakravarti A (1987) A global test of linkage disequilibrium. *Amer J Hum Genet* 41:A250.

Chakraborty R, Schull WJ (1976) A note on the distribution of the number of exclusions to be expected in paternity testing. *Amer J Hum Genet* 28:615–618.

Chakraborty R, Smouse PE, Neel JV (1988) Population amalgamation and genetic variation: observations on artificially agglomerated tribal populations of Central and South America. *Amer J Hum Genet* 43:709–725.

Deka R, Chakraborty R, Ferrell RE (1991) A population genetic study of six VNTR loci in three ethnically defined populations. *Genomics* 11:83–92.

Emigh TH (1983) On the number of observed classes from a multinomial distribution. *Biometrics* 39:485–491.

Feller W (1968) *An Introduction to Probability Theory and its Applications.* New York, Wiley.

Hanash SM, Boehnke M, Chu EHY, Neel JV, Kuick RD (1988) Nonrandom distribution of structural mutants in ethylnitrosourea treatment of cultured human lymphoblastoid cells. *Proc Natl Acad Sci USA* 85:165–169.

Schull WJ, Neel JV (1965) *The Effects of Inbreeding on Japanese Children.* New York, Harper and Row.

Schull WJ, Otake M, Neel JV (1981) Genetic effects of the atomic bombs: a reappraisal. *Science* 213:1220–1227.

Schull WJ, Rothhammer F (1990) *The Aymara: Strategies in Human Adaptation to a Rigorous Environment.* Amsterdam, Kluwer Academic Publishers.

Wainscoat JS, Hill AVS, Boyce AL, Flint J, Hernandez M, Thein SL, Old JM, Lynch JR, Falusi AG, Weatherall DJ, Clegg JB (1986) Evolutionary relationships of human populations from an analysis of nuclear DNA polymorphisms. *Nautre* 319:491–493.

13

The Maintenance of Genetic Variation in Bacterial Populations

BRUCE R. LEVIN

It seems appropriate that the maintenance of genetic variation is the topic I have been assigned for this festschrift. Estimating the amount and accounting for the persistence of genetic variation in populations was the main focus of my training as a graduate student under Jack Schull's mentorship, in the mid 1960s. Composing and validating stories about genetic diversity in populations—tales illustrated by equations, graphs, and tables of gene and genotype frequency data—was what most population geneticists did then, and still do (although now spots on gels and sequences of A's, T's, C's, and G's are included among the illustrations). The mechanisms responsible for maintaining genetic variability in populations continue to be a major focus of my research and that of the students, postdocs and collaborators working with me. However, our current stories are, I am pleased to say, different from those I learned as a student (or so they seem to me). Then my research was with sexually reproducing eukaryotes. For most of my post-student career, I have worked with bacteria.

In the following, I review some of what is known about the nature and magnitude of genetic variation in bacteria, emphasizing how the population genetics of bacteria is different from that of eukaryotes. I then consider current ideas about the mechanisms responsible for maintaining genetic variability in prokaryotic populations. I conclude with a discussion of some of the major, unsolved (unresolved) theoretical and experimental problems of the population genetics of bacteria, i.e., where action is. In this chapter, I give greater than random consideration to my own contributions and personal views of these subjects. I learned a great deal from Jack Schull, but he was never able to teach me the humility he practices so well.

THE POPULATION GENETICS OF BACTERIA IS DIFFERENT FROM THAT OF EUKARYOTES

Reproduction in bacteria is by binary fission; recombination is rare, restricted to the non-reciprocal exchange of small segments of the genome at a time, and achieved through the infectious transfer mediated by plasmid or phage vectors (conjugation or transduction) or by sequestering free DNA (transformation). As a result of the low rate of recombination, extinction and recolonization (Maruyama and Kimura, 1980) and periodic selection (Koch, 1974; Levin, 1981) are effective in purging genetic variability accumulating within populations due to mutation and occasional recombination. Multilocus genotypes are maintained for extensive periods and over wide geographic areas (see, e.g., Porras et al., 1986; Selander and Levin, 1980). This long-term persistence of multilocus genotypes has become known as the "clone concept" (Ørskov and Ørskov, 1983) and is fundamental to the typing of bacteria for epidemiological and clinical purposes.

A good deal of the genetic variation in bacterial populations involves genes carried on accessory genetic elements, plasmids, prophage, and transposons, rather than chromosomes. The average *E. coli* (which is, of course, your average bacterium) carries somewhat more than two plasmids, of generally of unknown function, "cryptic" (Caugant et al., 1981), a variety of functional and non-functional prophage (Campbell, 1981), and a plethora of insertion sequences and other transposable elements (Hartl and Sawyer, 1988). Many of the adaptations that make bacteria interesting—antibiotic, ultraviolet light, and heavy metal resistance; virulence determinants; fermentation and oxidation of "atypical" carbon sources; restriction–modification; and bacteriocin production—are coded for by genes carried on plasmids, prophage, and transposons.

Because of the low rates of recombination, the concepts of species and population in bacteria are different from those for sexually reproducing eukaryotes. Electrophoretic estimates of genic diversity in organisms classified as *E. coli* or *Haemophilus influenzae* are approximately 50% (Porras et al., 1986; Selander and Levin, 1980). Does this mean that these bacterial "species" have five times the genetic diversity of most higher organisms? Or, is this difference an artifact of the way these bacterial species are defined (primarily biochemical criteria)? Bacteria have nothing equivalent to the "Mendelian (panmictic) population" upon which most population genetic theory is founded. Indeed, bacteria may not have anything equivalent to a eukaryotic "species." Accessory elements and chromosomal genes can be exchanged among phylogenetically distant bacteria.

Because of the clonal structure of bacterial populations—"strong linkage disequilibria" in population genetics jargon—one can't treat selection at a locus as if it were independent of selection at other loci, even in the absence of epistasis. At any time, selection operating on the most favored gens(s) will affect the fate of selectively neutral genes and genes under weaker selection. Furthermore, because bacteria are haploid, that old chestnut heterosis cannot be invoked as a general explanation for the maintenance of genetic variability.

Because reproduction is asexual and recombination is through the infectious transfer of genes, mathematical models of the population genetics of bacteria are different from those of sexually reproducing eukaryotes. The rate of recombination depends on the absolute densities of genotypes, not just their relative frequencies (Levin, 1981). Models of the population dynamics of accessory genetic elements are more akin to the compartment models of epidemiology (e.g., Anderson and May, 1979) than the traditional p–q–N equations of eukaryotic population genetics.

MECHANISMS MAINTAINING GENETIC DIVERSITY IN BACTERIAL POPULATIONS

With combinations of chromosomal- and accessory-element–borne alleles of the same gene, one could have partial heterozygotes: hemizygotes (e.g., see appendix to Caugant et al., 1981). Thus, in theory some allelic variation in bacteria could be maintained by selection favoring individuals carrying different alleles of the same gene. However, there is no evidence for hemizygotes being at all common. Consequently, it seems reasonable to assume that heterosis is unlikely to play much, if any, role in maintaining the considerable chromosomal gene variability observed in bacterial species like *E. coli* and *H. influenzae*.

On the other hand, there are a plethora of other mechanisms that can account for the maintenance of allelic variation in bacteria, such as balances between opposing mutation rates or between recurrent mutation and selection. Neutral or near neutral genetic variation would also be anticipated due to recurrent mutation and genetic drift. Because of periodic (associated linkage) selection, individual populations of bacteria would be virtually monomorphic for neutral or near neutral genes. Over all populations, however, periodic selection would not affect the average genic diversity of neutral alleles or the average rate of substitution of genes that are under selection.

Clonal and therefore allelic variation in bacteria can be also maintained by a variety of ecological processes such as environmental heterogeneity (Helling et al., 1987; Levin, 1972; Stewart and Levin, 1973), interactions with phage (Chao et al., 1977; Levin et al., 1977), or different kinds of stabilizing frequency-dependent selection such as that resulting from the association between bacteria and antibiotics (Lenski and Hattingh, 1986) or bacteria with restriction-modification and phage (Levin, 1986b; Levin, 1988).

In theory, plasmids, phage, and transposons can be maintained as either symbionts or parasites. In the former case, these accessory elements carry genes that increase the fitness of their host cell or clone (e.g., those for antibiotic resistance in the presence of antibiotics). As long as the occasions where these accessory-element–borne genes are under positive selection are adequately frequent and/or the magnitude of the fitness advantage great, these elements will be maintained in the bacterial populations.

Parasitic accessory elements have to be maintained by infectious transfer overcoming the cost in host fitness associated with their carriage and their losses by vegetative segregation. Elucidating and evaluating the conditions under which this obtains has been a dominant theme of our research for the past 15 years. We have done theoretical and experimental studies of these "existence conditions" (Campbell, 1961) for lytic phage, conjugative plasmids, non-conjugative plasmids, temperate phage, and transposable elements (for reviews, see Levin, 1981, 1986b; Levin and Lenski, 1983, 1985; Stewart and Levin, 1984; and Condit et al., 1988; Simonsen, 1992). Lytic phage must be maintained as parasites (predators). However, the results of the studies we have done to date suggest that it is unlikely that plasmids, transposons, or temperate phage are maintained as pure parasites in natural populations of bacteria.

In our theoretical studies we have found conditions under which accessory elements of each type could be maintained as parasites. Our experiments, on the other hand, suggest that these conditions are unlikely to be met for plasmids and transposons. Based on our models and estimates of rates of infectious transfer of conjugative plasmids, mobilizable nonconjugative plasmids, and transposons, we have argued that it is unlikely that these accessory genetic elements are maintained as pure parasites in populations of realistic densities (for a possible exception see Lundquist and Levin, 1986). Although there is no reason to assume that the rate of infectious transfer of temperate phage is less than that of lytic, the a priori conditions for temperate phage to overcome competition with lytic mutants are restrictive. Although temperate phage could be maintained by infectious transfer alone, our theoretical results suggest that in the absence of positive selection for the carriage of the prophage, under most conditions they will be replaced by lytic mutants (Stewart and Levin, 1983).

SOME UNSOLVED PROBLEMS: WHERE THE ACTION SEEMS TO BE

What determines the distribution of multilocus genotypes that comprise a bacterial species? The combination of low rates of recombination and random extinction and/or periodic selection can account for the long-term persistence of multilocus genotypes, "clones." However, these mechanisms do not explain the total number of extant clones within a bacterial "species" or the amount of genetic difference between clones. Can, for example, the distribution of multilocus electrophoretic genotypes of *E. coli* (Selander et al., 1987) be attributed to genetic drift, recurrent mutation, and recombination of selectively neutral or near neutral alleles? At this juncture, the theory that predicts this distribution has not been developed for bacterial populations.

How important is horizontal transfer of chromosomal genes to the adaptation of bacteria to their environment and the evolution of bacterial genes and genomes? Although bacteria maintain a clonal population genetic structure, it is clear that evolution is not strictly within lineages. Not only is there interclone transfer of accessory genetic elements, there is also evidence for interclone exchanges of chromosomal regions. Bacteria of very different enzyme electrophoretic types have been found with identical restriction fragment patterns for the P-associated-pyelonephritis (Pap) region, a "virulence determinant" coded for by a series of chromosomal genes with considerable sequence diversity (Plos et al., 1989). There is also evidence for between-species movement of the genes responsible for coding penicillin binding proteins between species of *Neisseria* (Spratt et al., 1989, and review by Maynard Smith et al., 1991). Studies using DNA sequence data suggest that different segments of the *E. coli* chromosome and even portions of *E. coli*'s genes have different ancestries (DuBose et al., 1988; Dykhuizen and Green, 1986; Milkman and Stoltzfus, 1988). Are these genetic changes due to the random fixation of selectively neutral gene transfers (recombination), or are these recombinational events adaptive and established by selection?

Are the conditions for maintaining parasitic accessory elements greater than those currently believed? While, for image management, we may present the "no parasitic plasmids, transposons, and temperate phage hypotheses" with conviction, the theoretical and empirical basis of these hypotheses are, at best, modest. By and large they are founded on studies with mass-action models and their empirical analog, well agitated liquid cultures of bacteria, almost always laboratory strains of *E. coli.* Are the conditons for maintaining parasitic accessory elements in surface and other habitats where bacteria grow as colonies broader than those in mass, liquid, culture? At this juncture, there have only been a couple of quantitative studies of the surface population dynamics of plasmids (Goguen, 1980; Simonsen, 1990) and I don't know of any with phage and transposons. While these investigations suggest that the existence conditions for parasitic plasmids in surface culture are no greater than those in liquid, they provide little general theory and have been restricted to experiments with *E. coli* K-12 and plasmids of only one incompatability group, FII. Could it be that bacteria from natural sources are better donors and recipients of plasmids and transposons than laboratory strains? The results of a recent study with *E. coli* from natural sources and the plasmid R1 suggest that on average, wild strains transmit and receive plasmids at substantially lower rates than laboratory strains (Gordon, 1992), but is this the case for other plasmids? Are naturally occurring plasmids and transposons more infectious than their spoiled laboratory counterparts?

What is the origin of accessory genetic elements? It seems reasonable to assume that plasmids, phage, and transposons had their ancestry as cellular DNA or RNA or as other accessory elements. But, what are the selective pressures responsible for the evolution of these different kinds of accessory elements? Can one elucidate these selective pressures experimently, with extant plasmids, phage, and transposons?

Why are genes that augment bacterial fitness borne on plasmids, phage, and transposons rather than chromosomes? For vertically transmitted accessory elements, selection is anticipated to favor the evolution of symbiosis (Levin and Lenski, 1983) and this evolution has been observed to occur in studies with experimental populations (Bouma and Lenski, 1988). However, it is not clear why the genes coding for host-adaptive characters, like antibiotic resistances, continue to abide on accessory elements rather than chromosomes, where they are less prone to loss by vegetative segregation. Is the association between these genes and accessory elements transitory, a successional state rather than the end-product of this evolution? Or, is there something intrinsic to accessory-element–borne genes, or the nature of the selection pressure, which make their carriage on extrachromosomal elements more advantageous than being associated with chromosomes?

CODA

In this chapter, I have tried to demonstrate that the adage (inductive inference principle?), "What is true for *E. coli* is also true for Elephants, only more so," (attributed to J. Monod), is not entirely valid for population genetics. Because gene transfer is infectious and because of the role of accessory elements in bacterial adaptation, the population genetics of bacteria is different from that of sexually reproducing higher organisms. To be sure, the mechanisms responsible for maintaining genetic diversity in bacterial populations come down to variations on a theme by natural selection and genetic drift. However, I hope I have shown that those variations are sufficiently different to warrant separate treatment for the population genetics of bacteria.

On the other hand, with respect to some of the mechanisms of gene and molecular evolution, prokaryotes and eukaryotes are, I believe, sufficiently similar for inductive inference to be justified. More specifically, one can use the ability to manipulate bacteria and their genes in vitro and to measure small fitness differences to address general questions of the evolution of genes and gene action experimentally. Included among these "general questions" are (1) Are contemporary structural genes chimeras composed of modules, "domains," derived from other genes? (2) Is codon usage bias in structural genes the result of selection to maximize rates of translation? and (3) What are the constraints natural selection imposes on the products of structural genes? Much of what we know about molecular biology comes from investigations with bacteria. I believe that the contribution of studies with bacterial models will be equally important for our understanding of molecular population genetics and evolution.

ACKNOWLEDGMENTS

I wish to thank Judy Mongold, Margaret Riley, and Lone Simonsen for helpful comments. This enterprise has been supported by a grant from the National Institutes of Health GM 33782.

REFERENCES

Anderson RM, May RM (1979) The population biology of infectious disease Part I. *Nature* 280:361–367.

Bouma, JE, Lenski RE (1988) Evolution of a bacteria-plasmid association. *Nature* 335(6188):351–352.

Campbell A (1961) Conditions for the existence of bacteriophage. *Evolution* 15:153–165.

Campbell A (1981) The evolutionary significance of accessory DNA elements in bacteria. *Ann Rev Microbiol* 35:55–83.

Caugant DA, Levin BR, Selander RK (1981) Genetic diversity and temporal variation in the *E. coli* populations of a human host. *Genetics* 98:467–490.

Chao L, Levin BR, Stewart FM (1977) A complex community in a simple habitat: an experimental study with bacteria and phage. *Ecology* 58:369–378.

Condit R, Stewart FM, Levin BR (1988) The population biology of transposons: a priori conditions for maintenance as parasitic DNA. *Am Nat* 132:129–147.

DuBose RF, Dykhuizen DE, Hartl DL (1988) Genetic exchange among natural isolates of bacteria: recombination within the *phoA* gene of *Escherichia coli. Proc Nat Acad Sci US* 85:7036–7040.

Dykhuizen DE, Green L (1986) DNA sequence variation, DNA phylogeny and recombination in *E. coli. Genetics* 113:s71.

Gordon DM (in press) The rate of plasmid trnasfer among *Escherichia coli* strains isolated from natural populations. *J Gen Micro.*

Gougen JD (1980) The population dynamics of bacteria and their plasmids; some effects of surface populations. PhD thesis, University of Massachusetts, Amherst.

Hartl DL, Sawyer SA (1988) Why do unrelated insertion sequences occur together in the genome of *Echerichia coli*? *Genetics* 118(3):537–541.

Helling RB, Vargas CN, Adams J (1987) Evolution of *Escherichia coli* during growth in a constant environment. *Genetics* 116(3):349–358.

Koch A (1974) The pertinence of the periodic selection phenomenon to prokaryotic evolution. *Genetics* 77:127–142.

Lenski RE, Hattingh SE (1986) Coexistence of two competitors on one resource and one inhibitor: a chemostat model based on bacteria and antibiotics. *J Theor Biol* 122:83–93.

Levin BR (1972) Co-existence of two asexual strains on a single resource. *Science* 175:1272–1274.

Levin BR (1981) Periodic selection, infectious gene exchange and the genetic structure of *E. coli* populations. *Genetics* 99:1–23.

Levin BR (1986a) The maintenance of plasmids and transposons in natural populations of bacteria. In Novick R, Levy S (eds) *Antibiotic Resistance Genes: Ecology, Transfer and Expression.* Cold Spring Harbor, Cold Spring Harbor Laboratory, pp 57–70.

Levin BR (1986b) Restriction-modification and the maintenance of genetic diversity in bacterial populations. In Nevo E, Karlin S (eds) *Proceedings Conference of Evolutionary Processes and Theory.* New York, Academic Press, pp 669–688.

Levin BR (1988) Frequency dependent selection in bacteria. *Phil Trans R Soc Lond* B 319:472–549.

Levin BR, Lenski RE (1983) Coevolution in bacteria and their viruses and plasmids. In Futuyma DJ, Slatkin M (eds) *Coevolution.* Sunderland, MA, Sinauer Associates, pp 99–127.

Levin BR, Lenski RE (1985) Bacteria and phage: a model system for the study of the ecology and co-evolution of hosts and parasites. In Rollison D, Anderson RM (eds) *Ecology and Genetics of Host-Parasite Interactions.* New York, Academic Press, pp 227–242.

Levin BR, Stewart FM, Chao L (1977) Resource limited growth, competition and predation: a model and some experimental studies with bacteria and bacteriophage. *Amer Nat* 111:3–24.

Lundquist PA, Levin BR (1986) Transitory derepression and the maintenance of conjugative plasmids. *Genetics* 113:483–497.

Maruyama T, Kimura M. (1980) Genetic variability and effective population size when local extinction and recolinization of subpopulations are frequent. *Proc Natl Acad of Sci US* 77:6710–6714.

Maynard Smith J, Dowson CG, Spratt BG (1991) Localized sex in bacteria. *Nature* 349:29–31.

Milkman R, Stolzfus A (1988) Molecular evolution of the *E. coli* chromosome. II. Clonal segments. *Genetics* 120:359–366.

Ørskov F, Ørskov I (1983) Summary of a workshop on the clone concept in epidemiology, taxonomy and evolution of *Enterobacteriacae* and other bacteria. *J Infect Disease* 148:346–357.

Plos KS, Hull R, Hull S, Levin, BR, Ørskov I, Ørskov F, Svanborg-Edén C (1989) The distribution of the P-associated pili (Pap) region among *Escherichia coli* from natural sources: evidence for horizontal gene transfer. *Infection and Immunity* 57:1604–1611.

Porras O, Caugant DA, Gray B, Lagerard T, Levin BR, Svanborg-Edén C (1986) Difference in structure between type B and nontypable *Haemophilus influenzae* populations. *Infect and Immun* 53:79–89.

Selander RK, Caugant DA, Whittam T (1987). Genetic structure and variation in natural populations of *Escherichia coli.* In Neidhart F (ed) *Escherichia coli and Salmonella typhimurium.* Washington, DC, American Society of Microbiology, pp 1625–1648.

Selander RK, Levin BR (1980) Genetic diversity and structure in *Escherichia coli* populations. *Science* 210:545–547.

Simonsen L (1990) Dynamics of plasmid transfer on surfaces. *J Gen Microb* 136:1001–1007.

Simonsen L (1992) The existence conditions for bacterial plasmids: theory and reality. *J Microb Ecol* 22:187–205.

Spratt BG, Zhang Q-Y, Jones DM, Hutchinson A, Brannigan JA, Dowson CG (1989) Recruitment of a penicillin-binding protein gene from *Neisseria flavenscens* during emergence of penicillin resistance in *Neisseria meningitidis. Proc Natl. Acad Sci. USA* 86:8988–8992.

Stewart FM, Levin BR (1973) Partitioning resources and the outcome of interspecific competition: a model and some general consideration. *Amer Nat* 107:171–198.

Stewart FM, Levin BR (1977) The population biology of bacterial plasmids: a priori conditions for the existence of conjugationally transmitted factors. *Genetics* 87:209–228.

Stewart FM, Levin BR (1983) The population biology of bacterial viruses: why be temperate? *Theor Pop Biol* 26:93–117.

14

Some Theoretical Predictions for Electrophoretic Polymorphisms Maintained by Balancing Selection

PETER E. SMOUSE
RANAJIT CHAKRABORTY

With the development of electrophoretic assay methods for proteins in the 1960s, it became clear that genetic variation was more widespread in natural organisms than we had imagined. The neo-Darwinian predilection was to invoke widespread balancing selection, but the theoretical implications of overdominant genetic load for a large fraction of the genome (Hubby and Lewontin, 1966; Lewontin and Hubby, 1966) quickly led to the conclusion that most of the variation must be selectively neutral (Kimura, 1968; King and Jukes, 1969). The neutralist/selectionist debate is still with us two decades later (e.g., Clarke, 1975; Gillespie, 1986a,b; Nei et al., 1976), and while the genetic load argument has lost some of its apparent theoretical force in the interim, there is almost no compelling direct evidence for widespread selective maintenance of these electrophoretic polymorphisms (e.g., Mukai, 1964; Mukai and Yamazaki, 1980). The obvious alternative to neutrality is balancing selection, possibly heterosis of some sort, although frequency-dependent selection would also be adequate to explain the results (Clarke, 1964; Clarke and O'Donald, 1964; Cockerham and Burrows, 1971; Cockerham et al., 1972; Kojima and Yarbrough, 1967; Lewontin, 1958; Schutz et al., 1968; Wright, 1948). There is a large agricultural literature on heterosis (reviewed in Bush, 1988), but attempts to find it in natural populations have met with only mixed success.

There are some observations suggesting that selection may be involved in the maintenance of electrophoretically assayed genetic variation in natural populations:

1. When heterozygosity is measured by the number of loci at which an individual is heterozygous, increased heterozygosity is often associated with decreased phenotypic variability.

2. The frequency of the modal phenotype and symmetry are positively correlated with the heterozygosity level of an individual.
3. Parental fitness is positively correlated with the expected degree of heterozygosity among offspring (reviews in Livshits and Kobyliansky, 1985; Mitton and Grant, 1984; Zouros and Foltz, 1987).

Although these phenomena are not universal (Chakraborty, 1981; Gottlieb, 1977; Handford, 1980; Knowles and Grant, 1981; Mitton et al., 1981; Knowles and Mitton, 1980), the correlation between heterozygosity and fitness is deserving of some serious theoretical and empirical attention.

In the absence of specifically assayable genetic markers, it was never possible to choose between a pair of explanations: (a) overdominance, with heterozygotes having a growth, productivity, or homeostatic advantage, and (b) dominance, where hybrids among inbred lines escaped the consequences of homozygosity for deleterious recessives. The age of electrophoresis yielded one major advantage, the ability to assay individual genotypes at particular polymorphic loci. Measuring fitness on individuals remains very difficult, however, and the statistical power available to demonstrate selective differences among single-locus genotypes is small (Lewontin and Cockerham, 1959). Given the difficulties of measuring single-locus fitness effects with any real precision, it has become popular to look for the adaptive significance of multiple-locus individual genotypes.

If polymorphism is maintained by balancing selection, a general increase in heterozygosity will be roughly correlated with an increase in fitness (Turelli and Ginzburg, 1983). If the dominance hypothesis were valid, assayed genotypes would serve merely as convenient markers for the homozygosity level of the genome, and fitness should not vary across the choice of loci or alleles. If it turns out, however, that the loci in question are the causative loci themselves, then merely tallying the number of homozygous loci is not very good practice and does not make optimal use of the resolving power of the measured genotypes. There are several empiric demonstrations that it is better to treat each locus separately (Bush and Smouse, 1991; Bush et al., 1987; Chakraborty et al., 1986; Koehn et al., 1988; Shea, 1989). One could begin to assess the impact of different genotypes, without the necessity of relying on the averaging features of a classical cross-breeding design. The use of genetic markers to augment quantitative genetic analysis has elsewhere been called the *measured genotype* approach (Boerwinkle et al., 1986; Boerwinkle and Utterman, 1988; Sing et al., 1988). Our purpose here is to describe an alternative theoretical paradigm that formalizes the expected relationship in quantitative genetic terms and that makes better use of the available data.

OVERDOMINANCE AND THE ADAPTIVE DISTANCE MODEL

In a diploid population with K unlinked, polymorphic loci, the usual practice in relating a phenotypic score to the heterozygosity levels of individuals is to stratify individuals into $(K + 1)$ classes corresponding to the number of heterozygous loci (Mitton and Grant, 1984; Zouros and Foltz, 1987). No distinctions are made with regard to which loci are heterozygous or for which alleles. This convention is based on the premise that individuals who score high on the scale are genomically more highly heterozygous. The convention also weights each locus and each homozygote/heterozygote equally. For classical dominance theory, with assayed markers being mere neutral indicators of the genomic condition, that is appropriate; it has been shown, however, that heterozygosity for a small number of loci (less than a dozen in routine practice) is necessarily a poor indicator of total genomic heterozygosity for an individual (Chakraborty, 1981; Mitton and Pierce, 1980). With overdominance, with the assayed markers being those of direct interest, a heterozygosity count across loci is not an ideal choice of weights, because locus-specific effects range over four orders of magnitude (Koehn et al., 1988).

In order to circumvent such limitations, we have proposed an alternative scoring method, showing that it is useful to assign separate fitness contributions to each multiple-locus genotype in such a way that the scale measures the fitness decrement from the optimum K-locus genotype (Smouse, 1986). On the supposition that marginal overdominance (at each locus) is a general equilibrium *consequence* of multiple-locus polymorphism (Li, 1978), the genotypes $A_{1k}A_{1k}$, $A_{1k}A_{2k}$, and $A_{2k}A_{2k}$ for the kth locus can be assigned fitness values $\exp\{-S_k\} \approx (1 - S_k)$, 1, and $\exp\{-T_k\} \approx (1 - T_k)$, respectively, where S_k and T_k are assumed to be positive and small. At polymorphic equilibrium, the allele frequencies are given by

$$P_k = T_k/(S_k + T_k) \text{ and } Q_k = S_k/(S_k + T_k). \qquad [14.1]$$

In an attempt to construct a straight-line relationship between fitness and genotype, we have elsewhere (Smouse, 1986) described a scale transformation called "adaptive distance," formally defining (X_{ik}) for the kth locus as $1/P_k$, 0, and $1/Q_k$, corresponding to the three genotypes $A_{1k}A_{1k}$, $A_{1k}A_{2k}$, and $A_{2k}A_{2k}$, values chosen to linearize logarithmic fitness (Y_i), relative to the adaptive distance scale, according to

$$Y_i = \log(W_i) = -\beta_k X_{ik}, \qquad [14.2]$$

where $\beta_k = S_k T_k/(S_k + T_k)$, the polymorphic (segregational) genetic load measure of traditional theory (Morton et al., 1956). Under the multiplicative overdominance extension (Smouse, 1986), the mean fitness of a multiple-locus genotype can be written as

$$Y_i = \log(W_i) = -\sum_{i=1}^{K} \beta_k X_{ik} + \varepsilon_i, \qquad [14.3]$$

where X_{ik} is defined as $1/P_k$, 0, or $1/Q_k$ for the kth locus and ε_i is an environmental deviation for the ith individual, assumed independent of genotype. Eqn [14.3] makes a predictive statement about the relationship

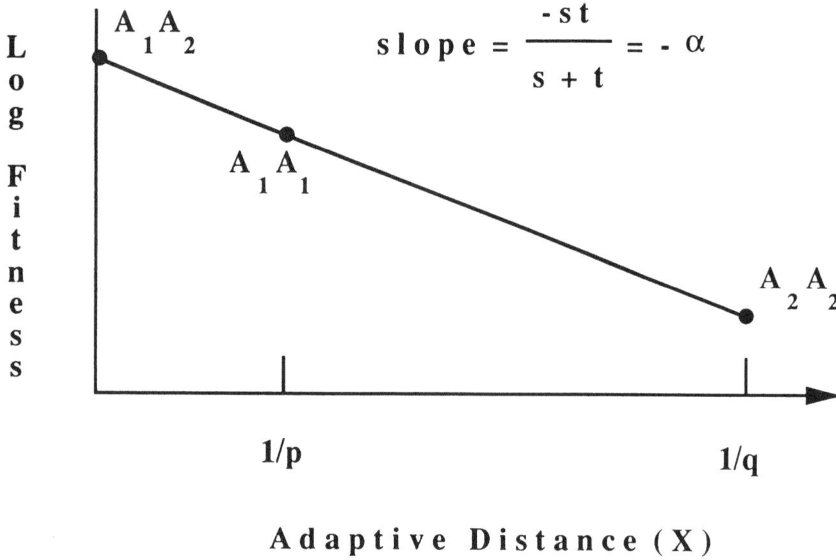

Figure 14.1. The relationship between adaptive distance and the natural logarithm of fitness for an overdominant fitness trait.

between fitness and multiple-locus genotype that can be evaluated in multiple regression fashion, without the necessity of lumping genotypes by the number of heterozygous loci. It develops that the dominance model is a degenerate special case, with all β's the same and $X_{ik} = 1$ for all homozygotes and $X_{ik} = 0$ for all heterozygotes. We have reduced the problem to a standard multiple regression/correlation form, as shown in Figure 14.1.

We have used this scoring convention in a study of *Pinus rigida* (Bush et al., 1987), with growth rate used as a fitness surrogate. We show that: (a) some loci have a much larger impact on growth than do others, (b) that a set of eight polymorphic loci accounts for a substantial portion of the variation in Y, and (c) heterotic selection is apparent in populations suffering from heavy competition, all arguing that the assayed loci themselves are those of interest for fitness. Mitton et al. (1986) have used the same convention in studies of the tiger salamander, as has Shea (1989) in populations of Engelmann spruce, both showing that the model provided useful resolution on the multiple-locus heterosis question. Marginal overdominance, as a selective consequence of multiple-locus balanced polymorphism, is a hypothesis that is compatible with data from natural populations.

QUANTITATIVE GENETICS

Multiple-locus (polygenic) traits are more usually studied in quantitative genetic fashion, and the adaptive distance scoring convention is unusual. We will now show that there is a convenient and particularly telling translation between the two scoring conventions. Recall that fitness, treated as a polygenic trait, has zero additive variance at equilibrium, as has been shown by Ewens and Thomson (1977); all of the additive × additive interaction variances are also zero at equilibrium. With overdominance, however, there is permanent dominance variance at equilibrium. Instead of scaling the fitness values of the genotypes $A_{1k}A_{1k}, A_{1k}A_{2k}$, and $A_{2k}A_{2k}$ as a_k, d_k, and $-a_k$ for the kth locus (Figure 14.2), the usual quantitative genetic convention (Falconer, 1981), we reorder and rescale for the adaptive distance analysis (Figure 14.2), using the convenient translation

$$a_k = \frac{(S_k^2 - T_k^2)}{2S_k T_k} = \frac{(S_k - T_k)}{2\beta_k} = \frac{(Q_k - P_k)}{2P_k Q_k} \qquad [14.4a]$$

and

$$d_k = \frac{-(S_k + T_k)^2}{2S_k T_k} = \frac{-(S_k + T_k)}{2\beta_k} = \frac{-1}{2P_k Q_k}, \qquad [14.4b]$$

with the additive effect of an allele substitution at the kth locus obtained by substituting equations [14.1], [14.4a] and [14.4b] into [14.5] below.

$$\alpha_k = [a_k + d_k(Q_k - P_k)] = 0 \qquad [14.5]$$

The additive and dominance variance components of adaptive distance (X_{ik}) at the kth locus become (after some manipulation)

$$V_A(X_k) = 2P_k Q_k \alpha_k^2 = 0, \qquad [14.6a]$$

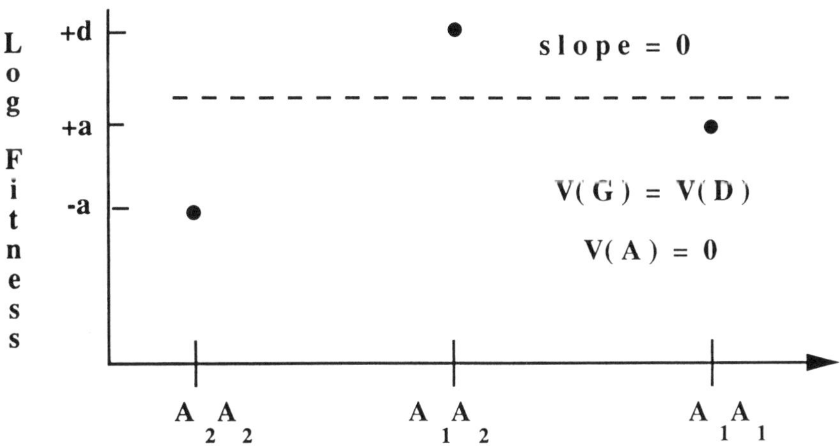

Figure 14.2. The classical quantitative genetic representation of an overdominant fitness trait.

and

$$V_D(X_k) = [2P_kQ_kd_k]^2 = 1. \qquad [14.6b]$$

Since the variance in adaptive distances (X's) at the kth locus can be partitioned into additive and dominance components, the variance of log-fitness (Y) in an equilibrium population yields a corresponding partition. Since the additive variances of the X-variables are zero at fitness equilibrium, the additive variance of log-fitness will also be zero; this is the defining condition for selectively maintained equilibrium of fitness (Ewens and Thomson, 1977). We have only dominance variance at equilibrium; allowing for the additional environmental variance $V_E(Y)$, we can write the total variance in Y as

$$V_T(Y) = \sum_{k=1}^{K} \beta_k^2 V(X_k) + V_E(Y), \qquad [14.7a]$$

and since X_{ik} takes the values $1/P_k$, 0, and $1/Q_k$ with probabilities P_k^2, $2P_kQ_k$, and Q_k^2 at equilibrium, we have $V(X_k) = 1$, reducing eqn [14.7a] to

$$V_T(Y) = \sum_{k=1}^{K} \beta_k^2 + V_E(Y) = V_D(Y) + V_E(Y). \qquad [14.7b]$$

This formulation also allows a translation of the fitness variance explained by the multiple-locus genotypes into expected familial correlations of log-fitness in an equilibrium population. The expected correlations between family members can be expressed in terms of the additive and dominance components of genetic variance alone (Falconer, 1981). The correlations for a general quantitative trait are completely standard, of course, but those for Y take particularly interesting values. With adaptive distance as our X-axis, the expected correlations for parent and offspring (PO), half sibs (HS), full sibs (FS), and double first cousins (DFC), respectively, become

$$r_{\mathrm{PO}} = \frac{1}{2}\cdot\frac{V_A(Y)}{V_T(Y)} = 0 = \frac{1}{2}\cdot\frac{V_A(Y)}{V_T(Y)} = r_{\mathrm{HS}}$$

$$r_{\mathrm{FS}} = \frac{1}{4}\cdot\frac{V_D(Y)}{V_T(Y)} \quad \text{and} \quad \frac{1}{16}\cdot\frac{V_D(Y)}{V_T(Y)} = r_{\mathrm{DFC}}. \qquad [14.8]$$

The relationships involving only additive components are identically zero at equilibrium, since $V_A(Y) = 0$. Studies involving parent–offspring or half-sib correlation (Falconer, 1981) should yield no evidence of additive heritability of fitness; those involving full-sibs or double first cousins should yield evidence of broad sense (overdominance) heritability. The argument can easily be extended to the case where there is interaction among the loci, in the same vein as described by Ewens and Thomson (1977), probably leading to non-zero dominance × dominance interaction variances at equilibrium, but because we have here assumed multiplicative overdominance across loci, all of the higher order (dominance × dominance) variances are zero, as long as linkage is loose enough to avoid selective disequilibria among the loci (Karlin and Liberman, 1979a,b).

At first glance, the reader might be inclined to conclude that we will require elaborate breeding studies to obtain estimates of these correlations, as is usual in quantitative genetics work, and that is certainly the preferred method of estimation, but it turns out that the familial components in eqn [14.8] can be roughly predicted even in the absence of family data, once we note that an estimate of $h_B^2 = \{V_D(Y)/[V_D(Y) + V_E(Y)]\}$ is obtainable from population survey. The key is to adopt the philosophy of the measured-genotype approach (Boerwinkle et al., 1986; Boerwinkle and Utterman, 1988; Sing et al., 1988). All we need to do is to assay the loci in question and obtain a credible measure of fitness. Given Y and the X's, we simply regress the former on the latter. The proportion of the variance in Y attributable to the X-values from eight polymorphic loci in the *Pinus rigida* study (Bush et al., 1987) ranged from $0.15 < h_B^2 < 0.50$ in eight populations.

DISCUSSION

The difficult task, of course, is to measure fitness for an individual. Ideally, we would need the full life cycle performance of each individual, measured (retrospectively) as its future reproductive value at birth. The sheer data requirements are such that we cannot realistically hope to achieve this goal; in routine practice, we can do no more than examine some convenient surrogate measure of fitness. Many studies, notably those among amphibians, trees, and marine bivalves, have used either size or growth rate as a conveniently measurable surrogate (Mitton and Grant, 1984; Zouros and Foltz, 1987). Other studies have examined some feature of metabolic performance or efficiency instead (Koehn et al., 1988). The literature is replete with success stories even with such crude measures of fitness (see reviews by Bush, 1988; Mitton and Grant, 1984; Zouros and Foltz 1987), suggesting that better assessment of fitness could only lead to a stronger case for the selective maintenance of genetic variation assayed by electrophoresis.

Lerner (1954) suggested that developmental homeostasis, operationally measured as bilateral symmetry, could also provide a surrogate fitness measure for individuals. Many authors have since surveyed the relationship between heterozygosity and symmetry on the premise that more heterozygous individuals will be developmentally more homeostatic than those who are less heterozygous (see review in Mitton and Grant, 1984). With a few exceptions (Hagen 1973; Sing 1970), most family studies on bilateral symmetry have found that the narrow sense (additive) heritability is either very small or not significantly different from zero (Bailit et al., 1970; Bradley, 1980; Leary et al., 1985; Mason et al., 1976; McGrath et al., 1984; Potter and Nance, 1976; Reeve, 1960; Singh, 1970; Sumner and Huestis, 1921). If Lerner's conjecture is valid, then we should see no additive variance at equilibrium.

There is some evidence that bilateral symmetry increases with heterozygosity (Beardmore and Shami, 1979; Fleischer and Johnston, 1983; Leary et al., 1983), but a formal genetic assessment of overdominance for bilateral symmetry (scaled as described above) lies in the future. Dermatoglyphic asymmetry in humans (Jantz and Webb, 1980) would seem to be an obvious candidate for careful examination.

Much has been made of the fact that so large a fraction of the genome is polymorphic, and hence a problem for genetic load, most of it based on the premise that the loci assayed are a random sample from the genome, chosen merely because convenient assay procedures are available. The truth is that there is a steady technical bias toward developing assay methods that detect polymorphisms; one cannot do genetics without variation! Extrapolation from half a dozen or a dozen (non-random) polymorphic loci to the genome as a whole is inappropriate. Having enriched our sample of the genome for the polymorphic portion, there is no theoretical reason why we should not at least consider the possibility that we have also enriched the sample for that part of the genome under balancing selection.

One lingering quibble with the overdominance model is that each of the genetic markers assayed marks a segment of chromosome; our marker locus may be a neutral gene to which we are attributing the fitness effects of a linked segment of chromosome (Kimura and Ohta, 1971). The overdominance we see may be *associative overdominance*, indirect selection on a flanking overdominant locus or on a large number of closely linked dominant loci with deleterious recessives, collectively behaving as a single overdominant locus. While there is ample simulation support for this explanation of the phenomenon as a theoretical possibility, the argument is based on the philosophy that we should invoke a locus (or loci) that (by the nature of the argument) we cannot observe in order to explain away a correlation between fitness and some locus which we can observe. While that strategy always "lets the observable marker locus off the selective hook," we gain nothing in this fashion. Whether we are describing overdominance or associative overdominance, we still have to account for a considerable amount of balancing selection. There is really nothing to be gained by invoking a phantom locus or set of loci. We might as well deal with the possibility that at least some of these electrophoretic polymorphisms are themselves (directly or indirectly) maintained by balancing selection.

DEDICATION AND ACKNOWLEDGMENTS

We would like to dedicate this chapter to Dr W. Jack Schull. We would like to thank Drs Robin Bush and Tom Meagher for helpful comments on earlier versions of the manuscript. PES is supported by NIH-R01-GM32589, the New Jersey Agricultural Experiment Station (NJAES)-32102 and USDA; RC is supported by NIH-R01-GM-20293 and NSF-BSR-8807910.

REFERENCES

Bailit HL, Workman PL, Niswander JD, McLean CJ (1970) Dental asymmetry as an indicator of genic and environmental conditions in human populations. *Hum Biol* 42:626–638.

Beardmore JA, Shami SA (1979) Heterozygosity and the optimum phenotype under stabilizing selection. *Aquilo Ser Zool* 20:100–110.

Boerwinkle E, Chakraborty R, Sing CF (1986) The use of measured genotype information in the analysis of quantitative phenotypes in man. I. Models and analytical methods. *Ann Hum Genet* 50:181–194.

Boerwinkle E, Utterman G (1988) Simultaneous effects of the apolipoprotein E polymorphism on apolipoprotein E, apolipoprotein B, and cholesterol metabolism. *Am J Hum Genet* 42:104–112.

Bradley BP (1980) Developmental stability of *Drosophila melanogaster* under artificial and natural selection in constant and fluctuating environments. *Genetics* 95:1033–1042.

Bush RM (1988) *The Relationship between Heterozygosity and Growth Rate in Pitch Pine* (Pinus rigida *Mill.) and Loblolly Pine* (Pinus tacda *L.*). PhD dissertation, University of Michigan, Ann Arbor.

Bush RM, Smouse PE (1991) The impact of electrophoretic genotype on fitness in *Pinus taeda* L. *Evolution* 45:481–498.

Bush RM, Smouse PE, Ledig FT (1987) The fitness consequences of multiple-locus heterozygosity: the relationship between heterozygosity and growth in pitch pine (*Pinus rigida*, Mill.). *Evolution* 41:787–798.

Chakraborty R (1981) The distribution of the number of heterozygous loci in an individual in natural populations. *Genetics* 98:461–466.

Chakraborty R, Ferrell RE, Barton SA, Schull WJ (1986) Genetic Polymorphism and fertility parameters in the Aymara of Chile and Bolivia. *Ann Hum Genet* 50:69–82.

Clarke B (1964) Frequency-dependent selection for the dominance of rare polymorphic genes. *Evolution* 18:364–369.

Clarke B (1975) The contribution of ecological genetics to evolutionary theory: detecting the direct effects of natural selection on particular polymorphic loci. *Genetics* 79:101–113.

Clarke B, O'Donald P (1964) Frequency dependent selection. *Heredity* 19:201–206.

Cockerham CC, Burrows PM (1971) Populations of interacting autogenous components. *Am Natur* 105:13–29.

Cockerham CC, Burrows PM, Young SS, Prout T (1972) Frequency-dependent selection in randomly mating populations. *Am Natur* 106:493–515.

Ewens WA, Thomson G (1977) Properties of equilibria in multi-locus genetic systems. *Genetics* 87:807–819.

Falconer DS (1981) *Introduction to Quantitative Genetics* (2nd ed) New York, Wiley.

Fleischer RC, Johnston RF (1983) Allozymic heterozygosity and morphological variation in house sparrows. *Nature* 304:628–630.

Gillespie JH (1986a) Variability of evolutionary rates of DNA. *Genetics* 113:1077–1091.

Gillespie JH (1986b) Natural selection and the molecular clock. *Mol Biol Evol* 3:138–155.

Gottlieb LD (1977) Genotypic similarity of large and small individuals in a natural population of the annual plant *Stephanomeria exigua* ssp. *coronaria* (compositae). *J Ecol* 65:127–134.

Hagen DW (1973) Inheritance of number of lateral plates and gillrakers in *Gasterosteus aculeatus. Heredity* 30:303–312.

Handford P (1980) Heterozygosity at enzyme loci and morphological variation. *Nature* 286:261–262.

Hubby JL, Lewontin RC (1966) A molecular approach to the study of genetic heterozygosity in natural populations. I. The number of alleles at different loci in *Drosophila pseudoobscura. Genetics* 54:577–594.

Jantz RL, Webb RS (1980) Dermatoglyphic asymmetry as a measure of canalization. *Ann Hum Biol* 7:489–493.

Karlin S, Libermann U (1979a) Central equilibria in multilocus systems. I: Generalized nonepistatis selection regimes. *Genetics* 91:777–798.

Karlin S, Libermann U (1979b) Central equilibria in multilocus systems. II: Bisexual generalized nonepistatis selection models. *Genetics* 91:799–816.

Kimura M (1968) Evolutionary rate at the molecular level. *Nature* 217:624–626.

Kimura M (1976) Population genetics and molecular evolution. *Johns Hopkins Med J* 138:253–261.

Kimura M, Ohta T (1971) Protein polymorphism as a phase of molecular evolution. *Nature* 229:467.

King JL, Jukes JH (1969) Non-Darwinian evolution. *Science* 164:788–798.

Knowles P, Grant MC (1981) Genetic patterns associated with growth variability in ponderosa pine. *Amer J Bot* 68:942–946.

Knowles P, Mitton JB (1980) Genetic heterozygosity and radial growth variability in *Pinus contorta. Silvae Genetica* 29:114–118.

Koehn RK, Diehl WJ, Scott TM (1988) The differential contribution by individual enzymes of glycolysis and protein catabolism to the relationship between heterozygosity and growth rate in the coot clam, *Mulinia lateralis. Genetics* 118:121–130.

Kojima K, Yarbrough KM (1967) Frequency-dependent selection at the esterase-6 locus in *Drosophila melanogaster. Proc Natl Acad Sci US* 57:645–649.

Leary RF, Allendorf FW, Knudsen KL (1983) Developmental stability and enzyme heterozygosity in rainbow trout. *Nature* 31:71–72.

Leary RF, Allendorf FW, Knudson RL (1985) Inheritance of meristic variation and the evolution of developmental stability in rainbow trout. *Evolution* 39:308–314.

Lerner IM (1954) *Genetic Homeostasis.* Edinburgh, Oliver and Boyd.

Lewontin RC (1958) A general method for investigating the equilibrium of gene-frequency in a population. *Genetics* 43:419–434.

Lewontin RC, Cockerham CC (1959) The goodness-of-fit test for detecting natural selection in random mating populations. *Evolution* 13:561–564.

Lewontin RC, Hubby JL (1966) A molecular approach to the study of genetic heterozygosity in natural populations. II. Amount of variation and degree of heterozygosity in natural populations of *Drosophila pseudoobscura. Genetics* 54:595–609.

Lewontin RC, Ginszburg LR, Tuljapurkar SD (1978) Heterosis as an explanation for large amounts of genic polymorphisms. *Genetics* 88:149–170.

Li CC (1978) *First Course in Population Genetics.* Pacific Grove, CA, Boxwood Press.

Livshits G, Kobyliansky E (1985) Lerner's concept of developmental homeostasis and the problem of heterozygosity level in natural populations. *Heredity* 55:341–353.

Mason LG, Ehrlich PR, Emmel TC (1976) The population biology of butterfly, *Euphydryas editha*. V. Character clusters and asymmetry. *Evolution* 21:85–91.

McGrath JW, Cheverud JM, Buikstra JE (1984) Genetic correlation between sides and heritability of asymmetry for nonmetric traits in rhesus macaques on Cayo Santiago. *Am J Phys Anthropol* 64:401–411.

Mitton JB, Carey C, Kocher TD (1986) The relation of enzyme heterozygosity to standard and active oxygen consumption and body size of tiger salamanders, *Ambystoma tigrinum*. *Physiol Zool* 59:574–582.

Mitton JB, Grant MC (1984) Association among heterozygosity, growth rate and developmental homeostasis. *Ann Rev Ecol Syst* 15:479–499.

Mitton JB, Knowles P, Sturgeon KB, Linhart YB, Davis M (1981) Associations between heterozygosity and growth rate variables in three western forest trees. In Conkle MT (ed) *Isozymes of North American Forest Trees and Forest Insects*. Genet Tech Rep PSW-48, USDA Forest Service, Pacific Southwest Forest Range Experimental Station, pp 27–34.

Mitton JB, Pierce BA (1980) The distribution of individual heterozygosity in natural populations. *Genetics* 95:1043–1054.

Morton NE, Crow JF, Muller HJ (1956) An estimate of the mutational damage in man from data on consanguineous marriages. *Proc Natl Acad Sci USA* 42:855–863.

Mukai T (1964) The genetic structure of natural populations of *Drosophila melanogaster*. I. Spontaneous mutation rate of polygenes controlling viability. *Genetics* 50:1–19.

Mukai T, Yamazaki T (1980) Test for selection on polymorphic isozyme genes using the population cage method. *Genetics* 96:537–542.

Nei M, Fuerst PA, Chakraborty R (1976) Testing the neutral mutation hypothesis by distribution of single locus heterozygosity. *Nature* 262:491–493.

Potter RH, Nance WE (1976) A twin study of dental dimension. I. Discordance, asymmetry, and mirror imagery. *Am J Phys Anthropol* 44:391–396.

Reeve ECR (1960) Some genetic tests on asymmetry of sternopleural bristle number in *Drosophila*. *Genet Res* 1:151–172.

Schutz WM, Brim CA, Usanis SA (1968) Inter-genotypic competition in plant populations. I. Feedback systems with stable equilibria in populations of autogamous homozygous lines. *Crop Sci* 8:61–66.

Shea KL (1989) The relationship between heterozygosity and fitness in Engelmann spruce. *Am J Bot* 76 (suppl):153.

Sing CF, Boerwinkle E, Moll PP, Templeton AR (1988) Characterization of genes affecting quantitative traits in humans. In Weir BS, Goodman MM, Eisen GJ, Namkoong G (eds) *Proceedings of the Second International Conference on Quantitative Genetics*. Sunderland, MA, Sinauer, pp 250–269.

Singh S (1970) Inheritance of asymmetry in finger ridge counts. *Hum Hered* 20:403–408.

Smouse PE (1986) The fitness consequences of multiple-locus heterozygosity under the multiplicative overdominance and inbreeding depression models. *Evolution* 40:946–957.

Sumner JL, Hucstis RR (1921) Bilateral asymmetry and its relation to certain problems in genetics. *Genetics* 6:445–485.

Turelli M, Ginzburg L (1983) Should individual fitness increase with heterozygosity? *Genetics* 104:191–209.

Van Valen L (1962) A study of fluctuating asymmetry. *Evolution* 16:125–142.

Wright S (1948) On the roles of directed and random changes in gene frequency in the genetics of natural populations. *Evolution* 2:279–294.

Zouros E, Foltz DW (1987) The use of allelic isozyme variation for the study of heterosis. *Isozymes Curr Top Biol Med Res* 15:1–60.

15

Application of Our Understanding of Genetic Variation in Native North America

EMÖKE J. E. SZATHMARY

What are important risk factors for the development of non-insulin-dependent diabetes mellitus (NIDDM) by Native Americans? The most likely response would have two components: one focusing on obesity, the other, on the amount of Amerindian ancestry. The higher the proportion of genes of Amerindian origin in a population, the greater the prevalence of NIDDM (Brosseau et al., 1979; Chakraborty et al., 1986; Gardner et al., 1984; Lee et al., 1985). The longer the duration of obesity in an individual (Knowler et al., 1981; Pettitt et al., 1983), or the type of obesity—for example, upper body fatness rather than just a generalized excess adiposity (Lee et al., 1985; Szathmary and Holt, 1983)—the more likely that an adult will develop diabetes. The organizing theory behind these observations is that in Indians (Knowler et al., 1983; Weiss et al., 1984) natural selection has favoured the spread of "thrifty genes" (Neel, 1962, 1982), postulated to be advantageous under conditions of alternating feast and famine but deleterious in times of continuous and ample food supply. Accordingly, NIDDM occurs in genetically susceptible individuals whose lifestyle differs radically from those of their ancestors. Because obesity is believed to have been a rare condition until recently, the NIDDM phenotype either did not emerge in the past, or emerged so rarely that the disease remained virtually unknown among Native Americans until the 1940s (West, 1974, 1978).

The evidence in support of the impact of obesity on the development of NIDDM is compelling. In West's opinion (1978) the genetic argument was not nearly as strong, although views on this have shifted (Weiss et al., 1984; Szathmary, 1987). In conjunction with this altered perspective, it needs to be acknowledged explicitly that Native individuals may differ in their degree of genetic risk for diabetes, as may Native populations. Without evidence to the contrary, variability is to be expected within and between groups for disease-influencing genes, whether the disease is simply inherited or is complex.

Variability among individuals *and* populations in the degree of lifestyle change that is conducive to the development of obesity should also be expected. Marked sociocultural differences occurred aboriginally in North America, and the duration and intensity of European contact conducive to lifestyle change have not been uniform across the continent.

Whether these expected genetic and cultural differences have actually influenced variation in the rates of NIDDM across *all* of Native North America remains to be demonstrated. However, in the Canadian context, Young et al. (1990b) have been reasonably successful in explaining the pattern of diabetes morbidity in Indians and Inuit (Eskimos) on the basis of geographic location, and known linguistic and cultural diversities. This chapter considers the rationale for the approaches employed in that study, and reviews its findings. An extension of its methods is also employed in an attempt to explain variation in diabetes prevalence among the Indian populations of the province of British Columbia.

ANTHROPOLOGICAL CLASSIFICATION OF LINGUISTIC AND CULTURAL DIVERSITY IN NORTH AMERICA, AND ATTRIBUTION OF THEIR ORIGINS

Linguistic relationships and culture areas

Language and culture area are fundamental to the way in which nonbiological variation in Native America has been classified. The most conservative view, based on the evidence from detailed comparative studies (Campbell and Mithun, 1979), is that there are 33 confirmed language families and 32 language isolates north of Mexico. Only three higher-order groupings of related language families are "solidly supported" (Campbell and Mithun, 1979, p 49), in contrast to the eight phyla containing 36 language families and 31 language isolates noted by Voegelin and Voegelin (1965). Currently the most radical perspective on linguistic relationships is Greenberg's (1987). He groups all the languages of North and South America into three stocks. Amerind, Na-Dene, and Eskaleut. Each has its own internal groupings, some of which were proposed earlier (e.g., by Sapir, 1929; Voegelin and Voegelin, 1965). Greenberg's model is undergoing vigorous criticism (e.g., Campbell, 1988; Chafe, 1987; Goddard, 1987), but it is not without support (e.g., Hymes, 1987).

In addition to groupings based on language, at least 10 different culture areas are recognized north of Mexico (Sturtevant and Heizer, 1978), although earlier works give as many as 16 (Murdock and O'Leary, 1975). Culture areas are geographic regions within which populations might differ in language, but otherwise share a variety of cultural attributes. The most important of these was the aboriginal subsistence pattern, because the methods employed to ensure a continuous food supply have ramifications for level of technological sophistication in the group, its sociopolitical organization, and population size.

Whatever the differences between the "lumpers" and the "splitters" in identifying the *exact* number of language families, language phyla, and culture areas in North America, it is fact that geographic boundaries circumscribe language distributions and culture areas, respectively. Do these phenomena tell us anything about the biological characteristics of the people themselves? This is a question worth asking because NIDDM does have a genetic base. The answer is associated with the origins of Native Americans, because the aboriginal inhabitants of the continent did not evolve in situ.

Origins

That the first migrants to North America were hunting-gathering people who came from Asia is not disputed by any serious scholar. There is ample genetic and morphological evidence to support the Asian connection (Kirk and Szathmary, 1985; Laughlin and Harper 1979). The continuing disputed questions concern the timing of human entry into North America, as well as the location of that entry. These issues have bearing on the origin of the linguistic and cultural diversity on the continent by the time of European colonization. If only one group of people entered North America overland from Asia during the last glaciation (20,000 to 14,000 years Before the Present [BP]) (Hopkins, 1979), this group is ancestral to all of the Native peoples of the Americas. The genetic, linguistic, and cultural diversity that is manifest today had then to have taken place in the Americas, as descendants of the original group moved southward; differentiated genetically, linguistically, and culturally; and eventually occupied both continents. However, if there were two routes of entry during the same time period, then it is possible that the ancestral populations differed genetically as well as linguistically at the onset. The variability seen in the descendant populations would have two different sources, and a single explanatory model, for example, concerning genetically based disease susceptibility acquired in an arctic environment (e.g., Ritenbaugh and Goodby, 1989; Weiss et al., 1984), could be inappropriate. The third scenario, not only of different routes of entry, but of different times of entry would have similar ramifications for variation in the distribution of a disease such as NIDDM.

Archaeological evidence shows that during the last glaciation humans were present on Beringia. The latter was a vast land mass of which more than 1,000,000 sq km was unglaciated (Morlan, 1983). Beringia extended from eastern Siberia, included the sea floor between the continents that became exposed when the Bering Sea receded during the glacial advances, encompassed Alaska, and ended near the eastern border of the Yukon Territory. At the height of the glacial advance, Beringia was bounded by ice on its North American perimeter. The narrow unglaciated "corridor" that extended out of the Yukon east of the Rocky Mountains apparently was a blind appendage at this time. It was blocked by ice not only between 18,000 and 16,000 years ago in the Athabasca Valley of Alberta (Reeves, 1983), but also at other times before 12,000 BP, further northwest, in the Peace River region

(White et al., 1985). The area was habitable by 11,700 ± 260 years BP (White et al., 1985), as were more southerly parts of the corridor, given the rapid melting of the Cordilleran and Laurentide ice shields after 14,000 years BP.

On the Asian side of Beringia, the oldest occupation at Lake Ushki on the Kamchatka Peninsula has been dated (C^{14}) to 14,300 ± 200 years ago, but may be 1,500 to 2,000 years older on the basis of archeomagnetic dating (Dikov, 1988). On the North American side, artifacts were present in horizons containing megafauna dated to 15,500 ± 130 BP and 12,900 ± 100 BP at the Bluefish Caves site in the northern Yukon Territory (Morlan, 1983). Assessment of these and more recently dated sites in Asian Beringia and in the Nenana Valley of Alaska (Powers and Hoffecker, 1989) has concluded that human occupation of Beringia had occurred by 15,000 BP. Powers and Hoffecker (1989) contend that differences in tool types, and differences in proportions of artifact mixtures in tool assemblages between 15,000 and 10,500 BP in Beringia suggest the presence of peoples of different origin or a rapid spread of tools of different type from different Asian foci. They favour the population movement notion. Climatic change after 15,000 BP may have been responsible for the migration of peoples with a bifacial-point technology along the Pacific rim, given the presence of bifacial projectile points in Late Glacial deposits in Japan, Sakhalin, Kamchatka, Alaska, and the Yukon. These may have been the forerunners of the Clovis fluted points widely distributed in North America between 11,500 and 11,000 BP. On the other hand, the microblade tradition typical of Holocene Alaskan and north Pacific assemblages, and also present at some Beringian sites dominates the collections recovered from interior Siberia (Powers and Hoffecker, 1989). In sum, these archaeologists articulate the majority view on the peopling of the Americas: entry overland from Asia, and dispersal of the occupying bands southward as the ice sheets receded. Their variation on the theme is the suggestion that migrants from the Pacific rim may have been the forerunners of the Paleo-Indians. The migrants from interior Siberia, on the other hand, did not move out of the northwest region of North America, as judged by the distribution of their distinctive wedge-shaped microblades in Alaska and northwestern British Columbia (Powers and Hoffecker, 1989). Another interpretation of the same set of data (Dikov, 1988; Dumond 1987) does not suggest that the blade-makers were ancestral to the Clovis people. Rather, the bearers of the mixed traditions of Beringia contributed to the formation of the cultures of the early Eskimo-Aleuts, and the Indian peoples of the Northwest Coast, including Alaska.

Arguments have also been presented for two alternate models of the occupation of North America. One involves entry during the period 17,000 to 13,000 BP, but not overland. Rather, movement along the seashore, over water, from one ice-free refugium to another (Fladmark, 1983) was a possibility. There is no archaeological evidence for such a migration, although Gruhn has argued (1988) that linguistic data suggest it occurred. Indeed, she emphasizes that such entry would have had to have taken place much earlier than Fladmark suggested, perhaps by 35,000 BP to permit sites dated to

33,000 BP at Monte Verde, Chile (Dillehay and Collins, 1988), or to more than 32,000 BP at Sao Raimundo Nonato, Brazil (Guidon and Delibrias, 1986). Gruhn (1988) noted that the greatest diversity of Amerindian languages occur along the Pacific northwest, in California, and along the Pacific side of Central America. Linguistic theory holds that the area of greatest linguistic diversity is also the oldest in terms of time depth of occupation.

The second alternate, and much disputed model for the peopling of the Americas also argues for a greater time depth of occupation than 14,000 years BP. There are a few well-dated, stratified sites in North and South America that indicate human presence south of the ice boundary formed by the coalescence of the Laurentide and Cordilleran glaciers around 18,000 BP. In addition to the Chilean site already noted, the Meadowcroft rock shelter just outside of Pittsburgh remains a serious contender for proof of early entry. At Meadowcroft a charcoal deposit without cultural associations yielded a date of 19,430 ± 800 BC (Adovasio et al., 1983). If the six oldest dates, all from the deepest layers (albeit differing layers) of the site, and all with cultural associations are averaged, then humans were present at Meadowcroft between 13,955 and 14,555 BP (Adovasio and Carlisle, 1988). The earliest possible occupation date, again based on an average, was 16,770 BP (Adovasio and Carlisle, 1988). Adovasio et al. (1983) postulate that the logical interpretation of their findings means that the ancestors of the Meadowcroft people had to have crossed Beringia 25,000 to 20,000 years ago, that is, before the glacial maximum.

Implications

There is no clear resolution in sight for the archaeological debate concerning the timing of the entry of humans into North America. Greenberg's (1987) tripartite linguistic hypothesis has generated considerable attention also among anthropologists who are not linguists, because he has linked it with dental and genetic data (Greenberg et al., 1986a). He and his co-authors (1986a) do not provide a single set of dates for each of the claimed migrations (Paleo-Indian; Na-Dene; Aleut-Eskimo), although they do say that from the linguistic perspective Paleo-Indian entry has to be older than 11,000 BP. Time depths of separation based on dental evidence yielded 14,000 years for the Macro-Indian (claimed descendants of Paleo-Indians) versus Northeast Asian comparison. The 20,000 year split between Macro-Indian and Aleut-Eskimo was attributed to an initial divergence in Asia. This is consistent with the view that biologically differentiated groups entered North America via Beringia more recently than 20,000 years ago (Greenberg et al., 1986b).

Although Greenberg et al. (1986a) felt that genetic evidence corroborated the dental and linguistic evidence for three migrations into North America, they also thought that the genetic evidence was not strong, and should be regarded as "supplementary rather than primary" (Greenberg et al., 1986a, p 487). However, they also noted that "the best direct genetic support" for their theory arises from the immunoglobulin Gm system. It had been suggested (Williams et al., 1985) that the current complement of Gm haplotypes in the descendants of the original occupying populations can be explained by differences among the founders in the retention and loss of particular haplotypes. Specifically, if $Gm^{1;21}$, $Gm^{1;2;21}$, and $Gm^{1;11;13}$ were present in northeastern Siberia, the ancestral Paleo-Indians could have lost $Gm^{1;11;13}$, and retained $Gm^{1;21}$ and $Gm^{1;2;21}$; the ancestral Na-Dene could have retained all three haplotypes while the ancestors to modern Eskimos could have retained $Gm^{1;21}$ and $Gm^{1;11;13}$, but lost $Gm^{1;2;21}$. In Williams et al.'s opinion (1985), this tripartite distribution of Gm haplotypes in Native Americans does not support the claim (Szathmary and Ossenberg, 1978) that Na-Dene are genetically closer to Eskimos than are other Indians.

Hypotheses about what genes were present and what genes were absent from ancient populations are based on deductions made from modern data. Whether the hypotheses have substance can be tested in various ways, including examination of the pattern of genetic distances among the descendant populations. Szathmary (1986a) undertook genetic distance analysis of Gm data corrected for European gene flow (from Table 18, Williams et al., 1985), and displayed the results in a dendrogram. Athapaskan-speaking Indians (members of the Na-Dene phylum) were always most closely linked to Eskimos. Ten of the 15 Eskimo and Athapaskan groups were clustered in one major branch of the dendrogram. Another four were in one sub-branch which was linked to another sub-branch that contained only Paleo-Indian groups, and the last (St Lawrence Island Eskimos) was linked to a branch that contained only Asian populations (Szathmary, 1986a; Fig 2; Szathmary, forthcoming, a). The Gm haplotypes certainly did not cluster as the Williams et al. (1985) model predicted: one tree branch for Eskimos, another tree branch for Na-Dene, and a third branch for speakers of Amerind. Contrary to Greenberg et al.'s opinion (1986a), Gm data do not provide support for a tripartite division of Native Americans.

Szathmary's (1993a) findings also emphasize that one locus, even as important a locus as immunoglobulin Gm, cannot singly reconstruct the biological history of a continent. Although much hope has been attached to evidence from mitochondrial DNA (mtDNA), to date, mtDNA restriction fragment length polymorphisms (RFLPs) do not provide evidence for three migrations from Asia. Rather, they suggest a close relationship among modern Beringian peoples, among them, Eskimos and Athapaskan Indians (Shields et al., 1992). Furthermore, sequence analysis in the control region of mtDNA indicates considerable genetic heterogeneity among the original migrants (Paabo et al., 1990), with "substantial molecular evolution" occurring after their arrival in North America (Paabo et al., 1990, p 277). Sequence analyses also show less mtDNA diversity between Eskimos and Na-Dene peoples than between "Paleo-Indian" descendants (Ward et al., 1992). The best genetic evidence on population origins is still based on many loci (Szathmary, 1993a, b), and this shows that Athapaskan-speaking peoples are genetically closer to Eskimos than are other Indians. The closeness reflects either common origin or the effects of ancient gene flow between their ancestors who occupied Beringia during the glacial maximum (Szathmary and Ossenberg, 1978). Until it is possible to identify which of these factors produced the extant genetic closeness, solid support for the migrations of three biologically distinct populations will remain elusive.

Although based on genetic and skeletal evidence, and a much scantier set of archaeological data available in 1978, Szathmary and Ossenberg's view is remarkably similar to Dikov's (1988) and Dumond's (1987). Szathmary (1984) also suggested that Indians without Na-Dene affiliation may be the descendants of people who were south of the glaciers, as shown by the dates accruing from Meadowcroft rock shelter. It is worth emphasizing that the "southern derived" people do not necessarily reflect an early migration in the sense of Greenberg et al. (1986a). Rather, if people were on the continent before the glacial maximum, it is possible that the advance of the glaciers simply separated far flung hunting bands, so that some remained in the northwest, on Beringia, while others were near the southern margin of the ice, as at Meadowcroft. Rogers, focusing on Indian linguistic data only, has presented a very similar scenario (1985).

What implications—if any—do the latter views have for a biological meaning for the distributions of language families and culture areas in North America? If there truly was an early migration (25,000–20,000 BP) into the Americas over Beringia, available evidence suggests that people south of the glaciers were separated from people northwest of the glaciers for at least 3000 and perhaps 6000 years (Szathmary 1993b). All of the incomers passed through a "cold filter" of arctic climate. After the ice receded, people northwest of the ice could move south, just as people south of the ice could move north. Such northeastern expansion, for example, has been hypothesized to be responsible for the movement of Plano Culture people from the Great Plains into the central and eastern subarctic (Szathmary, 1984; Wright, 1972).

As a population moves into an uninhabited area it will gradually increase in size and expand its territory (if successful). Over time, linguistic changes will not only occur but will also increase as bands move away from each other. There is a strong analogy between the process of linguistic differentiation and genetic differentiation, although the two processes are governed by very different mechanisms. Barring incidents of direct or indirect conquest, which might lead to the language of one group replacing the language of the other, the scenario of occupation outlined for the peopling of North America suggests that people speaking related languages should be biologically related to each other. Certainly native speakers of indigenous languages today tend learn their language from people to whom they are genetically related—their parents and grandparents. Spuhler (1972) tested these notions for Native American populations north of Mexico, and demonstrated that genetic distances were smaller within language families than between them. Indeed, genes predicted language family affiliation slightly better than they predicted culture area affiliations (Spuhler, 1979). The latter is not completely unexpected, as there is considerable overlap, especially in northern North

America, between language distribution and culture area membership. The association does not mean that there are genes coding for specific cultural attributes! Accordingly, although the possibility remains that the northeast Asiatic roots of Native North Americans are bifurcate, or trifurcate as Greenberg et al. (1986a) claim, genetic distances do not show a huge amount of genetic separation between the claimed "very different" populations. Whatever the truth of the ancient history, it does not prevent a currently demonstrable association between Indian genes and indigenous language north of Mexico.

LINGUISTIC AND CULTURAL DIVERSITY IN CANADA AND THE DISTRIBUTION OF NIDDM

Overview of diabetes in Native Canadians

Young (1990) has summarized the epidemiological information on the prevalence of diabetes in Canadian Native people. Briefly, few studies are available. With the exception of Szathmary's 1979 and 1985 oral glucose tolerance test field surveys, the studies that have been done rely on physicians' diagnoses, which vary in the diagnostic criteria employed. Furthermore, the only diabetes prevalence data for the nation as a whole are in the form of self-reports in the Canada Health Survey of 1978 (CHS) (Statistics Canada, 1981) and the General Social Survey of 1985 (GSS) (Statistics Canada, 1985). It is therefore difficult to know, with the type of precision and accuracy that is desirable, whether Canadian Indian and Inuit diabetes rates are elevated compared to a national average.

Known Native Canadian prevalence rates vary widely (Young, 1993); Young et al., 1990b). The age-adjusted rate of NIDDM in 1987 in Native Canadians ranged from a high of 87 per 1000 (provinces of New Brunswick, Nova Scotia, and Prince Edward Island) to a low of 4 per 1000 among the Inuit of the Northwest Territories (NWT). Prevalence exceeded 30/1000 in all regions except British Columbia (BC), the Yukon Territory (YT), and the NWT (16/1000 in BC, 12/1000 in the YT, 8/1000 in Dene [Athapaskan-speakers] and 4/1000 in the Inuit of the NWT). Rates in the latter three regions are below the age-adjusted morbidity rate (all races) in the United States, and are therefore likely to be below the Canadian national average (17/1000, in the CHS; 24/1000 in the GSS). The best comparison to date has been done by Evers et al. (1987) who found that in southwestern Ontario the age-adjusted prevalence of diabetes in Indians was 6.7 times higher than in Caucasians using the same clinic.

Although there is anecdotal material connecting obesity and NIDDM among Canadian Indians, there is little documentation available. Szathmary and Holt (1983) found no association between overall fatness and hyperglycemia (Hr-2 plasma glucose $\geqslant$ 1.11 mmol/L) among the Dogrib of the NWT. The body mass index (BMI) was not significantly higher among hyperglycemics, although the average indices in hyperglycemic men and women were 27.2 and 29.1, respectively (Szathmary, 1986b). Both Evers et al. (1987) and Young et al. (1985) found that there was a high rate of obesity in their samples, but neither was able to show, with the analytic methods they employed, that obesity was a risk factor for NIDDM. Among diabetic Kahnawake Mohawks, matched for age and sex with normal controls, BMIs were significantly higher in the diabetics. However, the normal controls were themselves obese by the usual standards (Adelson, 1987; Montour et al., 1989).

Studies of fat patterning and NIDDM in Native Canadians, although done on few populations, are more definitive. While Adelson (1987) could not demonstrate a real difference in her case-control study, upper-body obesity was demonstrable in hyperglycemic Dogrib (Szathmary and Holt, 1983). Obesity in Cree-Ojibwa (northeastern Manitoba, northwestern Ontario) is predominantly of the central type (Young and Sevenhuysen, 1989), and the waist-hip ratio in this group is a determinant of glycosylated hemoglobin, as is the BMI (Young et al., 1990a). The BMI in the Cree-Ojibwa also predicts fasting plasma glucose level.

The genetics of diabetes in Native Canadians is largely unexplored. A review of the evidence for genetic mechanisms in Indians (Szathmary, 1987) found contradictory claims about patterns of genetic transmission (e.g., Elston et al., 1974; Steinberg et al., 1970; Szathmary, 1985; Yamashita et al., 1984). In the Dogrib, as in the non-Indian population of the state of Michigan (Beaty et al., 1982) the most likely model for regulation of plasma glucose level is multifactorial, and it includes an intra-generational environmental component (Szathmary, 1985). No difference in the amount of Indian ancestry between diabetic and nondiabetic Kahnawake Mohawks could be demonstrated with genetic markers, including the immunoglobulin Gm system (Adelson, 1987). This resembles findings in other groups: Amerindian admixture estimates in individual Mexican Americans were also unrelated to diabetic status (Hanis et al., 1986).

Available records suggest that Native Canadians who have NIDDM have developed the disease recently. Young (1993) noted that of 1500 Indian examinations carried out in Saskatchewan in the 1930s, not one case of diabetes was found, although the age-adjusted rate in the province today is 39/1000. Of known cases among the subarctic Cree-Ojibwa (Young et al., 1985) and the Akwesasne Mohawks (St. Regis, Ontario) (Selikoff et al., 1986) half of the existing cases were diagnosed in the last 5 years of a 25-year period (Young, 1993). The average duration of diabetes in Kahnawake Mohawks (Quebec) is less than 7 years (Montour et al., 1989). Unpublished data also indicate that prevalence has increased between 1980 and 1985 in Saskatchewan (Young, 1993).

The risk of death from diabetes among Canadian Indians was 4.1 times higher in Indian women than other Candian women, and the corresponding risk in Indian men was 2.2 times higher than in their non-Indian counterparts (Mao et al., 1986). The risk of end-stage renal disease due to diabetes in Native Canadians was 3 to 5 times higher than in non-Native Canadians

(Young et al., 1989). Hypertension in the Cree-Ojibwa is associated with the duration of their diabetes (Young et al., 1985), while among diabetic Kahnawake Mohawks over 60% had at least one major complication (e.g., ischemic heart disease, stroke, peripheral vascular disease) (Montour et al., 1989).

This brief overview of diabetes in the Canadian Native population indicates that its epidemiology is similar to that noted in the United States (Gohdes, 1986). The appearance of the disease is even more recent than in the United States. There are considerable regional differences in prevalence rates. Lastly, known disease associations with diabetes and diabetes mortality rates indicate that NIDDM is not a benign disease in Canadian Native people.

Variation in diabetes morbidity by language, culture area, and other predictors: The Canadian national survey

In 1987 the Medical Services Branch of Health and Welfare Canada, the agency responsible for health care of Native Canadians, collected information on known cases of diabetes in all its field units (nursing stations, health centers). The 240,629 individuals surveyed through this reporting represented 76% of the registered[1] Indian and Inuit population of Canada. The diagnostic criteria for diabetes used by MSB physicians was not reported. The only additional information available on each diabetic case was age, sex, and residence—insufficient for assessment of the role played by obesity, or dietary differences, or physical activity levels, all of which are implicated in the onset of diabetes.

To get a picture of what does influence variation in the rate of Native diabetes, Young et al. (1990b) decided to examine the impact of factors that are responsible for extant cultural diversity: tribal, linguistic, and culture area affiliations (Figures 15.1 and 15.2). Language was used as an index of genetic relationship, while culture area membership was the index for aboriginal lifestyle that may persist today.

As a first step, Young et al. (1990b) determined the crude NIDDM prevalence for each community, which was then indentified according to tribe, language family, and culture area. The likelihood of a modern lifestyle was scored according to distance from a census metropolitan area, i.e., urban, rural, and remote. Communities were also located on a coordinate map, and degrees of longitude and latitude recorded. The data were then subjected to multiple regression analyses, with crude prevalence as the dependent variable.

[1]The term "registered" or "status" Indian under Candian law implies legal Indian status; that is, descent from Indian men whose chiefs signed treaty with the crown. According to the 1981 Canadian census (Gottesman, 1985) 80% of respondents who declared themselves as Indian were status Indians. Until recently, the child of a status Indian man and a white mother was considered a registered Indian, but offspring of unions of the reverse combination, were not. Indian women who married white men could not reside on reserves, and neither they nor their children were eligible for any benefits to which registered Indians are entitled. Canadian Indian and Inuit bands do not maintain rolls listing quantum of Native ancestry for each person on the band list, as do US tribes.

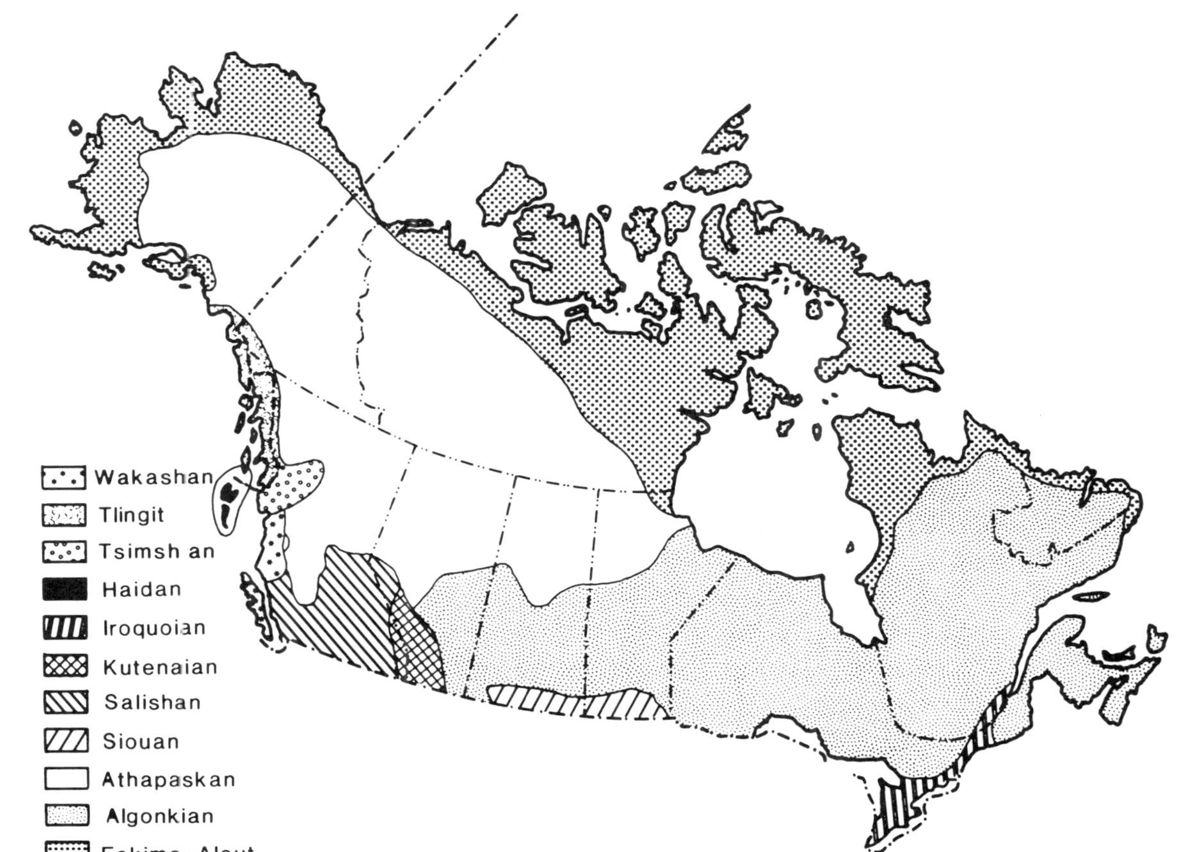

Figure 15.1. Geographic distribution of aboriginal language families in Canada. Copyright 1990; Pergamon Press.

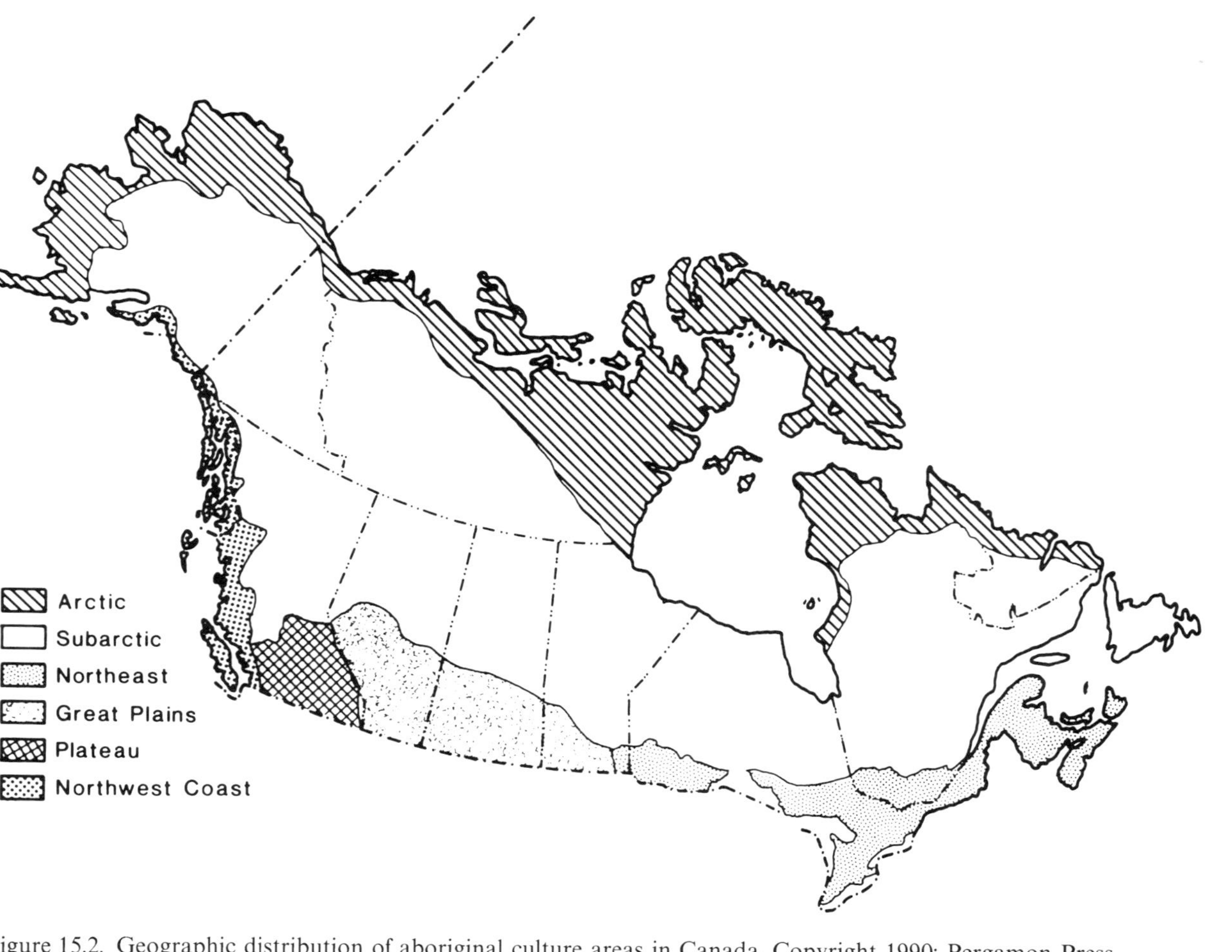

Figure 15.2. Geographic distribution of aboriginal culture areas in Canada. Copyright 1990; Pergamon Press.

Latitude, longitude, distance from a metropolitan area, and either composite language phylum–culture area variables (i.e., language family within a particular culture area), or separate language family, and culture area variables served as independent predictors (Young et al., 1990b).

When considered independently, each predictor had significant impact on the rate of NIDDM among the 421 communities. For example, urban residence was a more significant predictor of NIDDM morbidity than was rural residence, although both had significant impact on the diabetes rate. Similarly the Northeastern, Plains, and Subarctic culture areas were significantly better predictors of diabetes rate than the remaining three (Szathmary, 1988). However, when all variables were entered in hierarchical stepwise regression, only some proved to have significant impact on NIDDM morbidity. Table 15.1 shows that the most important predictor was latitude (adjusted $R^2 = 30.8\%$), and the further north a community, the lower the diabetes prevalence. Latitude, in combination with five language-culture area predictors explained over 48% of the variability in the rate of diabetes.

That some groups may indeed have greater genetic risk for diabetes than others was suggested by the language family variables in the equation. Algonkian-speaking peoples, whether in the Northeastern, Plains, or Subarctic culture areas, all had significantly higher rates of NIDDM than the reference group of Eskimos. After examining the beta ratios, Young et al. (1990b) concluded that language family membership was more important than culture area in influencing morbidity. For example, Algonkian-speakers in the Northeast were 2.3 times as likely as Iroquoians to have elevated rates of diabetes. Algonkians in the Plains were also more likely to have NIDDM than Siouans.

Table 15.1. Results of the hierarchical stepwise regression of crude NIDDM prevalence on the set of predictors[a]: Canadian national survey.[b]

Variables[c] Entered	$R^2_{(adj)}$	F	df	P	Variables in the Equation: Beta (β)	SE[d] of β	Reference Categories
Latitude					−0.090	0.042	Arc-Eskimo
NE-Algonkian					0.610	0.043	Remote
NE-Iroguoian					0.261	0.036	
SU-Algonkian					0.358	0.038	
PN-Siouan					0.212	0.036	
PN-Algonkian	0.48	66.61	6,414	0.00			

[a]The independent variables on which stepwise regression of crude prevalence was attempted included urban, rural, longitude, latitude, Northeastern-Algonkian, Subarctic-Algonkian, Plains-Algonkian, Plateau-Algonkian, Northeast Iroquoian, Subarctic-NaDene, Plains-NaDene, Plateau-NaDene, Northwest Coast-NaDene, Plains-Siouan, Plateau-Mosan, Northwest Coast-Mosan, Northwest Coast-Tsimshian (Penutian). The abbreviations in the table are: NE = Northeastern; SU = Subarctic; PN = Plains.

[b]The data set included 421 communities. Only the final results of the regression analysis are shown. Source: Young et al., 1990b.

[c]Only those variables that entered the regression are shown.

[d]SE = standard error

On the other hand, Young et al. (1990b) did not discount the influence of the environment in influencing diabetes rate. They found that when regression analyses were carried out within language families that were spread over different culture areas (e.g., Algonkians: Northeastern, Subarctic, Plains; Athapaskans: Subarctic, Plains, Plateau), some culture areas did have significant impact on morbidity. Young et al. reasoned that the influence of culture area within a language family distribution may be due to differences in the procurement of food from the wild. In some parts of Canada, Native peoples still rely on fish, migratory fowl, and wild ungulates to provide a significant amount of their diets. In some of these instances, food procurement is arduous, and this may have led to systematic influences in particular culture areas on physical activity level and caloric intakes, each of which is believed to play a role in the onset of obesity.

Did the results support the notion proposed by Niswander (1968) that particular cultures (e.g., horticulturalists compared to hunters and gatherers) have higher diabetes rates because their lifeways were more easily affected by colonizing Europeans than were other lifestyles? Iroquoian peoples were horticulturalists aboriginally, and their diabetes morbidity is the highest of any single language family in Canada (Young et al., 1990b). These results suggest a "yes," but cannot be regarded, without more detail, as definitive.

Aboriginal lifestyles, whether hunting–gathering or horticultural, have all been disrupted by the Euro-Canadian colonizers. Culture change conducive to the development of obesity and other possible diabetogenic factors is believed to be the basic reason for the emergence of NIDDM in Native peoples. The negative association of diabetes morbidity across Canada with latitude lends credence to this perspective. In Young et al.'s opinion, the inverse relationships could be due to diminishing degrees of Euro-Canadian lifestyle influence with increasingly northern residence, as the bulk of the Canadian population hugs the US-Canadian border, and their distribution thins northward. Alternatively, the relationship could also reflect duration of Euro-Canadian contact, rather than intensity of contact (and culture change). It is worth adding, that in the Arctic culture area, a different pattern emerged than in the country as a whole: Diabetes rate was positively associated with longitude (Young et al., 1990b). In the Arctic, the intensity of European contact, and the longest duration of contact emanate from the West, from the north Pacific, where European and American whaling ships were present before 1850. The whalers are known to have affected the populations on the shore, at minimum by making foreign goods and foods available. Young et al. (1990b) noted that the trend towards higher morbidity in the western Arctic was even seen in the Alaskan data of Schraer et al. (1988).

The national survey of diabetes among Native Canadians certainly indicates that variables that are responsible for aboriginal linguistic and cultural diversity also influence diabetic morbidity (Figure 15.3). Given that language family membership can be used as a surrogate measure of genetic relationship, some groups seem to be more prone than others to develop NIDDM. As Young et al. (1990b) noted, Algonkians, Iroquoians, and Siouans within the

Table 15.2. Age-adjusted[a] NIDDM rates in the Indian population of British Columbia.

	Men			Women		
Characteristic	Group Size	Rate	(95% Confifenc Interval)	Group Size	Rate	(95% Confidence Interval)
All Province	20,899	1.44	1.42–1.46	19,020	1.72	1.69–1.74
Location						
Remote	4,927	1.71	1.67–1.76	4,492	1.51	1.46–1.56
Rural	11,717	1.16	1.13–1.18	10,295	1.79	1.76–1.83
Urban	4,255	1.63	1.58–1.68	4,233	2.99	2.22–2.36
Culture Area						
Subarctic	633	1.27	1.16–1.38	575	2.26	2.08–2.44
Plateau	8,306	0.74	0.71–0.76	7,295	1.65	1.60–1.69
Northwest Coast	11,960	1.83	1.80–1.86	11,150	1.93	1.89–1.96
Language Family						
Algonkian	65	0	0	73	2.03	1.66–2.39
Athapaskan	3,892	0.61	0.58–0.64	3,553	1.29	1.23–1.34
Haida Isolate	485	1.06	0.97–1.16	379	0.83	0.74–0.93
Kootenai Isolate	201	0	0	171	3.26	2.92–3.61
Salishan	9,792	1.47	1.44–1.50	9,033	1.86	1.81–1.90
Tsimshian	3,225	1.89	1.83–1.94	2,904	1.84	1.77–1.91
Wakashan	3,239	1.79	1.73–1.86	2,907	1.77	1.70–1.84

[a] Age-Adjustment was by the direct method. The standard population was the Canadian national population, 1986. Rates are reported per 100 population.

borders of the United States are among the groups that are known to have high rates of diabetes.

Canadian Na-Dene and the Inuit have low rates of diabetes, as do their counterparts in the United States (West, 1974). However, in Canada, there is one group of Athapaskan-speakers who is different in this regard: the Sarcee of southern Alberta. West (1974) thought diabetes was "probably high" in the Kiowa-Apache of Oklahoma, but the group was not listed in the "top 10" tribes affected by NIDDM in the state (Lee et al., 1985). Both groups reside in the Plains culture area, and both groups are believed to have undergone some acculturation (and possibly admixture) with non-Athapaskan neighbors—Algonkians in the case of the Sarcee, and Kiowa (Uto-Aztecan language, phylum to which the Piman languages also belong) in the case of the Kiowa-Apache. Somewhat of a mystery, however, is the generally low rate of diabetes in the Indian population of British Columbia. In this province, none of latitude, longitude, remote, rural or urban location, nor language family affiliation, or culture area occupied, appear to have impact on crude diabetes prevalence. What might be the explanation?

The case of British Columbia

The registered Indian population of British Columbia numbers over 66,000 individuals, of whom 60% continue to live on reservations. Only 1.4% live on crown land, and the remainder are off reserve. The 247 Native communities in the province range from those that are tiny (less than 10 people resident) to large, containing over 1500 people (Cowichan, BC). However, the majority of the communities are small, with 84% including fewer than 500 residents. To eliminate potential bias because of very small population size, Young et al. (1990b) grouped the tiny communities with larger ones when geographic proximity, ethnicity, and band affiliation permitted this. This convention was followed here, so that a total of 132 communities was available for investigation. MSB field data apply to the on-reserve portion of the Native population only, that is, to 60% of known status Indians in British Columbia.

The National Survey did not attempt to adjust crude prevalence rates for age by community, because age-sex breakdowns of diabetic cases were not available for all communities. However, the Pacific Region (i.e., British Columbia) did collect the relevant data in virtually all instances. Age adjustment in this paper employed the 1986 year's end age–sex breakdowns for each community recorded by the Department of Indian and Northern Affairs (DIAND). The Canadian national population data for 1986 (Ministry of Supply and Services, 1987) were used for the standard population. To avoid the problems of age-adjustment in very small populations, adjustment was done for the province as a whole, and by sex, tribe, language family, culture area, and remote, rural or urban residence. Direct standardization and calculation of the 95% confidence intervals followed the methods of Armitage and Berry (1987). Regression analysis used the same protocol as in the national survey, except that age-adjusted diabetes per tribe was the dependent variable. When a tribe extended over more than one culture area, or more than one location (urban, rural, remote), age adjustment was undertaken for each portion of the tribe in the given categories (e.g., Tsimshian tribe in remote location; Tsimshian in rural location).

Table 15.2 provides information on some selected traits in British Columbia Indians. In general, age-adjusted rates of NIDDM are higher in women, and they conform to increase in NIDDM prevalence as nearness to a census metropolitan area increases. The reverse is not true for the men. Otherwise, the only thing that is striking about the contents of Table 15.2 is the low rates of diabetes in the various categories. The rates are lower than that reported for the United States national average (2.4 per 100: Freeman et al., 1989), and are well below the rates for related Indian populations of the US Pacific Northwest. Freeman et al. (1989) had adjusted their data to the 1980 United States population, but standardization of their reported crude rates to the 1986 Canadian national population figures did not change the findings.

The pattern of NIDDM morbidity in Canada also seems different than in Washington state. The United States rates for Plateau and Nothwest Coast culture areas were 7.2 and 4.5, respectively, but the Canadian age-sex–adjusted rate was only 1.14 and 1.89, respectively. When the United States data are grouped to reflect language family, (Salishan: Spokane, Colville, Quinault, Lummi; Wakashan: Makah), their age-adjusted diabetes rates are still twice as high as observed in British Columbia.

One possibility for the difference in rates north and south of the border is under-reporting of diabetes in British Columbia. However, because diabetic cases are known from the most remote communities, and from all of the eight MSB districts in the Pacific Region, one would not expect MSB physicians to be unaware of the presence of NIDDM. The disease is known to be familial, and has received a great deal of publicity in various media outlets. If under-reporting is not the answer, one possible explanation for a low rate of NIDDM among British Columbia Indians is recency of disease onset. The latter appears to be the case in general, in Canada.

The regression of age-adjusted diabetes on all of the predictors employed by Young et al. (1990b) proved futile. However, when the rate was regressed on tribal affiliation, whether in the province as a whole (Table 15.3), or within each of the three culture areas of British Columbia, variation in the age-adjusted rates was partially explained. Of the six tribes that entered into the provincial regression, three are Salishan, and the others are Athapaskan, Tsimshian, and Wakashan, respectively (Table 15.4). Too much cannot be read into these observations, but it is possible that the pattern of NIDDM morbidity is identifying a language group more at risk for diabetes than are the others. The presence of significantly higher rates of NIDDM in remote, linguistically unrelated Northwest Coast locations also suggests that British Columbia may just be beginning to manifest the tip of what may become a provincially distributed diabetes problem.

Is the current distribution of NIDDM in the Indians of British Columbia

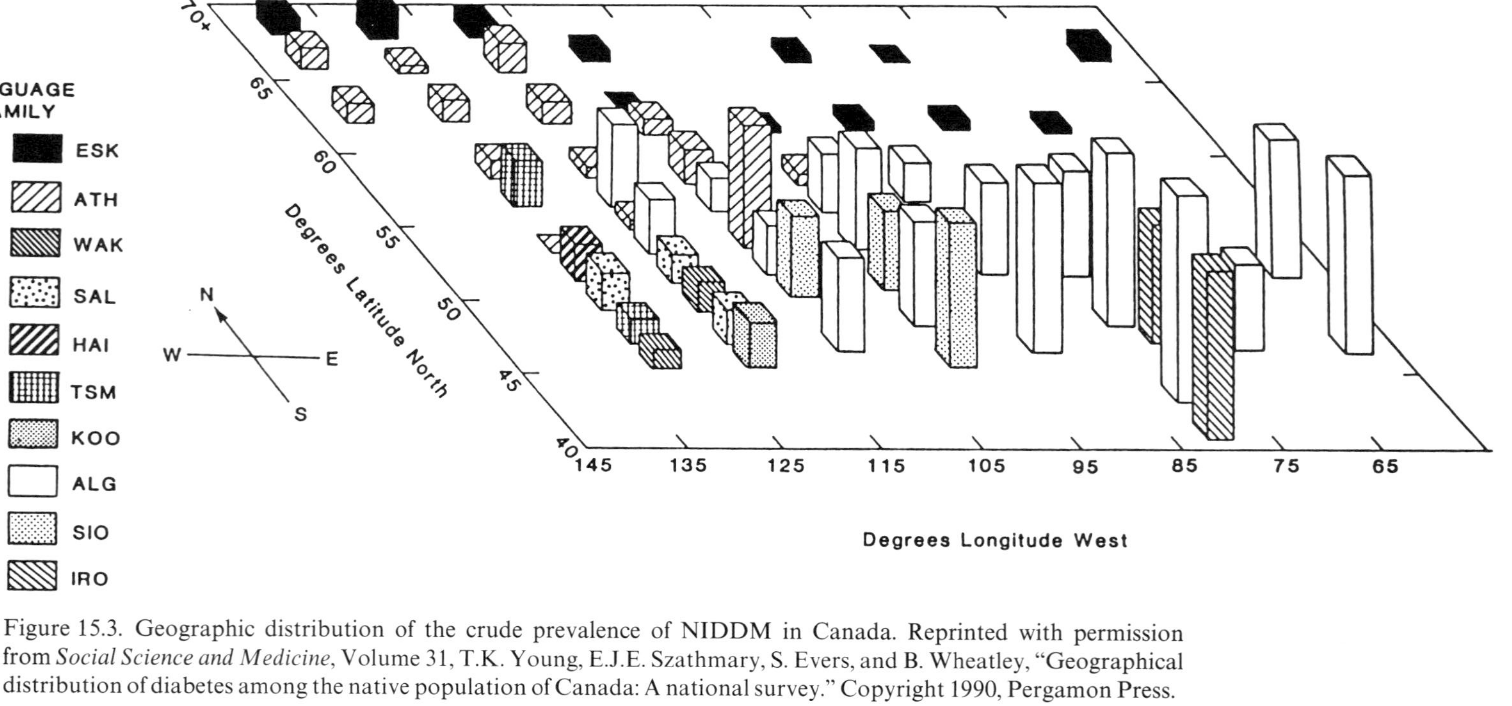

Figure 15.3. Geographic distribution of the crude prevalence of NIDDM in Canada. Reprinted with permission from *Social Science and Medicine*, Volume 31, T.K. Young, E.J.E. Szathmary, S. Evers, and B. Wheatley, "Geographical distribution of diabetes among the native population of Canada: A national survey." Copyright 1990, Pergamon Press.

Table 15.3. Results of hierarchical stepwise regression of age-adjusted NIDDM prevalence on all British Columbian tribes.[a]

Variables Entered[b]	$R^2_{(adj)}$	F	df	P	Variables in the Equation: Beta (B)	Variables in the Equation: Standard Error of B	Reference Group
Sechelt					0.56	0.56	Semiahmoo
Bella Coola					0.41	0.13	
Tahltan					0.35	0.11	
Niska					0.35	0.10	
Lillooet					0.23	0.09	
Kwakiutl	.79	22.6	6,28	.00	0.22	0.08	

[a]There are 28 tribes in British Columbia. Because some of them spread over different culture areas, or into remote, rural, and urban regions, the total number of tribal units on which regression was done is 35.
[b]Only the variables that entered the equation are shown.

in keeping with any expectations that might arise from population history and known genetic relationships? The answer is "maybe." Linguists have grouped Wakashan and Salishan families into the Mosan phylum, and Sapir (1929) placed the Mosan division into his Macro-Algonquian group. Campbell and Mithun (1979) consider such aggregations unwarranted without better evidence than is available. However, if the second order classification is correct (i.e., Wakashan and Salishan in the Mosan phylum), and if indigenous language can be used as an index of genetic relationship, then it is not surprising that 66% of the tribes with significantly higher rates of diabetes than others can be placed into one language phylum.

Historically, it is known that several of the Northwest Coast societies carried on warfare with each other, and abducted people to work as slaves in their class-structured communities. Some married into the abducting group, and produced children who might, or might not, remain in the slave class. Anthropologists have argued that because of this, gene flow across the Northwest Coast was extensive (e.g., Turner, 1985). If so, then NIDDM susceptibility genes may also have undergone a mixing across whatever genetic boundaries may have been there "in the beginning." However, genetic distance analysis that has been done with Northwest Coast data (e.g., Nootka; Haida; Tlingit) does not show evidence of this. Rather, the Haida and Tlingit group with Northern Athapaskans from the subarctic interior in the dendrograms based on the genetic distances, and the Nootka are distinct from these Na-Dene and from the northern Algonkians (Szathmary, 1979). It is not likely that the low NIDDM morbidity in British Columbians is due to relatively recent events of gene flow.

Evidence from Beringian prehistory has led different archeologists to suggest (Dikov, 1988; Dumond, 1987) that the makers of the blade/microblade assemblages of interior Beringia, Alaska, and the Northwest Coast contributed to the cultures of Aleuts, Eskimos, and Northwest Coast Indians. Indeed, Dumond says explicitly (1987) "These would form the remote ancestors of Na-Dene-speakers and, probably, other Indians of the Northwest Coast, as well as one portion of the ancestors of at least some later Eskimo-Aleuts."

Table 15.4. Age-adjusted rate[a] of NIDDM in six tribes of British Columbia.

Tribe	Group Size	Culture Area	Location	NIDDM Rate	(95% Confidence Interval)
Athapaskan					
Tahltan	160	Subarctic	Rural	4.42	4.04–4.80
Salishan					
Sechelt	439	Northwest Coast	Urban	6.43	6.12–6.76
Bella Coola	638	Northwest Coast	Rural	5.03	4.80–5.25
Lillooet	1,942	Plateau	Rural	3.15	3.04–3.23
Tsimshian					
Niska	1,581	Northwest Coast	Remote + Rural	2.62	2.52–2.73
Wakashan					
Kwakiutl	2,001	Northwest Coast	Remote	3.03	2.92–3.14

[a] Rate is reported per 100 population.

Because Eskimos tend to have the lowest rates of NIDDM of any North American peoples, and northern Athapaskans are similar, the prehistory argument would predict low NIDDM prevalences in British Columbian Indians generally. Whatever one's opinion of the thesis, it is worth noting that the 1985 age-adjusted prevalence rates (per 100) of Alaskan interior Athapaskans, "coastal Indians," Aleuts, and Eskimos, were 1.7, 2.7 to 3.1, 2.7, and .88, respectively (Schraer et al., 1988). These rates are not out of line with the age-adjusted NIDDM prevalences calculated for each of the language families of British Columbia's Indians.

CONCLUSIONS

The Canadian national study (Young et al., 1990b) and to some degree, the detailed analysis of the province of British Columbia, show that linguistic and cultural variables do influence the pattern of diabetes morbidity in aboriginal North America. Because linguistic relationship north of Mexico has been shown to be a good predictor of underlying biological relationship, the patterning of NIDDM morbidity suggests that all Native American populations are not equally susceptible to this diseases.

Can we attribute differences in NIDDM morbidity in Native Americans to different origins of their ancestors? Some have tried to do so (e.g., Wendorf, 1989), or asserted that it was so (e.g., Sievers and Fisher, 1985). This paper has tried to show that the evidence for three distinct migrations across Beringia during the last glaciation is not as strong as is commonly assumed by non-specialists. If any ancient history argument has any bearing on modern NIDDM patterning, then in my opinion, the separation of ancestral groups already on the continent during the glacial maximum has greater explanatory power. Certainly this might account for the Canadian national survey results, and the findings in British Columbia. However, what is clearly more important than the ancient genetic connections are the more recent ones. Indigenous ethnicity manifest through tribal and language family affiliations does influence NIDDM morbidity. Ultimately, even these phenomena are less important than the degree of culture change and disruption of aboriginal lifestyles brought about by the invaders, who have become the dominant peoples of Canada and the United States.

ACKNOWLEDGMENTS

I am grateful for the following individuals for their help with material examined in this paper: Dr David Martin, MSB Pacific Region, for making the British Columbia data available to me; Dr T.K. Young, University of Manitoba, and Ms Carol DeVito, McMaster University, for data preparation.

REFERENCES

Adelson N (1987) Mohawk diabetes, obesity and fat patterning. MA thesis, McMaster University, Hamilton, Ontario, Canada.

Adovasio JM, Carlisle RC (1988) The Meadowcroft rock shelter. *Science* 239:713–714.

Adovasio JM, Donahue J, Cushman K, Carlisle RC, Stuckenrath R, Gunn JD, Johnson WC (1983) Evidence from Meadowcroft rock shelter. In Shutler R Jr (ed) *Early Man in the New World*. Beverly Hills, CA, Sage Publications, pp 163–190.

Armitage P, Berry G (1987) *Statistical Methods in Medical Research*. Blackwell Scientific Publications. Oxford,

Beaty TH, Neel JV, Fajans SS (1982) Identifying risk factors for diabetes in first degree relatives of non-insulin dependent diabetic patients. *Am J Epidemiol* 115:380–397.

Brosseau JD, Elkema RC, Crawford AC, Abe TA (1979) Diabetes among the three affiliated tribes: correlation with degree of Indian inheritance. *Am J Public Health* 69:1277–1278.

Campbell L (1988) Review of *Language in the Americas*, by JH Greenberg. *Language* 64:591–615.

Campbell L, Mithun M (eds) (1979) *The Languages of Native America*. Austin, University of Texas Press.

Chafe W (1987) Review of *Language in the Americas*, by JH Greenberg. *Current Anthropol* 28:652–653.

Chakraborty R, Ferrell RE, Stern MP, Haffner SM, Hazuda HP, Rosenthal M (1986) Relationships of prevalence of noninsulin dependent diabetes mellitus to Amerindian admixture in the Mexican Americans of San Antonio, Texas. *Genet Epidemiol* 3:435–454.

Dillehay TD, Collins MB (1988) Early cultural evidence from Monte Verde in Chile. *Nature* 332:150–152.

Dikov NN (1988) On the road to America. *Natural History* 97(1):10–14.

Dumond DE (1987) A re-examination of Eskimo-Aleut prehistory. *Am Anthropol* 89:32–56.

Elston RC, Namboodiri KK, Nino HV, Pollitzer WS (1974) Studies on blood and urine glucose in Seminole Indians: implications for segregation of a major gene. *Am J Hum Genet* 26:13–34.

Evers S, McCracken E, Antone T, Deagle G (1987) Prevalence of diabetes in Indians and Caucasians in southwestern Ontario. *Can J Pub Health* 78:240–243.

Fladmark KR (1983) Times and places: environmental correlates of mid-to-late Wisconsonian human population expansions in North America. In Shutler R Jr (ed) *Early Man in the New World*. Beverly Hills, CA, Sage Publications, pp 13–42.

Freeman WL, Hosey GM, Diehr P, Gohdes D (1989) Diabetes in American Indians of Washington, Oregon, and Idaho. *Diabetes Care* 12:282–288.

Gardner LI, Stern MP, Haffner SM, Gaskill SP, Hazuda HP, Relethford JH, Eifler CW (1984) Prevalence of diabetes in Mexican Americans. Relationship to percent of gene pool derived from Native American sources. *Diabetes* 33:86–92.

Goddard I (1987) Review of *Language in the Americas*, by JH Greenberg. *Current Anthropol* 28:656–657.

Gohdes DM (1986) Diabetes in American Indians: a growing problem. *Diabetes Care* 9:609–613.

Gottesman D (1985) Native people, Demography. In *The Canadian Encyclopaedia.* (vol 3) Edmonton, Alberta, Canada, Hurtig Publishers, pp 1212–1213.

Greenberg JH (1987) *Language in the Americas.* Stanford CA, Stanford University Press.

Greenberg JH, Turner CG, Zegura SL (1986a) The settlement of the Americas: a comparison of linguistic, dental and genetic evidence. *Current Anthropol* 27:477–488.

Greenberg JH, Turner CG, Zegura SL (1986b) Reply to Comments of the Settlement of the Americas. *Current Anthropol* 27:492–497.

Gruhn R (1988) Linguistic evidence in support of the coastal route of earliest entry into the New World. *Man* 23:77–100.

Guidon N, Delibrias G (1986) Carbon-14 dates point to man in the Americas 32,000 years ago. *Nature* 321:769–771.

Hanis CL, Chakraborty R, Ferrell RE, Schull WJ (1986) Individual admixture estimates: disease associations and individual risk of diabetes and gallbladder disease among Mexican-Americans in Starr County, Texas. *Am J Phys Anthropol* 70:433–442.

Hopkins DM (1979) Landscape and climate of Beringia during late Pleistocence and Holocene time. In Laughlin WS, Harper AB (eds) *The first Americans: Origins, Affinities and Adaptations.* New York, Gustav Fischer, pp 15–42.

Hymes D (1987) Review of *Language in the Americas,* by JH Greenberg. *Current Anthropol* 28:659–663.

Kirk R, Szathmary E (eds) (1985) *Out of Asia. Peopling the Americas and the Pacific.* Canberra, The Journal of Pacific History.

Knowler WC, Pettitt DJ, Bennett PH, Williams RC (1983) Diabetes mellitus in the Pima Indians: genetic and evolutionary considerations. *Am J Phys Anthropol* 62:107–114.

Knowler, WC, Pettitt DJ, Savage PJ, Bennett PH (1981) Diabetes incidence in Pima Indians: contributions of obesity and parental diabetes. *Am J Epidemiol* 113:144–156.

Laughlin WS, Harper AB (eds) (1979) *The first Americans: Origins, Affinities, and Adaptations.* New York, Gustav Fischer.

Lee ET, Anderson PS, Bryan J, Bahr C, Coniglione T, Cleaves M (1985) Diabetes, parental diabetes, and obesity in Oklahoma Indians. *Diabetes Care* 8:107–133.

Mao Y, Morrison H, Semenciw R, Wigle D (1986) Mortality on Indian reserves. *Can J Pub Health* 77:263–268.

Ministry of Supply and Services (1987) *Canada Year Book.* Ottawa, Ontario.

Montour L, Macauley AC, Adelson N (1989) Diabetes mellitus in Mohawks of Kahnawake, PQ: A clinical and epidemiological description. *Can Med Assoc J* 141:549–552.

Morlan RE (1983) Pre-Clovis occupation north of the ice sheets. In Shutler R Jr (ed) *Early Man in the New Wold.* Beverly Hills, CA, Sage Publications, pp 47–64.

Murdock GP, O'Leary TJ (1975) *Ethnographic Bibliography of North America.* New Haven, Human Relations Area Files.

Neel JV (1962) Diabetes mellitus: a "thrifty" genotype rendered deterimental by progress? *Am J Hum Genet* 14:353–362.

Neel JV (1982) The thrifty genotype revisited. In Kobberling J, Tattersall R (eds) *The Genetics of Diabetes Mellitus.* New York, Academic Press, pp 243–250.

Niswander J (1968) Discussion, medical problems of American Indian groups. In Pan American Health Organization, *Biomedical Challenges Presented by the American Indian.* Sci Pub No 165, Washington, DC, pp 133–136.

Paabo S, Dew K, Frazier BS, Ward RH (1990) Mitochondrial evolution and the peopling of the Americas. *Am J Phys Anthropol* 81:277.

Pettitt DJ, Baird HR, Aleck KA, Bennett PH, Knowler WC (1983) Excessive obesity in offspring of Pima Indian women with diabetes during pregnancy. *The New Eng J Med* 308:242–245.

Powers WR, Hoffecker JF (1989) Late Pleistocene settlement in the Nenana Valley Valley, Central Alaska. *Amer Antiquity* 54:263–287.

Reeves, BOK (1983) Bergs, barriers and Beringia: reflections on the peopling of the New World. In Masters PM, Fleming NC (eds) *Quaternary Coastlines and Marine Archaeology: Towards to Prehistory of Land Bridges and Continental Shelves.* New York, Academic Press, pp 389–411.

Ritenbaugh, C, Goodby C-S (1989) Beyond the thrifty gene: metabolic implications of prehistoric migration into the New World. *Med Anthropol* 11:227–236.

Rogers RA (1985) Glacial geography and native North American languages. *Quat Res* 23:130–137.

Sapir E (1929) Central and North American languages. *Encyclopaedia Britannica* (14th ed). 5:138–141.

Schraer CD, Lanier AP, Boyko EF, Gohdes D, Murphy NJ (1988) Prevalence of diabetes mellitus in Alaskan Eskimos, Indians and Aleuts. *Diabetes Care* 11:693–700.

Selikoff IJ, Hammond EC, Levin SM (1986) Diabetes among the people of the St. Regis Reserve. (Unpublished report submitted to the Department of National Health and Welfare, Canada), Mount Sinai School of Medicine, New York (mimeo).

Shields GF, Hecker K, Voevoda MI, Reed JK (1992) Absence of Asian-specific region V mitochondrial marker in Native Beringians. Am J Hum Genet 50:758–765.

Sievers ML, Fisher JR (1985) Diabetes in North American Indians. In National Diabetes Data Group (eds) *Diabetes in America.* NIH Pub No 85–1468, Washington, DC, pp XI-1 to XI-20.

Spuhler JN (1972) Genetic, linguistic and geographical distances in native North America. In Weiner JS, Huizinga J (eds) *The Assessment of Population Affinities in Man.* Oxford, Clarendon Press, pp 72–95.

Spuhler JN (1979) Genetic distances, trees, and maps of North American Indians. In Laughlin WS, Harper AB (eds) *The First Americans: Origins, Affinities and Adaptations.* New York, Gustav Ficher, pp 135–184.

Statistics Canada (1981) Canada health survey. The health of Canadians. Cat No 82–538, Ottawa.

Statistics Canada (1985) General social survey analysis series: health and social support, 1985. Cat No 11–612, no 1, Ottawa.

Steinberg AG, Rushforth NB, Bennett PH, Burch TA, Miller M (1970) On the genetics of diabetes mellitus. In Cerasi E, Luft R (eds) *The Pathogenesis of Diabetes Mellitus.* New York, John Wiley & Sons, pp 237–260.

Sturtevant WC, Heizer RF (1978) Preface. In Sturtevant WC (gen ed) *Handbook of North American Indians* , Heizer RF (ed) Vol 8, California. Washington, DC, Smithsonian Institution, pp xiii–xv.

Szathmary EJE (1979) Blood groups of Siberians, Eskimos, and subarctic and northwest coast Indians: the problem of origins and genetic relationships. In Laughlin WS, Harper AB (eds) *The First Americans: Origins, Affinities and Adaptations.* New York, Gustav Fischer, pp 185–210.

Szathmary EJE (1984) Peopling of northern North America: clues from genetic studies. *Acta Anthropogenet* 8:79–110.

Szathmary EJE (1985) The search for genetic factors controlling plasma glucose levels in Dogrib Indians. In Chakraborty R, Szathmary EJE (eds) *Diseases of Complex Etiology in Small Populations: Ethnic Differences and Research Approaches.* New York, Alan R. Liss, pp 199–226.

Szathmary EJE (1986a) Comment on Greenberg JH, Turner CG, Zegura SL, *The Settlement of the Americas. Current Anthropol* 27:490–491.

Szathmary EJE (1986b) Diabetes in arctic and subaractic populations undergoing acculturation. *Collegium Anthropologicum* 10:145–158.

Szathmary EJE (1987) Genetic and environmental risk factors. In Young TK (ed) *Diabetes in the Canadian Native Population: Biocultural Perspectives.* Toronto, Canadian Diabetes Association, pp 27–66.

Szathmary EJE (1988) Unpublished calculations from the National Diabetes Survey.

Szathmary EJE (1993a, in press) Modelling ancient population relationships from modern populations genetics. In Bonnichsen R, Steele DG (eds) *Methods and theory for Investigating the Peopling of the Americas.* Corvallis, OR, Center for the Study of the First Americans.

Szathmary EJE (1993b, in press) Genetics of aboriginal North Americans. *Evolutionary Anthropology.*

Szathmary EJE, Holt N (1983) Hyperglycemia in Dogrib Indians of the Northwest Territories, Canada: Association with age and a centripetal distribution of body fat. *Hum Biol* 55:493–515.

Szathmary EJE, Ossenberg NS (1978) Are the biological differences between North American Indians and Eskimos truly profound? *Current Anthropol* 19:673–701.

Torroni A (1990) Personal communication. Departments of Biochemistry and Anthropology, Emory University School of Medicine, Atlanta, GA.

Turner CG II (1985) The dental search for Native American origins. In Kirk R, Szathmary E (eds) *Out of Asia. Peopling the Americas and the Pacific. The Journal of Pacific History*, Canberra, Australia, pp 31–78.

Voegelin CF, Voegelin FM (1965) Classification of North American Indian Languages. *Anthropol Linguistics* 7:121–150.

Ward RH, Redd A, Valencia D, Frazier BL, Matsumara B, Santas M (1992) Molecular evolution and linguistic differentiation in the Americas. *Am J Phys Anthropol (Supp)* 14:170–171.

Weiss KM, Ferrell RE, Hanis CL (1984) A new world syndrome of metabolic diseases with a genetic and evolutionary basis. *Yearbk of Phys Anthropol* 27:153–178.

Wendorf M (1989) Diabetes, the ice free corridor, and the PaleoIndian settlement of North America. *Am J Phys Anthropol* 79:503–520.

West KM (1974) Diabetes in American Indians and other native populations of the new world. *Diabetes* 23:841–855.

West KM (1978) Diabetes in American Indians. *Advan Metab Disord* 9:29–48.

White JM, Mathewes RW, Mathews WH (1985) Late Pleistocene chronology and environment of the "ice free corridor" of northwestern Alberta. *Quaternary Res* 24:173–186.

Williams RC, Steinberg AG, Gershowitz H, Bennett PH, Knowler WC, Pettitt DJ, Butler W, Baird R, Dowd-Rea L, Burch TA, Morse HG, Smith CG (1985) Gm allotypes in Native Americans: evidence for these distinct migrations across the Bering land bridge. *Am J Phys Anthropol* 66:1–20.

Wright JV (1972) The Shield Archaic. *Publications in Archaeology* No 3, National Museum of Canada, Ottawa.

Yamashita T, Mackay W, Rushforth N, Bennett PH, Houser H (1984) Pedigree analysis of non-insulin dependent diabetes mellitus (NIDDM) in the Pima Indians suggest dominant mode of inheritance. *Am J Hum Genet* 36:183S.

Young TK (1993, in press) Diabetes among Canadian Indians and Inuit: an epidemiological overview. In Joe J, Young RS (eds) *Culture change, Diabetes and Native Americans*. New York Mouton Press.

Young TK, Kaufert JM, McKenzie JK, Hawkins A, O'Neill J (1989) Excessive burden of end-stage renal disease among Canadian Indians: a national survey. *Am J Pub Health* 79:756–758.

Young TK, McIntyre LL, Dooley J, Rodriguez J (1985) Epidemiologic features of diabetes mellitus among Indians in northwestern Ontario and northeastern Manitoba. *Can Med Assoc J* 132:793–797.

Young TK, Sevenhuysen G (1989) Obesity among northern Canadian Indians: patterns determinants and consequences. *Am J Clin Nutr* 49:786–793.

Young TK, Sevenhuysen GP, Ling N, Moffatt MEK (1990a) Determinants of Plasma glucose level and diabetic status in a northern Canadian Indian population. *Can Med Assoc J* 142:821–830.

Young TK, Szathmary EJE, Evers S, Wheatley B (1990b) Geographic distribution of diabetes among the native population of Canada: a national survey. *Soc Sci Med* 31:129–139.

16

Genetic Variation and Evolution of Human Populations

MASATOSHI NEI
GREGORY LIVSHITS
TATSUYA OTA

Genetic variation within and between populations provides important information on the evolutionary history of the populations. To obtain a reliable picture of the evolutionary history, however, one must use gene frequency data from a large number of loci because gene frequency changes are known to be subject to large stochastic errors. Yet, experimentalists often use data from a single locus or a small number of loci. This is particularly so when DNA polymorphism data are used. For example, Wainscoat et al. (1986) constructed a phylogenetic tree for human populations using a DNA region that involves a part of the β-globin gene cluster. Phylogenetic trees from mitochondrial DNA are also often used for inferring the evolutionary history of human populations (Cann et al., 1987; Vigilant et al., 1989), though mtDNA is inherited as a single genetic entity and therefore is equivalent to a single genetic locus.

Before DNA polymorphism data became available, Cavalli-Sforza and Edwards (1964) and Edwards and Cavalli-Sforza (1964) analyzed gene frequency data for five blood group loci and concluded that human populations can be divided into two major groups: The Euroafrican (Europeans and Africans) and the Greater Asian (Asians, American Indians, Oceanians, and native Australians). From this observation and the fossil record available at that time, they proposed that *Homo sapiens* originated in western Asia. Later, Nei and Roychoudhury (1974, 1982) studied the genetic relationships of three major groups of humans (Europeans, Asians, and Africans) using 56 polymorphic loci and suggested that Europeans and Asians are genetically closer to each other than to Africans. This result implicitly supported the hypothesis of an African origin of *H. sapiens*. Recently, Cavalli-Sforza et al. (1988) reached a similar conclusion by using 42 polymorphic loci (120 polymorphic alleles) in their study of the differentiation of 42 world populations.

A similar result was also obtained by Bowcock et al. (1987) in an analysis of restriction-site data for nuclear DNA. However, the number of loci used in these studies is still too small to make any statistically significant conclusion about the evolutionary relationships of the three major groups of humans.

With this in mind, we recently conducted an extensive literature survey and compiled gene frequency data from various human populations for a large number of loci (Roychoudhury and Nei, 1988). This enabled reexamination of the evolutionary relationships of the three major groups of humans (Nei and Livshits, 1989, 1990). At the same time, we examined the extent of genetic variation within and between populations and the genetic relationships for various human populations. In this chapter, we present some of the results so far obtained.

GENETIC VARIATION WITHIN AND BETWEEN THREE MAJOR GROUPS OF HUMANS

We first examined the extent of genetic variation in the three major groups of humans to see if there is any difference in the amount of genetic variation. We computed the average heterozygosities for four different sets of genetic loci, i.e., those for proteins, blood groups, HLA and immunoglobulins (IMG), and DNA markers. The number of "loci" used for these four sets were 80 (62 polymorphic), 32 (19 polymorphic), 8 (all polymorphic), and 61 (59 polymorphic), respectively. The gene frequencies for these loci were taken from Roychoudhury and Nei (1988), except those for DNA markers given in Bowcock et al. (1987). Only loci that are in common for all the major groups of humans were used.

Europeans, Asians, and Africans were represented by northern Europeans (mainly British), eastern Asians (mainly Japanese), and west-central Africans (mainly Bantu from Nigeria and Cameroon), respectively. When gene frequency data for a particular locus were not available from the British, Japanese, or Bantu population, data from their neighboring populations or Americans whites, orientals, or blacks were used. The use of neighboring populations seems to be acceptable for the present purpose, becuase the gene frequency differences between them (e.g., British and Italians, Japanese and Chinese, or Bantu and Pygmies) are generally much smaller than the differences between the three major groups of humans (Roychoudhury and Nei, 1988). About 20% of the gene pool of American blacks is known to have come from whites (Reed, 1969). However, because gene frequency data from American blacks are used only for 26 loci, this procedure is unlikely to affect our conclusion. Furthermore, as will be seen later, there is not much difference in average heterozygosity between American blacks and Nigerians when the same set of loci are used.

The average heterozygosities (H) thus obtained are presented in Table 16.1. Protein and blood group loci show similar average heterozygosities for all three major groups of humans, whereas HLA + IMG show the highest

Table 16.1. Average heterozygosities for four different groups of genetic loci in three major groups of humans.

Set of Loci	No. of Loci	Europeans	Asians	Africans
Proteins	80	0.186 ± 0.22	0.164 ± 0.022	0.179 ± 0.021
Blood groups	32	0.179 ± 0.043	0.145 ± 0.038	0.164 ± 0.038
HLA and IMG	8	0.619 ± 0.113	0.700 ± 0.057	0.678 ± 0.084
DNA markers	61	0.359 ± 0.019	0.319 ± 0.025	0.271 ± 0.024
Total	181	0.262 ± 0.017	0.237 ± 0.017	0.229 ± 0.016

Note: The genetic loci used in this analysis are the same as those of Nei and Livshits (1989), except *DIA4*, *GSR*, *HPX*, *HK3*, and *NY*. These loci were later found to be inappropriate for the analysis. The same comment applies to Tables 16.2 and 16.3.

heterozygosity value, reflecting the well-known high degree of polymorphism at these loci. DNA markers show an intermediate H value, but this is because most of the marker loci included in this study were polymorphic. At each set of genetic loci, there are some racial differences in heterozygosity value, but none of the differences is statistically significant except that for DNA markers between Europeans and Africans. The racial differences for all loci are also not significant, though Europeans show the highest H value.

To examine the extent of genetic differences between the three major groups of humans, we computed estimates of Nei's (1972) genetic distance (D) for the four sets of genetic "loci." The results obtained are presented in Table 16.2.

All data sets except the HLA + IMG loci show the same genetic relationships among the three major groups of humans; the genetic distance between Europeans and Asians is smallest and that between Asians and Africans is largest. The t-test described by Nei (1987) shows that the distance between Europeans and Asians is significantly smaller than that between Asians and Africans in all data sets except in HLA and IMG loci. When all four data sets are combined, the distance between Europeans and Asians is significantly smaller than either that between Europeans and Africans or that between Asians and Africans. We can therefore conclude that Europeans and Asians are genetically closer to each other than to Africans. It should be noted that

Table 16.2. Estimates of genetic distance between three major groups of humans.

Set of Loci	No. of Loci	Europeans/ Asians	Europeans/ Africans	Asians/ Africans
Proteins	80	0.030 ± 0.009[a]	0.036 ± 0.010	0.053 ± 0.014
Blood groups	32	0.020 ± 0.010[a]	0.063 ± 0.034	0.087 ± 0.042
HLA and IMG	8	0.329 ± 0.122	0.701 ± 0.341	0.390 ± 0.169
DNA markers	61	0.061 ± 0.012[a]	0.082 ± 0.017	0.111 ± 0.025
Total	181	0.042 ± 0.007[b]	0.066 ± 0.011	0.083 ± 0.013

[a]The distance between Europeans and Asians is significantly smaller than that between Asians and Africans at either the 1% or the 5% level.

[b]This distance is significantly smaller than either that between Asians and Africans or that between Europeans and Africans.

this conclusion is not dependent on the use of Nei's distance measure, because in the range of genetic distances considered here most distance measures are highly correlated (Chakraborty and Tateno, 1976).

Our results for blood groups loci are somewhat different from those of Nei and Roychoudhury (1982), in which the smallest distance was obtained between Europeans and Africans and the largest distance between Asians and Africans. (None of the distance comparisons was statistically significant.) There seem to be two major reasons for the difference between the present results and the previous ones: (a) the number of loci used has increased considerably; and (b) the present analysis used the frequencies of three alleles (Fy^a, Fy^b, and Fy) rather than two ($Fy^a, Fy^b + Fy$) at the Duffy locus and the haplotype frequencies rather than the allele frequencies for the component loci at the MNS and RH gene complexes. Note that allele Fy is fixed in central African populations, whereas it is virtually absent in European and Asian populations and that haplotype frequencies are more informative for discriminating populations from one another than allele frequencies are.

At the HLA and IMG loci, the genetic distance between Asians and Africans is smaller than that between Europeans and Africans, though the difference is not statistically significant. This is partly due to the fact that unidentified (blank) alleles at the HLA-C and HLA-DR loci are higher in Asians (0.385 and 0.113) and Africans (0.467 and 0.263) than in Europeans (0.173 and 0.040). In the present study, these blank alleles were treated as single alleles. (The frequencies of the blank alleles at the HLA-B loci were small for all three populations.) Our results for the DNA markers are similar to those of Bowcock et al. (1987), though these authors used a smaller number of "loci" and did not reach any statistically significant conclusion.

At any rate, the results in Table 16.2 indicate that Europeans and Asians are genetically closer to each other than to Africans and suggest that Africans and the European-Asian group first separated and then Europeans and Asians diverged. This conclusion is the same as that of Nei and Roychoudhury (1974), but the fact that our results are now statistically confirmed gives strong support to this previous view.

Nevertheless, it is important to note that the amount of inter-group genetic variation is a small fraction of the total variation in humans despite the statistically significant differences. Table 16.3 shows the decomposition of

Table 16.3. Intragroup (H_S) and inter-group (D_{ST}) genetic variation among three major groups of humans.

Set of Loci	No. of Loci	H_T	H_S	D_{ST}	G_{ST}
Proteins	80	0.197	0.176	0.021	0.11
Blood groups	32	0.193	0.162	0.031	0.16
HLA and IMG	8	0.748	0.663	0.085	0.11
DNA markers	61	0.353	0.313	0.040	0.11
Total	181	0.273	0.241	0.032	0.12

the total genetic variation as measured by heterozygosity (H_T) into the within-population (H_S) and the between-population (D_{ST}) components for the four different sets of genetic loci (Nei, 1973). In all sets of loci, the proportion (G_{ST}) of genetic variation attributable to population differences is only 11% to 16%. This means that the differences among the three major groups of humans are much smaller at the gene level than those observed in such morphological characters as skin color and hair texture, where the variation attributable to population differences is often very high (Nei and Roychoudhury, 1972).

GENETIC VARIATION IN SMALL POPULATIONS

According to the neutral mutation theory, the amount of genetic variation is expected to be smaller in small populations than in large populations (Kimura, 1983). This prediction generally holds true when average heterozygosity is related to species population size for various groups of organisms (Nei and Graur, 1984). Human geneticists are aware that some small populations such as the Yanomama are less polymorphic than large populations, but there seems to be no systematic study on this subject. We therefore examined the average heterozygosities for small populations in comparison with those of large populations.

Table 16.4 shows the average heterozygosities of six small populations in comparison with those of their respective control populations. The controlled population used here is a well-defined, large population, which is relatively closely related to the small population under investigation and for which gene frequency data are available. Because the genetic loci used are not the same for all populations, the H value of a small population can be compared only with that of its control population. Table 16.4 shows that small populations tend to have a smaller H value than large populations. Particularly in the Yanomama population, the H for 27 protein loci is only 0.040, which is less than half the H value of the Japanese population. The H for 15 blood group loci is also smaller than that of the Japanese. The difference in H for all loci between the two populations is significant at the 1% level. The Hadza population in African and the Gainj population in Papua New Guinea also show a significantly smaller H value compared with their respective control population. However, other populations do not really show a smaller H value. The Tristan da Cunha population, which inhabits a small island located southwest of Africa in the Atlantic, has been studied extensively because of its possible inbreeding. Roberts (1980) has estimated that the inbreeding coefficient of this population is 0.04. Our results show that this population is an heterogeneous as the English population. This is probably because this population is composed of descendants of Africans, Europeans, and Chinese (Jenkins et al., 1985).

Table 16.5 shows another comparison of small and large populations. In this comparison, the same set of genetic loci (22 protein loci and 8 blood

Table 16.4. Average heterozygosities in some isolated populations and their control populations.

	No. of Loci	Isolate	Control	Difference
		Yanomama	*Japan*	
Proteins	27	0.040 ± 0.018	0.096 ± 0.031	−0.056
Blood groups	15	0.237 ± 0.065	0.272 ± 0.062	−0.035
Total	42	0.111 ± 0.029	0.159 ± 0.032	−0.048
		Hadza	*Nigeria*	
Proteins	10	0.118 ± 0.058	0.142 ± 0.056	−0.024
Blood groups	12	0.202 ± 0.073	0.233 ± 0.067	−0.031
Total	22	0.164 ± 0.048	0.192 ± 0.044	−0.028
		Gainj	*Japan*	
Proteins	32	0.068 ± 0.023	0.090 ± 0.028	−0.022
Blood groups	10	0.193 ± 0.068	0.304 ± 0.069	−0.111
Total	42	0.098 ± 0.024	0.141 ± 0.030	−0.043
		Atacameno	*Japan*	
Proteins	17	0.244 ± 0.053	0.257 ± 0.048	−0.013
Blood groups	6	0.258 ± 0.102	0.319 ± 0.100	−0.061
Total	23	0.248 ± 0.046	0.273 ± 0.043	−0.025
		Ojibwa	*Japan*	
Proteins	23	0.074 ± 0.029	0.078 ± 0.029	−0.004
Blood groups	9	0.366 ± 0.083	0.313 ± 0.081	0.053
Total	32	0.156 ± 0.039	0.144 ± 0.036	0.012
		Tristan da Cunha	*English*	
Proteins	20	0.116 ± 0.041	0.135 ± 0.043	−0.019
Blood groups	5	0.490 ± 0.130	0.469 ± 0.123	0.021
Total	25	0.191 ± 0.051	0.201 ± 0.049	−0.010

Note. The genetic loci (written in gene symbols; see Roychoudhury and Nei, 1988) used in this study are as follows. Yanomama: *ACP1, ADA, AK1, CA1, CA2, ESD, G6PD, GPI, IDH1, LDHA, LDHB, MDH1, NP, PEPA, PEPB, PGD, PGM1, PGM2, SOD1, TPI, AG, ALB, CP, GC, HBB, HPA*, TF, ABO, DI, FY, JK, K, LE, LU, MG, MNS, P, RH, SE, VW, WR, XG.* Hadza: *ACP1, AK1, LDHA, LDHB, MDH1, PGD, PGM1, PGM2, HPA*, TF, A1A2BO, DI, FY, HE, JK, JS, K, LU, MNS, RH, V, WR.* Gainj: *ACP1, ADA, AK1, CA1, CA2, ESD, GLO1, GOT1, G6PD, GPI, GPT1, IDH1, LDHA, LDHB, MDH1, NP, PEPA, PEPB, PEPC, PEPD, PGD, PGK, PGM1, PGM2, SOD1, TPI, ALB, CP, GC, HBB, HPA*, TF, ABO, DI, FY, GE, JK, K, LE, MNS, P, RH.* Atacameno: *ACP1, ADA, AK1, ALAD, CHE1, ESD, GLO1, GPT1, PGD, PGM1, BF, C3, GC, HPA*, PLG, PI, TF, ABO, DI, FY, K, MN, RH.* Ojibwa: *ACP1, ADA, AK1, DIA1, GOT1, G6PD, GPI, IDH1, LHDA, LHDB, NP, PEPA, PEPB, PEPC, PGD, PGK, PGM1, PGM2, ALB, CP, GC, HPA*, TF, A1A2BO, DI, FY, JK, K, LU, MNS, P, RH.* Tristan da Cunha: *ACP!, ADA, AK1, CA1, CA2, ESD, G6PD, GPT1, GPX1, IDH1, PEPA, PEPB, PEPC, PEPD, PGD, PGM1, PGM2, HBB, HPA*, TF, A1A2BO, HE, MNS, RH, SE.*

group loci) were used, so that the H values from different populations can be compared. Large populations show an H value between 0.19 and 0.23, whereas small populations uniformly fall below 0.19. Particularly, the native Australian and the Gainj, New Guinea, populations show an H value which is about half that of large populations. Table 16.5 also shows that English and U.S. whites have essentially the same heterozygosity, as do Nigerians and U.S. blacks.

Table 16.5. Average heterozygosities (H) and population sizes (N) of small populations in comparison with those of large populations.

Population/Country	H	$N(\times 10^3)$	Source of N
Small Populations			
1. Eskimos, St. Lawrence	0.181 ± 0.041	1	Crawford et al., 1981
2. Dogrib, Canada	0.133 ± 0.033	1.7	Helm, 1981
3. Macushi, Brazil	0.157 ± 0.041	3.5	Neel et al., 1977
4. Baniwa, Brazil	0.165 ± 0.039	1.5	Mestriner et al., 1980
5. Cayapo, Brazil	0.187 ± 0.043	1.5	Salzano et al., 1971
6. Guami, Costa Rica	0.148 ± 0.037	50	Barrantes et al., 1982
7. San, Southern Africa	0.167 ± 0.039	50	Nurse et al., 1985
8. Native Australians (N)[a]	0.109 ± 0.027	25	McNally, 1982
9. Native Australians (W)[a]	0.112 ± 0.031	25	McNally, 1982
10. Gainj, New Guinea	0.115 ± 0.031	1.5	Long et al., 1986
11. Solomon Is., Melanesia	0.166 ± 0.039	25	Lai and Bloom, 1982
12. Ainu, Japan	0.163 ± 0.039	12	Acad American Enc, 1981
Large Populations			
13. U.S. whites	0.210 ± 0.046	180,000	Acad American Enc, 1981
14. U.S. blacks	0.197 ± 0.041	20,000	Acad American Enc, 1981
15. Nigerians	0.190 ± 0.039	8,000	Acad American Enc, 1981
16. English	0.212 ± 0.046	54,000	Acad American Enc, 1981
17. Georgeans, USSR	0.199 ± 0.044	5,000	Acad American Enc, 1981
18. Iranians	0.228 ± 0.046	37,000	McNally, 1982

[a] N: Northern Australia. W: Western Australia. The genetic loci used are as follows. Polymorphic loci: *ACO1, ACP1, AK1, ESD, GLO1, G6PD, GPI, PGD, PGM1, PGM2, CP, GC, HBB, HPA*, TF, A1A2BO, DI, FY, JK, K, MNS, P, RH*. Monomorphic loci: *IDH1, LDHA, LDHB, MDH1, PEPB, PGK, ALB*.

GENETIC RELATIONSHIPS OF HUMAN POPULATIONS

In a previous investigation, we studied the genetic relationships of various human populations by using Nei's genetic distance (Nei and Roychoudhury, 1982). The method of phylogenetic construction used was the unweighted pair-group method with arithmetic means (UPGMA). The results obtained were generally reasonable from the archeological and historical records of human populations. However, there were several anomalous features in the phylogenetic trees when small isolated populations were included. It was speculated that these anomalies were caused by small effective sizes of some populations or bottleneck effects that occurred in the recent past. Bottlenecks of population size are known to increase genetic distance significantly (Chakraborty and Nei, 1977).

To examine whether this speculation is correct, we have now studied the relationship between genetic distance (D) and the mean ($\bar{H}$) of the average heterozygosities of the two populations involved. The rationale is that population bottlenecks are expected to reduce average heterozygosity but increase genetic distance. Therefore, if some of the populations studied were subjected to bottleneck effects, a negative correlation is expected between D and $\bar{H}$. Figure 16.1 shows the results obtained for six mongoloid populations

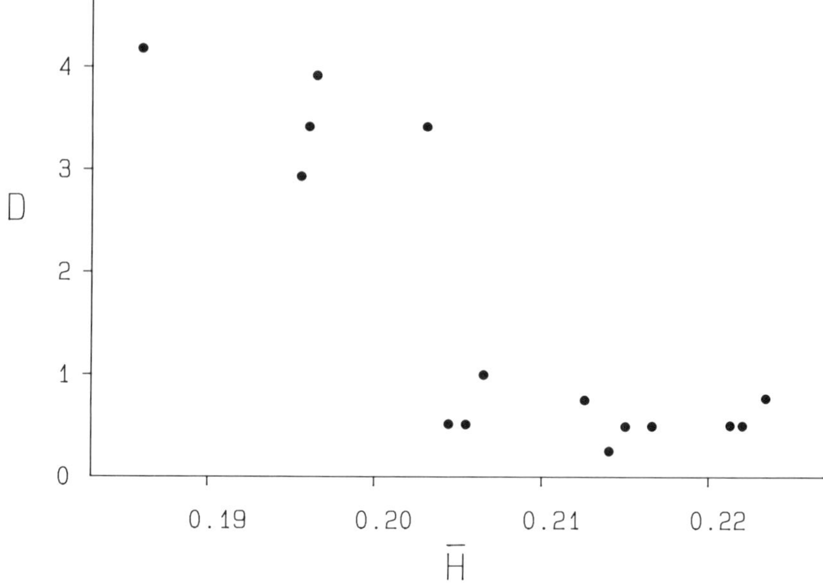

Figure 16.1. Correlation between genetic distance (D) and the mean ($\bar{H}$) of the average heterozygosities of the two populations involved for Ainu, Bhutanese, Chinese, Eskimo, Japanese, and Malays. Redrawn from Livshits and Nei (1990).

(Japanese, Chinese, Bhutanese, Malays, Ainu, and Eskimos). Clearly, there is a negative correlation between D and $\bar{H}$, the correlation coefficient being -0.844 ($P < 0.001$). Similar negative correlations have been observed for several other population sets (Livshits and Nei, 1990). Therefore, these results support Nei and Roychoudhury's previous speculation.

When genetic distances are affected by inbreeding effects, UPGMA is not an appropriate method for constructing phylogenetic trees, because it depends on assuming a constant rate of evolution. Several methods allow for variation of evolutionary rate among populations. Among these methods, Saitou and Nei's (1987) neighbor-joining method is known to be quite efficient in obtaining the correct phylogenetic trees and was employed here. In this case it is important to use a proper genetic distance measure as well. Nei et al. (1983) showed that for obtaining the correct topology of a tree the following distance is more powerful than Nei's D.

$$D_A = \sum_{k=1}^{r} \left(1 - \sum_{i=1}^{m_k} \sqrt{x_{ik} y_{ik}}\right) \Big/ r, \qquad [16.1]$$

where x_{ik} and y_{ik} are the frequencies of the ith allele at the kth locus in populations X and Y, respectively, whereas m_k and r are the number of alleles at the kth locus and the number of loci examined, respectively. We therefore used the neighbor-joining (NJ) method with this distance. However, to see the effect of the assumption of a constant ecolutionary rate, we also constructed a dendrogram using UPGMA with the D_A distance. In this approach there is no need to include monomorphic loci, so only polymorphic loci were used.

Our work on this project is still in progress. In this chapter we present only the results obtained for 18 representative populations from the world. The D_A distances for these 18 populations are presented in Table 16.6, and the UPGMA tree obtained is given in Figure 16.2. The general pattern of this tree is similar to that of Nei and Roychoudhury (1982) with African, Caucasian, and Asian Mongoloid populations forming separate clusters. As expected from the results given in Table 16.2, Caucasian and Asian Mongoloid populations are genetically much closer to each other than to Africans. However, the splitting patterns of Australians, Papua New Guineans, and American Indians are different from what anthropologists generally believe. That is, these populations are thought to be more closely related to Asian Mongoloids.

These rather anomalous splitting patterns are apparently caused by the effects of bottlenecks or small effective sizes these populations have experienced. We have therefore constructed another tree by using the neighbor-joining method (Figure 16.3). This tree now has three major clusters that correspond to Africans, Caucasians, and Greater Asians. Since the neighbor-joining method takes care of varying evolutionary rates, this tree seems to reflect the evolutionary pathways of human populations more adequately than the UPGMA tree.

According to the NJ tree, American Indians (Baniwa and Macushi) seem to have been the first to split from the rest of the Greater Asians. By contrast, native Australians and Papuans are closely related to Filipinos and Indonesians, suggesting that they are derived from the latter populations. Of course, the number of loci used in this study is rather small so that too much confidence cannot be given to this tree. Nevertheless, the results of this phylogenetic analysis are consistent with the view held by many anthropologists.

Both UPGMA and NJ trees show that English and Finns are genetically very close compared with other populations. This is consistent with Nei and Roychoudhury's previous findings that Western Europeans are genetically very close. By contrast, two Eskimo tribes in Alaska and Canada are much more different from each other than Europeans populations are. This suggests that these two small populations have been inbred or isolated for a considerable period of time. In general, small populations such as the Lapp, Ainu, Papuan, and Pygmy have a long branch, suggesting that they experienced bottlenecks or an extended period of small effective size.

DISCUSSION

We have seen that the extent of genetic variation is nearly the same for large European, Asian, and African populations but that small populations in some isolated areas of the world have considerably less genetic variation.

Table 16.6. Average heterozygosities and genetic distances ($D_A \times 10^2$) for 18 representative populations of the world.

	(1)	(2)	(3)	(4)	(5)	(6)	(7)	(8)	(9)	(10)	(11)	(12)	(13)	(14)	(15)	(16)	(17)	(18)
(1) Native Australian	18.5																	
(2) Papuan	4.7	17.5																
(3) Baniwa (S. America)	5.9	9.8	23.5															
(4) Macushi (S. America)	4.8	9.4	0.6	22.9														
(5) Eskimo (Alaska)	5.8	5.8	4.9	5.6	26.2													
(6) Eskimo (Canada)	5.9	8.2	3.2	3.9	1.5	29.1												
(7) Ainu	4.8	5.9	5.2	5.6	3.1	3.2	26.5											
(8) Japanese	3.6	5.3	3.8	3.9	2.2	1.7	1.1	28.1										
(9) Chinese	3.4	5.6	5.1	5.3	3.4	3.0	2.5	1.1	27.0									
(10) Filipino	2.8	4.3	3.5	3.2	3.0	2.6	2.5	1.2	1.6	25.2								
(11) Indonesian	3.2	4.5	3.7	3.6	2.8	3.2	2.2	1.1	1.9	1.1	27.0							
(12) English	7.3	8.7	4.7	5.1	4.4	4.0	4.7	3.4	5.5	4.1	4.3	33.2						
(13) Finn	7.7	8.7	4.6	5.4	4.0	3.7	4.4	3.1	5.2	3.7	4.0	0.4	33.8					
(14) Indian	7.6	8.8	4.7	5.4	4.3	4.0	3.9	2.8	4.6	3.1	3.0	1.3	1.1	34.7				
(15) Lapp	7.5	7.3	6.0	6.5	4.1	4.4	4.6	3.7	5.8	4.1	4.1	1.1	1.2	1.7	32.2			
(16) Yoruba (Nigeria)	14.7	15.4	14.5	14.0	13.5	15.3	15.7	13.4	13.5	12.9	13.2	11.3	11.6	12.6	12.4	23.6		
(17) Pygmy	18.6	20.4	16.2	15.7	16.7	18.1	17.9	16.4	17.4	16.2	16.3	14.0	14.2	14.8	15.3	2.6	23.1	
(18) San (Bushman)	10.0	12.6	10.9	10.5	9.2	10.6	11.7	10.0	11.6	9.8	10.1	7.0	7.2	8.4	7.9	6.0	7.0	29.4

Note. The genetic loci used (all polymorphic) are *ACPI, ADA, AK1, ESD, PGM1, PGM2, HPA*, GC, TF, ABO, MNS, RH, FY. K* and *P*.
The figures on the diagonal are the average heterozygosities per locus in percentages.

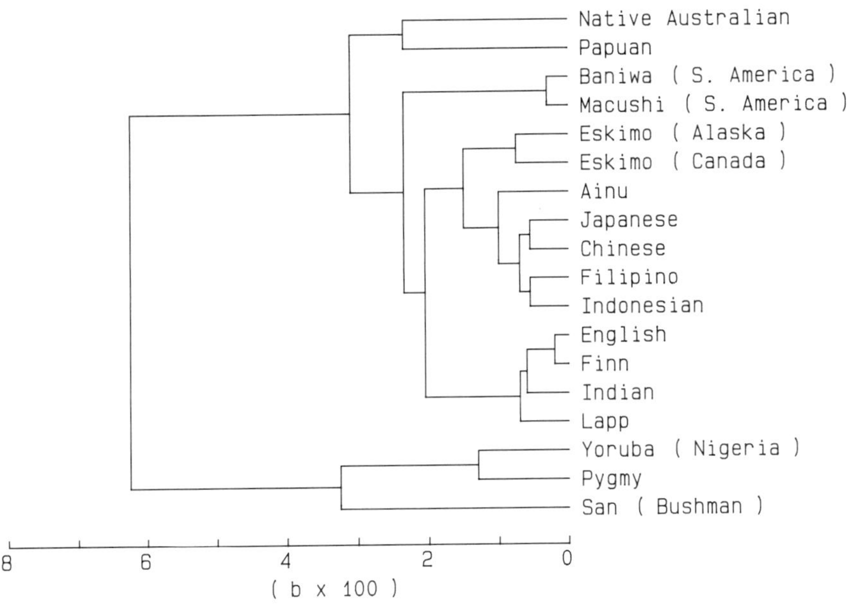

Figure 16.2. Dendrogram for 18 representative world populations obtained by UPGMA.

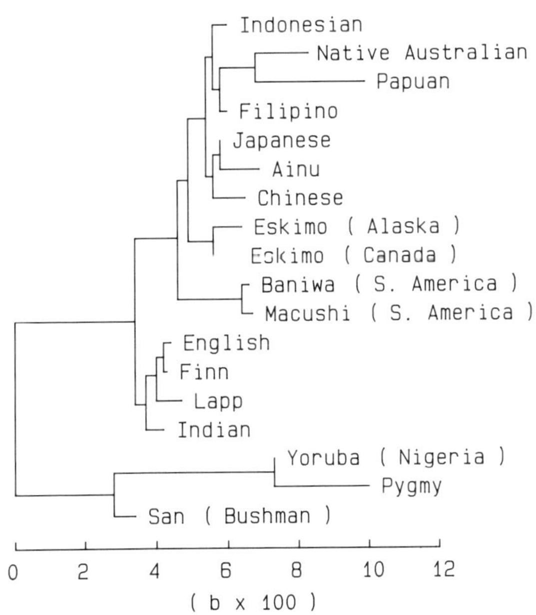

Figure 16.3. Dendrogram for 18 representative world populations obtained by the neighbor-joining method.

These small amounts of genetic variability are apparently caused by population bottlenecks or extended periods of small effective size (inbreeding effects).

In this connection it is interesting to note that in the case of American Indians in South America, population amalgamation does not significantly increase average heterozygosity (Chakraborty et al., 1988). This suggests that the entire South American Indian populations went through the same bottleneck relatively recently and that new mutations have not significantly increased average heterozygosity since then. This is quite reasonable because these populations are known to have migrated from Asia probably less than 20,000 years ago. It remains to be seen, however, whether the small populations in other parts of the world have experienced the same type of bottlenecks or not.

One of the implications of finding a bottleneck effect is that in the study of genetic relationships of human populations, proper care should be taken of this factor, because this affects any measure of genetic distance. When there is any possibility of inbreeding effects, genetic relationships of populations should be studied by using a method that allows variation of evolutionary rate in different populations.

We have also emphasized the importance of using a large number of polymorphic loci for the study of genetic relationships of populations. Only when a sufficient number of polymorphic loci are used, can we determine the relationships statistically. In recent years, mitochondrial DNA or a small segment of genomic DNA has been used for this purpose (Long et al., 1990; Vigilant et al., 1989). However, these DNAs are inherited essentially as a single entity, so that the reliability of the conclusions obtained from these studies is low (Nei and Livshits, 1989, 1990). While Vigilant et al.'s major conclusion about the splitting of three major groups of humans is supported by our study, the details of the genetic relationships they obtained (their Figure 1) do not agree with ours (Figure 16.3 in this chapter). Of course, the number of loci we used is not very large, so it is not yet clear which tree is really correct. It is hoped that gene frequency data for more polymorphic loci will be studied in the near future.

ACKNOWLEDGMENTS

We thank Professor Jim Neel for his comments on an earlier version of this chapter. This study was supported by research grants from NIH and NSF.

REFERENCES

Academic American Encyclopedia (1981) Danbury, CT, Grolier.

Barrantes R, Smouse PE, Neel JV, Mohrenweiser HW, Gershowitz H (1982) Migration and genetic infrastructure of the Central American Guaymi and their affinities with other tribal groups. *Am J Phys Anthrop* 58:201–214.

Bowcock AM, Bucci C, Hebert JM, Kidd JR, Kidd KK, Friedlaender JS, Cavalli-Sforza LL (1987) Study of 47 DNA markers in five populations from four continents. *Gene Geography* 1:47–64.

Cann RL, Stoneking M, Wilson AC (1987) Mitochondrial DNA and human evolution. *Nature* 325:31–36.

Cavalli-Sforza LL, Edwards AWF (1964) Analysis of human evolution. In Geerts SJ (ed) *Genetics Today*. Oxford, Pergamon Press, pp 923–933.

Cavalli-Sforza LL, Piazza A, Menozzi P, Mountain J (1988) Reconstruction of human evolution: bringing together genetic, archaeological and linguistic data. *Proc Natl Acad Sci USA* 85:6002–6006.

Chakraborty R, Nei M (1977) Bottleneck effects on average heterozygosity and genetic distance with the stepwise mutation model. *Evolution* 31:347–356.

Chakraborty R, Smouse PE, Neel JV (1988) Population amalgamation and genetic variation: observations on artificially agglomerated tribal populations of Central and South America. *Am J Hum Genet* 43:709–725.

Chakraborty R, Tateno Y (1976) Correlations between some measures of genetic distance. *Evolution* 30:851–853.

Crawford MM, Mielke JH, Devor EJ, Dykes DD, Polesky HF (1981) Population structure of Alaskan and Siberian indigenous communities. *Am J Phys Anthrop* 55:167–185.

Edwards AWF, Cavalli-Sforza LL (1964) Reconstruction of evolutionary trees. *Systemat Assoc Publ* 6:67–76.

Helm J (1981) Dogrib. In Helm J (ed) *Handbook of North American Indians* (vol. 6). Washington, DC, Smithsonian Institution, pp 291–309.

Jenkins T, Beighton P, Steinberg AG (1985) Serogenetic studies on the inhabitants of Tristan da Cunha. *Ann Hum Biol* 12:363–371.

Kimura M (1983) *The Neutral Theory of Molecular Evolution*. Cambridge, Cambridge University Press.

Lai LYC, Bloom J (1982) Genetic variation in Bougainville and Solomon islands populations. *Am J Phys Anthrop* 58:369–382.

Livshits G, Nei M (1990) Relationships between intrapopulational and interpopulational genetic diversity in man. *Ann Hum Biol* 17:501–513.

Long JC, Naidu JM, Mohrenweiser HW, Gershowitz H, Johnson PL, Wood JW, Smouse PE (1986) Genetic characterization of Gainj- and Kalam-speaking peoples of Papua New Guinea. *Am J Phys Anthrop* 70:75–96.

Long LC, Chakravarti A, Boehm CD, Antonarakis S, Kazazian HH (1990) Phylogeny of human β-globin haplotypes and its implication for recent human evolution. *Am J Phys Anthrop* 81:113–130.

McNally R (1982) *The Atlas of Mankind*. Chicago, Rand McNally & Company.

Mestriner MA, Simoes AL, Salzano FM (1980) New studies on the esterase D polymorphism in South American Indians. *Am J Phys Anthrop* 52:95–101.

Neel JV, Tanis RJ, Migliazza EC, Spielman RS, Salzano F, Oliver WJ, Morrow M and Bachofer S (1977) Genetic studies of the Macushi and Wapishana Indians. I. Rare genetic variants and a "private polymorphism" of esterase A. *Hum Genet* 36:81–107.

Nei M (1972) Genetic distance between populations. *Am Nat* 106:283–292.

Nei M (1973) Analysis of gene diversity in subdivided populations. *Proc Natl Acad Sci USA* 70:3321–3323.

Nei M (1987) *Molecular Evolutionary Genetics*. New York, Columbia University Press.

Nei M, Graur D (1984) Extent of protein polymorphism and the neutral mutation theory. *Evol Biol* 17:73–118.

Nei M, Livshits G (1989) Genetic relationships of Europeans, Asians, and Africans and the origin of modern *Homo sapiens. Hum Hered* 39:276–281.

Nei M, Livshits G (1990) Evolutionary relationships of Europeans, Asians and Africans at the molecular level. In Takahata N, Crow JF (eds) *Population Biology of Genes and Molecules.* Tokyo, Baifukan, pp 251–265.

Nei M, Roychoudhury AK (1972) Gene differences between Caucasians, Negro, and Japanese populations. *Science* 177:434–436.

Nei M, Roychoudhury AK (1974) Genic variation within and between the three major races of man, Caucasoids, Negroids and Mongoloids. *Am J Hum Genet* 26:421–443.

Nei M, Roychoudhury AK (1982) Genetic relationhip and evolution of human races. *Evol Biol* 14:1–59.

Nei M, Tajima F, Tateno Y (1983) Accuracy of estimated phylogenetic trees from molecular data. II. Gene frequency data. *J Mol Evol* 19:153–170.

Nurse GT, Weiner JS, Jenkins T (1985) *The Peoples of Southern Africa and their Affinities.* Oxford, Clarendon Press.

Reed TE (1969) Caucasian genes in American Negroes. *Science* 165:762–768.

Roberts DF (1980) Genetic structure and the pathology of an isolated population. In Eriksson AW, Forsius H, Nevenlinna HR, Workman PL, Norio PK (eds) *Population Structure and Genetic Disorders.* New York, Academic Press, pp 7–26.

Roychoudhury AK, Nei M (1988) *Human Polymorphic Genes: World Distribution.* Oxford, Oxford University Press.

Saitou N, Nei M (1987) The neighbor-joining method: a new method for reconstructing phylogenetic trees. *Mol Biol Evol* 4:406–425.

Salzano EM (1971) Demographic and gene relationship among Cayapo Indians of Brazil. *Soc Biol* 18:148–157.

Vigilant L, Pennington R, Harpending H, Kocher TD, Wilson AC (1989) Mitochondrial DNA sequences in single hairs from a southern African population. *Proc Natl Acad Sci USA* 86:9350–9354.

Wainscoat JS, Hill AVS, Boyce AL, Flint J, Hernandez H, Thein SL, Old JM et al. (1986) Evolutionary relationships of human populations from an analysis of nuclear DNA polymorphisms. *Nature* 319:491–493.

17

Nucleotide Diversity in Humans and Evidence for the Absence of a Severe Bottleneck during Human Evolution

WEN-HSIUNG LI
WEIJUN XIONG
SHYAN-WOEI LIU
LAWRENCE CHAN

How much genetic variability exists in human populations? This is an old question. Over 20 years ago electrophoretic data indicated that human populations, like many other outbreeding populations, contain considerable genetic variability (Harris, 1966). Although we have good knowledge about the variability at the protein level, the more basic question is how much variation exists at the DNA level. Even though recombinant DNA techniques have long provided the necessary tools for addressing this question, no extensive study has been conducted. Fortunately, we have been able to obtain adequate information in answer to this question from published DNA sequence data. Our analysis indicates that the variability in humans is rather low at the nucleotide level, compared to that in *Drosophila* populations.

Does this observation imply that a population bottleneck occurred in the human species in the recent past? Studying mitochondrial (mt) DNA variation in human populations, Brown (1980) suggests that the human species went through a severe bottleneck about 180,000 years ago, and Cann et al. (1987) suggest that all present-day human mtDNAs were derived from a "common mother" about 200,000 years ago. However, Klein and co-workers (Figuerosa et al., 1988; Mayer et al, 1988) argue that no severe bottleneck has occurred in the human lineage since its separation from the chimpanzee lineage because humans and chimpanzees share many common alleles at MHC loci. We have analyzed the apolipoprotein (apo) C-II gene structure of apoC-II deficiency patients: two homozygous siblings from a Japanese family and a homozygous patient from a Caucasian family in Venezuela. We argue below that our apoC-II data provide strong evidence for absence of severe bottlenecks during human evolution. Furthermore, our data are more consistent with the "Multiregional Evolution" model of modern human origins (Wolpoff et al., 1984) than with the Complete Replacement or "Out of Africa" model (Stringer and Andrews, 1988). Both models assume that modern humans originated in Africa, but the latter model postulates a rapid, complete replacement of indigenous populations in other regions by those Africans, whereas the former suggests a gradual transformation of archaic regional human populations into modern ones by gene flow and natural selection.

NUCLEOTIDE DIVERSITY IN HUMANS

Searching through the literature or data banks, one can find that many human genes, particularly their coding regions, have been sequenced at least twice. Unfortunately, most of these data are not suitable for estimating the level of genetic variation among individuals, because the sequences usually have not been checked carefully against each other. In fact, a large fraction of the observed differences between sequences may represent sequencing errors. A good example is provided by the apolipoprotein B gene sequences. We have compared one genomic sequence, four complete cDNA sequences, three partial cDNA sequences, and a partial mRNA sequence (Yang et al., 1989). Our comparison reveals 75 differences among these sequences: 1 three-codon deletion, 53 nonsynonymous changes, and 21 synonymous changes; the gene contains 4563 codons. Five of the differences are certainly bona fide variants, having been checked by direct peptide sequencing, detected as restriction site polymorphisms, or observed in both our laboratory and others. Further, 12 of the differences have been reported by more than one laboratory and so are probably genuine variants because the chance of making the same sequencing error at a nucleotide position in two or more laboratories is small. In all other cases, it is difficult to distinguish sequencing errors from authentic variants, but the fact that there were 53 nonsynonymous differences and only 21 synonymous differences suggests that many of the differences represent sequencing errors. In fact, there is circumstantial evidence supporting this possibility in many cases. The evidence is particularly strong in those cases where the differences involve rearrangements of two or three nucleotides, a situation that is unlikely to arise by point mutation. For example, one laboratory observed CGA at codon 766 whereas all others observed CAG at this codon.

Fortunately, we have been able to find 27 sequence pairs suitable for the present purpose (Table 17.1). Most of these sequence pairs have each been obtained by the same laboratory: they were usually one cDNA and one genomic sequence; the cDNA and genomic libraries were generally constructed from different individuals. In the other cases, one laboratory has compared their sequence (usually genomic) carefully with a published sequence. From these data, we have estimated the nucleotide diversity, which is defined as the number of differences per nucleotide site between two

Table 17.1. Number of nucleotide differences per site between two gene sequences randomly chosen from a human population.

	Noncoding sites		Sites in coding regions			
				2-fold deg.		4-fold
Gene	5′UT	3′UT	Nondeg.	Nonsyn.	Syn.	deg.
Acid glycoprotein	0/78	0/143	0/383	0/134	0/134	0/83
Angiogenin	NA	0/175	0/271	0/100	0/100	0/67
Angiotensinogen	0/39	0/602	1/919	0/268	0/268	0/265
Adenosine deaminase	0/72	0/311	0/701	0/212	0/212	2/173
α-Amylase, salivary	0/199	0/30	0/1017	0/301	0/301	0/215
Factor VIII	0/109	1/1800	0/4558	1/1527	0/1527	0/965
Factro IX	NA	0/1389	1/900	0/292	0/292	0/188
Factor X	NA	NA	0/927	0/300	0/300	0/201
ApoA-I	0/86	0/57	0/508	0/164	0/164	0/126
ApoA-II	0/9	0/112	0/188	0/60	0/60	0/49
ApoE	0/61	0/142	0/571	0/158	3/158	0/168
Calcitonin/ CGRP	0/74	NA	0/235	0/80	0/80	0/66
Erythropoietin	0/179	0/565	0/359	0/103	0/103	0/114
Ferritin H	0/208	0/160	0/348	0/132	0/132	0/66
Fibrinogen γ	0/80	0/241	1/858	0/288	1/288	0/162
Parathyroid hormone	0/74	0/348	0/221	0/72	0/72	0/49
Phosphoglycerate kinase	0/79	0/434	0/813	0/230	0/230	0/205
Prolactin	0/4	0/145	0/430	0/141	0/141	0/107
Protein C	0/98	0/360	0/899	0/274	0/274	0/207
Renin	0/42	0/172	0/795	0/217	0/217	0/203
Ribosomal Protein S14	0/33	0/45	0/294	0/74	0/74	0/85
Steroid 21-hydroxylase	0/32	0/492	0/932	0/287	0/287	0/260
SOD-1	NA	0/94	0/297	0/87	0/87	0/75
Thymidine kinase	1/57	0/659	0/450	0/133	0/133	0/116
Tumor necrosis factor	0/152	0/789	0/440	0/130	0/130	0/126
TPA	0/218	0/755	0/1077	0/358	1/358	1/248
VIP	0/176	7/767	0/325	0/108	0/108	0/74
Total	1/2159	8/1077	4/19716	1/6230	5/6230	3/4663
	0.0005	0.0007	0.0002	0.0002	0.0008	0.0006

Note: Taken from W.-H. Li and S.-W. Liu (unpublished).

randomly chosen sequences from a population (Nei and Li, 1979). We consider coding and noncoding regions separately (Table 17.1). In coding regions, a site is labeled nondegenerate if all changes at that site are nonsynonymous (amino acid-changing), twofold degenerate if one of the three possible changes is synonymous, and fourfold degenerate if all three possible changes are synonymous. The calculation can easily be done by using the computer program of Li et al. (1985).

Table 17.1 shows that the level of nucleotide diversity is only about 0.02% at nondegenerate sites, and is about 0.06% at fourfold degenerate sites and in the 5′ and 3′ untranslated (UT) regions. The observation that the synonymous component of the nucleotide diversity at twofold degenerate sites is slightly higher than the total diversity at fourfold degenerate sites is largely due to the contribution of two mutant apoE alleles, which differ by one nonsynonymous change and three synonymous changes. The results in Table 17.1 may be taken as representing the approximate level of diversity in the American white population, because most of the libraries were constructed from white Americans. However, it should be taken as an upper estimate because some of the libraries were from Japanese, Europeans, or other groups, and because a fraction of the differences may have arisen from sequencing errors.

The nucleotide diversity in humans is much lower than that found for various *Drosophila* species (see Aquadro, 1990). Table 17.2 presents a summary of the nucleotide diversity in three *Drosophila* species estimated from six-base recognition restriction enzymes. Although restriction enzymes do not detect all variation between sequences, the level of nucleotide diversity estimated by this method fror the Adh region is the same as that obtained by direct nucleotide sequencing (Kreitman, 1983). The regions shown in Table 17.2 contain mostly sequences that do not code for amino acids. The lowest level of diversity is observed in *D. melanogaster* (0.4%), which is also the value obtained from a larger data set (Aquadro, 1992). Yet this level is sevenfold higher than that in humans. The levels of diversity are even higher in *D. simulans* and *D. psuedoobscura* (Table 17.2). Thus, the level of nucleotide diversity in humans is about one order of magnitude lower than that in *Drosophila*.

There are two possible explanations for the low nucleotide diversity in humans: the effective size of the human species has been relatively small

Table 17.2. Nucleotide diversity in species of *Drosophila*.

	Drosophila Species		
Location	*D. melanogaster*	*D. simulans*	*D. pseudoobscura*
X chromosome			
per	0.001	0.007	—
Second chromosome			
Adh	0.006	0.015	0.026
Amy	0.006	—	0.019
Third chromosome			
rosy	0.005	0.018	—
Average	0.004	0.014	0.023
	(180 kb)	(165 kb)	(58 kb)

Note: Taken from Aquadro (1990).

in the past, or the species has gone through a severe population bottleneck in the recent past. In the following we provide evidence against the second possibility.

APOLIPOPROTEIN C-II DEFICIENT MUTANT GENES

Apolipoproteins C-II (apoC-II) is the physiological activator of lipoprotein lipase, an important enzyme in chylomicron and very low density lipoprotein metabolism. Patients with a deficiency of apoC-II have abnormalities in lipoprotein metabolism and present with elevated triglyceride, chylomicrons, and very low density lipoprotein. This syndrome is inherited as an autosomal recessive disorder.

In an effort to determine the mutational defects that cause the deficiency, we have analyzed the gene structure of two homozygous siblings from a Japanese family and a homozygous patient from a Caucasian family in Venezuela. We denote the mutant allele from the two Japanese siblings as apoC-II_{Jap} and that from the Venezuelan patient as apoC-II_{Ven}. These two mutant alleles share a deletion in exon 3 that causes a shift in the reading frame, but both contain several other mutations when compared to each other or to the normal human apoC-II allele (Table 17.3). For comparison, we have also analyzed the sequences of the normal Japanese allele and a chimpanzee apoC-II gene, finding that the former is identical to and the latter differs by only three nucleotides from the published normal Caucasian apoC-II sequence.

The fact that apoC-II_{Jap} and apoC-II_{Ven} share a frameshift deletion suggests that they were derived from an ancestral mutant gene that contained the deletion, because it is unlikely that the same deletion has occurred independently in the two alleles. However, the two alleles must have diverged a long time ago because they differ by five nucleotides, one single-base deletion and

Table 17.3. Degree of divergence between apolipoprotein C-II gene sequences.

Sequence pairs	Base Substitutions	Deletions: # of deletions	Deletions: Total # of bases involved	Total Divergence
ApoC-II_{Jap} vs Hum ApoC-II	1/965 (0.001)	3/965 (0.003)	4/965 (0.004)	0.005
ApoC-II_{Ven} vs Hum ApoC-II	4/965 (0.004)	1/965 (0.001)	1/965 (0.001)	0.005
ApoC-II_{Jap} vs Hum ApoC-II_{Ven}	5/965 (0.005)	2/965 (0.002)	3/965 (0.003)	0.008
Hum vs Chimp ApoC-II	3/965 (0.003)	0	0	0.003
Hum vs Chimp η globin pseudogene	32/2156 (0.015)	5/2156 (0.002)	10/2156 (0.005)	0.020

Note: Taken from Xiong et al. (Unpublished).

one two-base deletion, whereas the normal human and chimpanzee apoC-II genes differ by only three nucleotides. To obtain a minimum estimate of their separation time we assume that the two mutant alleles have been nonfunctional and have accumulated mutations as fast as pseudogenes. If we count each base deletion as a base substitution, then the degree of sequence divergence between apoC-II_{Jap} and apoC-II_{Ven} is 0.008 and that between human and chimpanzee η globin pseudogene is 0.020 (Table 17.3). Since $0.008/0.020 = 0.40$, the divergence time between apoC-II_{Jap} and apoC-II_{Ven} is estimated to be 40% of the divergence time between human and chimpanzee, which has been estimated to be between 5 to 10 million years (Myr) ago (Li and Tanimura, 1987; Sarich and Wilson, 1967). Therefore, the two mutant alleles could have been separated for 2 million years or longer.

The above conclusion does not depend on the assumption of the two alleles being derived from a common mutant allele because we can also treat each mutant allele separately. For example, using the chimpanzee sequence as a reference we can infer that all the divergence (0.005; Table 17.3) between apoC-II_{Jap} and the normal human sequence occurred in apoC-II_{Jap}. Since only half of the divergence (0.020) between human and chimpanzee η globin pseudogene can be attributed to the human lineage, the age of apoC-II_{Jap} is estimated to be $0.005/0.010 = 50\%$ of the divergence time between human and chimpanzee or at least 2.5 Myr. The same conclusion holds for apoC-II_{Ven}.

How the two defective alleles have persisted in human populations for such a long time is a puzzling question. We note that patients with apoC-II deficiency clearly have a selective disadvantage because they have major disturbances in lipid metabolism and often present with abdominal pain caused by repeated attacks of pancreatitis. This can be diagnosed at any age, but the symptoms have often been traced back to childhood or adolescence. Of course, the presentation is highly variable; patients can be totally asymptomatic or they could die of acute pancreatitis. Note, however, the two defective alleles under study each cause a severe form of deficiency in homozygotes.

One possible explanation is that the two defective alleles have been maintained by heterozygote advantage. However, the advantage must be very small, for two reasons. First, although heterozygotes of apoC-II deficiency are asymptomatic, they may have delayed clearance of alimentary fat. When compared as a group with a normal population of similar age and sex, they appear to have significantly elevated triglycerides and very low density lipoproteins, and decreased high density lipoprotein cholesterol. While it is unclear whether their lipid abnormalities result in any clinically significant health impairment, it is highly unlikely that the trait confers a considerable selective advantage to such individuals. Second, apoC-II deficiency is a rare disease. To date, the defect has been described in only 13 families, who have come from diverse geographic areas including Northern and Southern Europe, Asia, and North and South America; all the other patients have defects different from the two families reported here. The limited number of cases described worldwide and the failure of epidemiologic studies using apoC-II radioimmunoassays involving hundreds of normolipidemic or hyperlipidemic individuals to identify such patients indicate that it is a rare disorder with a frequency probably considerably less than 1/10,000. The frequency would have been higher if the heterozygote advantage is considerable.

It is possible that the two mutant alleles happened by chance to have accumulated more mutations than expected. To take this possibility into account, let p be the expected number of nucleotide differences per site between the two sequences compared. Then if L sites are compared, the variance of the estimated p value is $p(1-p)/L$ under the assumption of a binomial distribution. In the present case $p = 0.008$ and $L = 965$ (Table 17.3), so the standard error is about 0.003 and the 95% confidence interval is from 0.002 to 0.014. Based on the lower limit, we estimate that the two alleles have persisted for at least $2 \times 10^6 \times (0.002/0.008) = 500{,}000$ years.

We now show that a recessive defective allele can persist in a population for only a short time if a bottleneck occurs. The average extinction time of such an allele can be calculated by the following formula:

$$t = 4N_e p[-ln(p) - 0.5ln(2N_e s) + 0.5(0.5\pi/N_e s)^{1/2} + 0.71],$$

where N_e is the effective population size, p is the frequency of the allele, and s is the selective disadvantage of mutant homozygotes (Li and Nei, 1972). The two defective alleles under study both cause an absolute apoC-II deficiency in homozygotes and are likely to have a substantial effect on fitness; so, we assume $s = 0.01$. In humans, the effective population size is about 1/3 of the actual size N (Nei and Imaizumi, 1966) and so $N_e = N/3$. As mentioned above, the frequency of apoC-II deficient patients is probably less than 1/10,000 and so the sum of the frequencies of all apoC-II deficiency alleles is $<1\%$. We assume that the frequency of any particular alleles is $<0.5\%$. We note from Table 17.4 that if $N = 1000$ and $p = 0.5\%$, then $t = 36$ generations or $36 \times 20 = 720$ years. This is relatively short. Further, note that the mean time for one of the two alleles to become lost is only about half the t value for a particular allele. Thus, if $p \leqslant 0.001$ or $N \leqslant 250$, one of the two alleles would have become lost is several generations.

Brown's (1980) suggestion of a severe bottleneck during human evolution was based on the argument that the current level of mtDNA variation in

Table 17.4. Mean extinction time (t) for a recessive defective allele.

Population Size (N)	Allele Frequency	t (generations) $s = 0.05$	$s = 0.01$
1000	0.5%	29.4	36.0
1000	0.1%	8.0	9.3
500	0.5%	16.1	19.6
250	0.5%	8.8	10.8

humans could have been generated from a single pair of individuals that existed 180 to 360 × 10^3 years ago. This reasoning is not proper, because the mtDNA variation in any species could have been generated by a single pair of individuals that existed some time ago, but this does not imply that the species went through a severe bottleneck at that time. The results presented above and the existence of trans-species polymorphic MHC alleles in humans provide strong evidence for the absence of a severe bottleneck during hominid evolution and suggest that our genetic materials have been derived from many different ancestors in the past. The term "a common mother" is misleading because although all extant human mtDNA lineages could have been derived from a single sequence about 200,000 years ago, as suggested by Cann et al. (1987), mtDNA is but a small part of the total human DNA. It has also been suggested that the early Asian Homo (such as Java man and Peking man) contributed no surviving mtDNA lineages to the gene pool of our species. This may or may not be true, but the fact that the Japanese and the Venezuelan (Caucasian) apoC-II mutant alleles diverged at least 500,000 years ago suggests that the early Asian *Homo erectus* contributed a substantial part to the pool of human nuclear DNAs. If so, then the "Multiregional Evolution" model (Wolpoff et al., 1984) of modern human origins is more plausible than the "Out of Africa" model (Stringer and Andrews, 1988). (See the Chapter 18 by Spuhler in this volume for a more detailed explanation of the two models.) Both models assume that modern humans originated in Africa, but differ with respect to how modern humans spread to other part of the world. The latter model postulates a rapid, complete replacement with little or no hybridization so that, except in Africa, archaic *Homo sapiens* became extinct without contributing to modern human populations. On the other hand, the former model postulates a gradual transformation of archaic regional human populations into modern ones by gene flow and natural selection. This model allows regional (genetic) continuities and is therefore more compatible with our result.

ACKNOWLEDGMENTS

We thank Peter Smouse for many valuable comments. This study was supported by NIH grants to L. C. and W. H. L.

REFERENCES

Aquadro CF (1990) Contrasting levels of DNA sequence variation in Drosophila species revealed by "six-cutter" restriction map surveys. In Clegg M, O'Brien S (eds), *Molecular Evolution, UCLA Symposium on Molecular and Cellular Biology* (New Series, vol 122). New York, Alan R. Liss,, Inc., pp. 179–189.

Brown WM (1980) Polymorphism in mitochondrial DNA of humans as revealed by restriction endonuclease analysis. *Proc Natl Acad Sci USA* 77:3605–3609.

Cann RL, Stoneking M, Wilson AC (1987) Mitochondrial DNA and human evolution. *Nature* 325:31–36.

Figuerosa F, Günther E, Klein J (1988) MHC polymorphism pre-dating speciation *Nature* 335:265–271.

Harris H (1966) Enzyme polymorphisms in man. *Proc Royal Soc London* (Ser B) 164:198–310.

Kreitman M (1983) Nucleotide polymorphism at the alcohol dehydrogenase locus of *Drosophila melanogaster. Nature* 304:412–417.

Li W-H, Nei M (1972) Total number of individuals affected by a single deleterious mutation in a finite population. *Am J Hum Genet* 24:667–679.

Li W-H, Tanimura M (1987) The molecular clock runs more slowly in man than in apes and monkeys. *Nauture* 326:93–96.

Li W-H, Wu C-I, Luo C-C (1985) A new method for estimating synonymous and nonsynonymous rates of nucleotide substitution considering the relative liklihood of nucleotide and codon changes. *Mol Biol Evol* 2:150–174.

Mayer WE, Jonker M, Klein D, Ivanyi P, van Seventer G, Klein J (1988) Nucleotide sequences of chimpanzee MHC class I alleles: evidence for trans-species mode of evolution. *EMBO Journal* 7:2764–2774.

Nei M, Imaizumi Y (1966) Effects of restricted population size and increase in mutation rate on the genetic variation of quantitative characters. *Genetics* 54:763–782.

Nei M, Li W-H (1979) Mathematical model for studying genetic variation in terms of restriction endonucleases. *Proc Natl Acad Sci USA* 76:5269–5273.

Sarich VM, Wilson AC (1967) Immunological time scale for hominid evolution. *Science* 158:1200–1203.

Stringer CB, Andrews P (1988) Genetics and fossil evidence for the origin of modern humans. *Science* 239:1263–1268.

Wolpoff MH, Zhi WX, Thorne AG (1984) Modern *Homo sapiens* origins: a general theory of hominid evolution involving the fossil evidence from East Asia. In Smith FH, Spencer P (eds) *The Origins of Moder Human: A World Survey of the Fossil Evidence.* New York, Alan R. Liss, Inc, New York, pp 411–483.

Yang C-Y, Gu Z-W, Weng S-A, Kim TW, Chen S-H, Pownall HJ, Sharp PMS-W, Li W-H, Gotto Jr. AM, Chan L (1989) Structure of apolipoprotein B-100 of human low density lipoproteins. *Arteriosclerosis* 9:96–108.

18

Population Genetics and Evolution in the Genus *Homo* in the Last Two Million Years

JAMES N. SPUHLER

Two contradictory hypotheses are in contention today on the evolutionary origin of anatomically modern humans, populations belonging to the species *Homo sapiens sapiens*:

1. The *Multiregional Transformation Hypothesis* holds that archaic *Homo sapiens* living in Africa, Europe, Asia, and Indonesia evolved from *Homo erectus* by anagenic species transformation (sense of Sewall Wright, 1932, 1982a) in those regions and were ancestors of modern *Homo sapiens* in those regions.
2. The *African Replacement Hypothesis* holds that modern *Homo sapiens* evolved from *Homo erectus* by species splitting (for some investigators by Punctuated Equilibrium in the sense of Eldredge and Gould, 1972) only in Africa and spread rapidly to replace archaic *Homo sapiens* everywhere with little or no hybridization so that, except in Africa, archaic *Homo sapiens* became extinct without contributing genes to modern human populations.

The members of archaic human populations are sometimes assigned to different formal subspecies, for example, in Europe *Homo sapiens neanderthalensis* (King, 1864), in Africa *H.s. rhodesiensis* (Woodward, 1921), in East Asia, *H.s. daliensis* (Wu, 1981), and in the Indonesia *H.s. soloensis* (Oppenoorth, 1932); references and details are in Campbell (1965). Many biological anthropologists prefer to use geographical names to designate the major regional human populations (Garn, 1971).

Nearly all evolutionists agree with Darwin that hominids evolved in Africa where the living chimpanzees and gorillas are our closest biological relatives. Schwartz (1974) is nearly alone in claiming that the Asian orangutans are our closest living relatives. Although some disagree about phyletic details there is strong agreement that *Homo erectus* evolved first in Africa about 1.5 million years ago (mya) and that *Homo habilis* from east Africa dating about 2.5 mya is the ancestor of *Homo erectus*. The *Homo erectus* populations spread to Eurasia by about 1 mya.

Figure 18.1 represents the trellis or netlike phyletic relations postulated by the multiregional hypothesis. Over 50 specimens of archaic forms could be added to the middle of the trellis and some 40 specimens of *Homo erectus* could be included near the bottom. Further, about 50 specimens of subfossil anatomically modern humans showing regional morphological characters can be placed near the top of the trellis. The African replacement hypothesis removes all of these 100-plus fossil hominids from the ancestry of modern humans holding that only the African geographical lineage is ancestral.

In some part, acceptance of the recent African replacement hypothesis rests on continuation of a deep prejudice championed by Marcellen Boule in his 1908–1911 monographs on the La Chapelle aux Saints Neanderthal fossil skeleton (and since persistent among many western European paleoanthropologists) that demands removal of Neanderthals from ancestry of all modern human populations (Hammond, 1982) rather than treating Neanderthals as "a race of our own species, not an ancestor or any form of 'missing link'" as Gould (1987, p 41) once put it—a view currently supported by many east European, Chinese, Japanese, Australian, and American paleoanthropologists (see, for example, relevant papers in Smith and Spencer, 1984, and in Mellers and Stringer, 1989).

The Neanderthal problem is complicated by the fact that there may be a real discontinuity in western Europe, where local Neanderthal populations were not proximal ancestors of local Cro-Magnon populations that in fact intruded and replaced. Rather other, earlier Neanderthal populations (or their equivalent archaic populations of western Asia) were the proximal ancestors of Cro-Magnon predecessors when they lived further to the east. A similar situation exists in modern Greenland, where local Eskimo populations were not proximal ancestors of the first Scandinavian invaders, yet Eskimos and Danes (strikingly different in cranial morphology) represent two interfertile races of *Homo sapiens*.

Acceptance of the replacement hypothesis and exclusion of all non-African archaic humans from ancestral status requires postulation of an excessive amount of parallel evolution in all non-African regions to account for the independent origin of archaic humans. In eastern Europe, China, and Australasia several archaic fossil specimens reliably classified as *Homo sapiens* closely resemble in some metrical and nonmetrical characters both modern and middle Pleistocene *erectus* forms in the same region.

The two hypotheses differ in theoretical predictions that can be tested by the fossil record (Stringer and Andrews, 1988, Table 1). The model of multiregional evolution predicts:

1. Continuity of regional morphological patterns from middle Pleistocene to present,

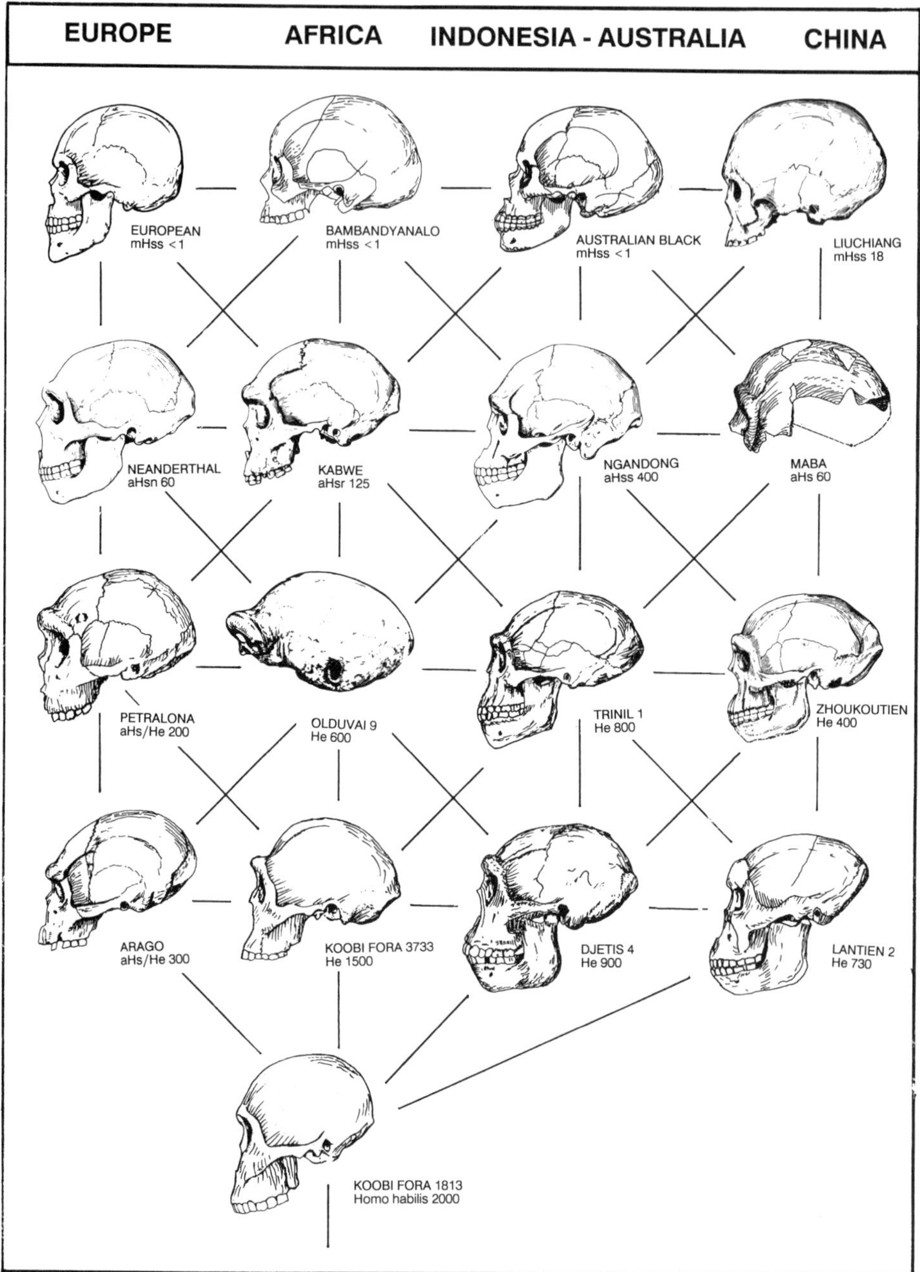

Figure 18.1. Phylogenetic reticulum of the hominids. The vertical divisions represent five evolutionary phases: *Homo habilis*, early *Homo erectus* (He), later *Homo erectus*, archaic *Homo sapiens* (aHs), and modern *Homo sapiens* (mHss). Vertical placement is not proportional to time, indicated in kiloyears before present. The horizontal groups correspond, on the whole, to four major geographical races of today. The vertical lines represent ancestry, the horizontal lines geographical distribution and morphological differentiation, and the diagonals genetic interchange.

2. Transitional fossils widespread in each region, and
3. Regional characters of high antiquity in non-African regions.

The Recent African evolution model predicts in contradiction:

1. Continuity of patterns only after the late Pleistocene appearance of anatomically modern man in Africa to the present,
2. Transitional fossils restricted to Africa, population replacement elsewhere, and
3. Regional characters of low antiquity in non-African regions.

Before presenting a new geometrical analysis of some archaic fossils that supports the multiregional transformation and rejects the African replacement hypothesis it is helpful to review some doubts and suspicions about current evolutionary theory applied to the origin of modern humans.

GENERAL THEORIES OF EVOLUTION

To the extent that theoretical issues are considered in their publications, most paleoanthropologists accept either Punctuated Equilibrium (PE) of Eldredge and Gould (1972) or the Synthetic Theory (ST) of Dobzhansky (1937); Mayr (1942); and Simpson (1944). I believe that the Shifting Balance theory (SB) of Sewall Wright (1931, 1932, 1977, 1978, 1982b) provides a better theoretical foundation for both neontological and paleontological study of hominid and other simian primate populations. Of course, Wright's Shifting Balance theory was of immense importance to Dobzhansky and to Simpson; much less so, to Mayr (Provine, 1986) in formulation of the Synthetic theory.

Treatment of the connection between the concept of morphological (and physiological) change and the concept of speciation is fundamentally different in PE and SB. PE defines speciation as a hypothetical, unified event; to an experimentalist a definition with very little empirical biological content; SB makes speciation many observed heterogeneous events, placed in a rich biological context.

Macroevolution (evolution above the species level) involves both major morphological change and reproductive isolation. Early taxonomists and paleontologists made morphological difference the primary criterion in naming species. Following the more extensive taxonomic studies of the early 1990s, especially by Osgood (1909) and Rensch (1929), reproductive isolation became the primary criterion (a criterion awkward for, but acceptable by paleontologists, including Carroll [1988] and Romer [1966]. SB holds that morphological change and speciation (defined by the isolation criterion) are wholly distinct phenomena that are usually associated but may occur separately (Wright, 1982a,b). Differences in total morphological pattern among breeds of horses, dogs, and geographical races of humans are greater than such differences among many Old World primate species. Yet, experimental studies of many genera, including *Drosophila*, reveal sibling species that are reproductively isolated in nature, but with minor or no morphological differences. Supporters of PE often quote the "hopeful monsters" and completely hypothetical "chromosomal repatterning" of Goldschmidt (1940) in support of a necessary connection between speciation and morphological change, but most geneticists insist that Goldschmidt had no concrete examples of such "speciation" (see Wright, 1941a). In fairness to Goldschmidt we should emphasize that he wrote before the discovery that DNA is the gene material and when very little was known about the fine structure of chromosomes.

PE denies or denigrates anagenic species transformation; SB supports both cladistic species splitting and anagenic species transformation. As early as 1932 Wright provided a theoretical, statistical, genic basis for anagenic species transformation. The Dobzhansky (1963) review of Carleton Coon's *The Origin of Races* (1961) is often incorrectly quoted as being against the multiregional transformation hypothesis, but, in fact, in that review Dobzhansky strongly rejected Coon's fundamentally different theory on the *independent*, no gene flow, polyphyletic origin of five major human races. Dobzhansky (1950, and later papers and books) explicitly accepted a multiregional transformation hypothesis in polytypic species held together by gene flow:

> Different populations (races) of a polytypic species may be descended largely from different races of the ancestral species and may differ in some genes in which these ancestral races differed. And yet, a polytypic species may still evolve as a single genetic system. Favorable mutants or gene combinations arrived at in one part (race) of such a species, may under the influence of natural selection, eventually spread to all other parts and thus become a common property of the entire species. Thus local autonomy of the gene pools of racial populations does not preclude retention of a basic unity of the species as a whole. I would like to point out that this view agrees quite well with the conclusions reached by the late Weidenreich on the basis of purely morphological analysis of pre-human populations....

The late Pleistocene origin of the polar bear (*Ursus arctos*), sufficiently different in morphology and behavior to be assigned by some to a different genus (*Thalarctos*), is a clearly demonstrated example of anagenic species transformation without species splitting (Hecht, 1974).

PE claims stasis in morphological phenotypes after rapid change during or soon after speciation; SB accounts both for lineages with stasis and those showing gradual change. The hominid fossil evidence shows that brain weight increased from about 500 grams in australopithecines to about 1000 grams in *Homo erectus* to nearly 1500 grams in *sapiens* populations (Tobias, 1971), and that this spectacular change was probably gradualistic (Wolpoff, 1984), requiring (as shown by Cavalli-Sforza and Bodmer, 1971, p 692) surprising small selection differentials. The brain increased in weight by about 500 grams over a period of 500,000 years or 25,000 generations averaging 20 years per generation. Using a standard deviation of 100 grams for brain weight, the change in units of standard deviations per generation is $500/(100 \times 25000) = 0.002$. With a heritability for brain weight of 0.50, the average selection is only 0.004 per generation, a very small value compared with selection differentials achieved under artificial selection in mammals. The gradualistic reduction and repatterning of the face and dentition is also fully documented (Brace et al., 1987; Turner, 1987). Gradualistic change in cranial measurements over periods of several thousands years has been demonstrated by Piontek (1979) and Schwidetsky (1967, 1972) for European and by Suzuki (1969) for Japanese populations.

PE does not account for widespread subspeciation in humans, primates, and other vertebrates. SB does (Wright, 1982b).

PE does not fit the evidence for fossil primates (Gingerich, 1978), fossil hominids (Wolpoff, 1980, 1084), modern human population genetics (Neel, 1984; Spuhler, 1984), nor for the empirical observation on the population genetics of some other species (Stebbins and Ayala, 1981; Charlesworth et al., 1982). SB does (Wright, 1982a,b).

SHIFTING BALANCE THEORY OF EVOLUTION

The basic concepts developed by Sewall Wright in his Shifting Balance theory cluster about the word *balance* in three diverse meanings which should be kept distinct:

First is the balance between heredity (factors of genetic homogeneity) and variation (factors for genetic heterogeneity) established in the genetic mechanism itself in all eukaryotic, bisexual diploid organisms. In his first major paper on evolution Wright (1931) asserted that "Evolution as a process of cumulative change depends on the proper balance of the conditions which at each level (gene, chromosome, cell, individual, local race) make for genetic homogeneity or genetic heterogeneity." To illustrate this premise he grouped the factors of evolution that tend toward genetic homogeneity or heterogeneity of the species in more or less definitely opposing pairs that give the basis for the shifting state of balance on which evolution depends:

gene duplication	Δ	gene mutation (u, v)
gene aggregation	Δ	random division of the aggregate
mitosis	Δ	chromosome aberration
conjugation	Δ	reduction (meiosis)
linkage	Δ	crossing over
restricted population size ($1/[2N]$)	Δ	hybridization (m)
environmental pressures (s)	Δ	individual adaptability
crossbreeding among groups (m_1)	Δ	subdivision of group ($1/[2N]$),
individual adaptability	Δ	local subgroup environment (s_1)

Second is the more literal sort of balance between the evolutionary pressures of mutation, gene flow, and selection.

Third is a more subtle sort of balance between these pressures as a group and random or stochastic processes which, in conjunction with selection, give rise to processes of trial and error that may lead to the creation of new ecological opportunities.

Both the second and third form of balance reach maximum significance where there is a finely divided population structure (Wright, 1950; p 279). But, Wright's theory is not restricted to subdivided populations; it covers large, panmictic populations as well as small inbred populations. Thus, Nei's (1987, p 420) list of premises for Wright's Shifting Balance theory applies only to that part of the theory (it is, of course, the most distinctive and innovative part) dealing with subdivided populations and not the whole theory that covers both panmictic and subdivided populations, large and small, including the case (Wright's "C") where selection is the major factor in important, large-scale evolutionary episodes such as the origin of the tetrapods or the radiation of South American mammals.

Wright's theory differs fundamentally from the evolution theories of Fisher (1928, 1930), Haldane (1924, 1932), and Kimura (1968) on the assumed relations of genotypes to phenotypes (illustrated diagrammatically in Figure 18.2). Figure 18.2A shows the 1 to 1 relation of Mendelian major gene and unit character assumed by early evolutionary geneticists, including Hardy, Weinberg, and Castle. Figure 18.2B adds the assumption of polygenic inheritance, uncomplicated by gene interaction and pleiotropy. Fisher (1930) held that only minor genes are likely to be selected favorably and that nonadditive relations between genotype and phenotype (dominance, epistasis) merely slow down progress by selection in panmictic species populations. Fisher brought

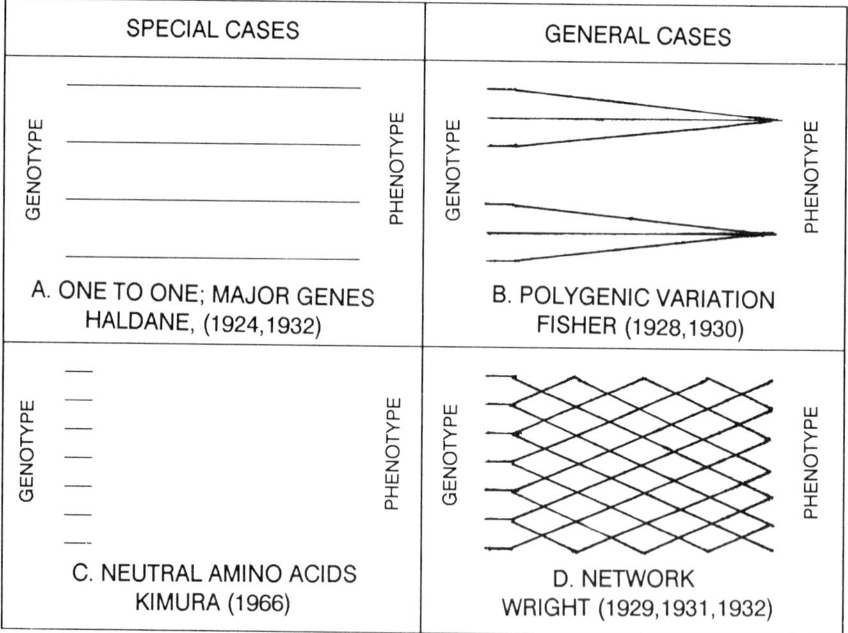

Figure 18.2. Four assumptions on the relationship of genotype to phenotype. Wright's Shifting Balance theory is based on Assumption D. (Modified from Wright, 1982, Fig 1).

all evolution under a single, simple formula, his Fundamental Theorem of Natural Selection that "the rate of increase in fitness of any organism at any time is equal to its (additive) variation in fitness at that time." It is clear that Fisher's theory does not hold for genes that modify a heterozygote and also have other pleiotropic effects on the homozygotes (Charlesworth, 1979; Wright, 1977). Figure 18.2C applies to neutral genes with no phenotypic effect that is subject to selection (Kimura, 1968) that account for the extensive polymorphism of proteins revealed by electrophoresis in many species. The network in Figure 18.2D shows each gene has multiple pleiotropic effects in general not additive. These interactions are inevitable results of complex biochemical and developmental reactions that occur between primary gene action and ultimate phenotypes subject to selection. Pleiotropy is universal in genes that control the variability of metrical and discrete characters of anthropological and primatological interest (Cavalli-Sforza and Bodmer, 1971; Sing, et al., 1988; Wright, 1982a).

Wright's 1932, largely non-mathematical paper on "The Roles of Mutation, Inbreeding, Crossbreeding and Selection in Evolution" is his most influential and widely quoted paper (Provine, 1986). His two-part figure (redrawn here as Figure 18.3) shows the random variability in the distribution of gene frequencies with respect to a single locus under various specified conditions regarding population number, selection coefficients, forward and back mutation rates, and population exchange with the rest of the species (gene flow), in (A) the whole species, and (B) a local race or deme. In the whole species the tendency toward random variation is balanced by mutation and selection resulting in a high, narrow distribution of probability densities when $4NU = 8$, $4NS = 80$, that is, very large; a lower, broader distribution when $4NU = 8$, $4NS = 20$, that is, medium; and a U-shaped distribution when $4NU < 1$, $4NS = 0$, that is very small. In a local race (B) subject to a small amount of crossbreeding with the rest of the species, the tendency toward random fixation is balanced by immigration (gene flow) instead of by mutation and selection. For example, when $4Nm = 16$, $q_m = 0.75$, and $4Ns = 0$, that is, medium, the distribution is low and broad as in the middle curve of (B). In later papers Wright made an important theoretical modification in the equations shown in Figure 18.3 by replacing the limited selection term (e^{4Nsx}) with the vastly richer term W^{2N} (Provine 1986, p 305).

Wright developed the idea of an adaptive landscape or fitness surface (Figure 18.4) as a visual aid in understanding his SB theory. With 10 alleles in each of 1000 loci the number of possible combinations in 10^{1000}, much larger than Eddington's estimate of the number of electrons and protons in the universe, 10^{67}. Any actual population is confined to an infinitesimal portion of the possible gene combinations. If the chance that a random combination is as adaptive as those characteristic of the species is 10^{-100}, room remains for 10^{800} separate peaks, each surrounded by 10^{100} more or less similar combinations. Depiction of the field of possible gene combinations graded with respect to adaptive value requires 9000 dimensions for adequate representation, so that the two dimensions of Figure 18.4 poorly represent

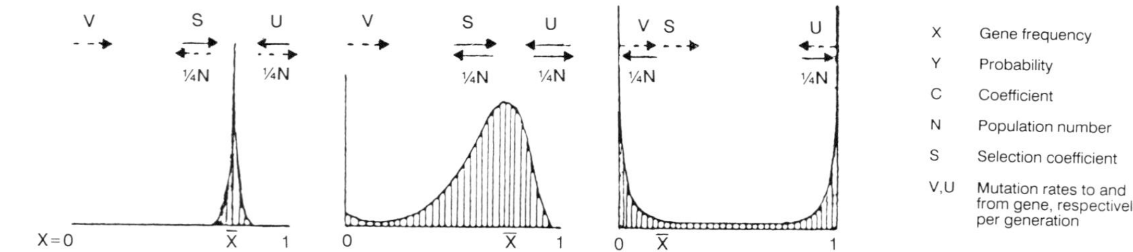

Figure 18.3. Random variability of a gene frequency in Mendelian populations under various specified conditions as expressed by Wright in 1932. The abscissas represent a scale of gene frequencies, 0% to the left, 100% to the right. (Redrawn from Wright, 1932, Fig 3).

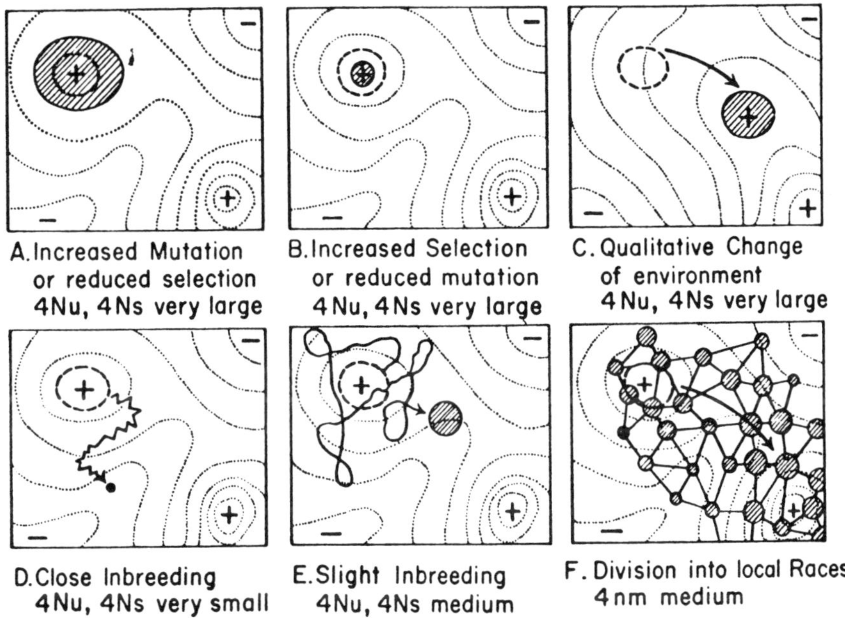

Figure 18.4. Wright's adaptive surface diagrams illustrating some possible modes of evolution in Mendelian populations. The multiple dimensions of gene frequency change for 1000s of heterallelic loci are reduced to two dimensions on which all genotypes are supposed to be located and the dimension of selective value is represented by contours. The heavy broken contour indicates an initially occupied field. Crosshatched areas (multiple subpopulations in F) indicate a later occupied field. Courses of evolutionary change are indicated by arrows. Effective population numbers, N (total), n (local); v (mutation), s (selection), m (migration). (From Wright, 1932, Fig. 4).

such a field. The contour lines represent the scale of adaptive value. For Wright (1932)

> The problem of evolution as I see it is that of a mechanism by which the species may continually find its way from lower to higher peaks in a field. In order for this to occur there must be some trial and error mechanism on a grand scale by which the species may explore the region surrounding the small portion of the field which it occupies. To evolve, the species must not be under strict control of natural selection....

Wright concluded that increased mutation or reduced selection of (A) and the increased selection or reduced mutation of (B) are possible but not adequate to explain evolution to an important extent. Qualitative change of environment when 4NU, 4NS are very large, as Fisher (1930) argued in his *The Genetical Theory of Natural Selection*, may be of great importance in macroevolution, as mentioned above. Wright concluded that the situations

(D) and (E) are not of general evolutionary importance, except that (D) may explain cases of extinction. In (F) Wright considers

> ... the case of a large species which is subdivided into many small, local races, each breeding largely within itself but occasionally outbreeding. The field of gene combinations occupied by each of these local races shifts continually in a nonadaptive fashion (except in so far as there are local differences in the conditions of selection). The rate of movement may be enormously greater than in the preceding case [E] since the conditon for such movement is that the reciprocal of the population number be of the order of the proportion of crossbreeding [1/N ? m] instead of the mutation rate. With many local races, each spreading over a considerable field and moving relatively rapidly in the general field about the controlling peak, the chances are good that one at least will come under the influence of another peak. If a higher peak, this race will expand in numbers and by crossbreeding with the others will pull the whole species toward the new position. The average adaptiveness of the species thus advances under intergroup selection, an enormously more effective process than intragroup selection. The conclusion is that subdivision of the species into local races provides the most effective mechanism for trial and error in the field of gene combinations.

In subdivided populations there are three phases in the shifting balance process of evolution:

1. *Phase of Random Drift.* The set of gene frequencies in each deme drift at random in a multidimensional stochastic distribution about the equilibrium set characteristic of a particular fitness peak. The set of equilibrium values are a result of pressures on the gene frequencies due to recurrent mutation, recurrent immigration from other demes, and selection. The fluctuation in the gene frequencies responsible for stochastic distribution (or random drift) may be due to accidents of sampling or to fluctuation in the coefficients measuring the various pressures.
2. *Phase of Mass Selection.* Occasionally, in one of the demes the set of gene frequencies drifts far enough to cross one of the many two-factor saddles in the surface of fitness values. There follows a period of relatively rapid change in this deme, dominated by selection among individuals (or families) until the set approaches the equilibrium associated with the newly controlled fitness peak, about which it now drifts at random and thus returns to the first phase, but in general at a higher level.
3. *Phase of Interdeme Selection.* A deme in which the set of gene frequencies comes under control of a fitness peak superior to those controlling the set in nearby demes tends to produce a greater surplus population and, by excess dispersion, systematically shifts the position of equilibrium of these towards its own position until the same saddle is crossed in them, and they all move autonomously to control by the same fitness peak. This process tends to spread through the species in concentric circles. Two such circles, spreading from different centers, may overlap and give rise to a new center that combines the two different favorable interaction systems and becomes a still more

active population source. The virtually infinite field of interaction systems may be explored in this way with only a small number of novel mutations, as alleles which have been rare come in time to displace the previously more abundant ones (nearly verbatim from Wright 1977, pp 454–455).

Periodic reductions in size of hominid breeding populations were significant during the first 99% of the existence of the genus *Homo* due to chance deviations in the genotypes of survivors at the time of least numbers. Many local populations were liable to frequent extinctions with restoration from the progeny of a few migrants from neighboring demes. In such regions the line of gene continuity of large subdivided populations may have passed through bottlenecks of extremely small numbers even though the whole species contained millions of individuals (Figure 18.5).

The mathematical theory of Phases 1 and 2 is well developed by Wright, but he did not publish the mathematical foundations of Phase 3 (Wright, 1982b). Crow et al. (1990) examined the third phase where Wright had postulated that favorable gene combinations from a fitter deme are exported by migration to the rest of the population. A number of workers had pointed out that, intuitively, the genotypes favorable in the local donor demes would be broken up by recombination in the recipient demes (Haldane, 1959; Nei, 1987; Crow, 1986). But Crow et al. (1990) show that, in the case of several fitness models, migration rates as small as one order of magnitude less than

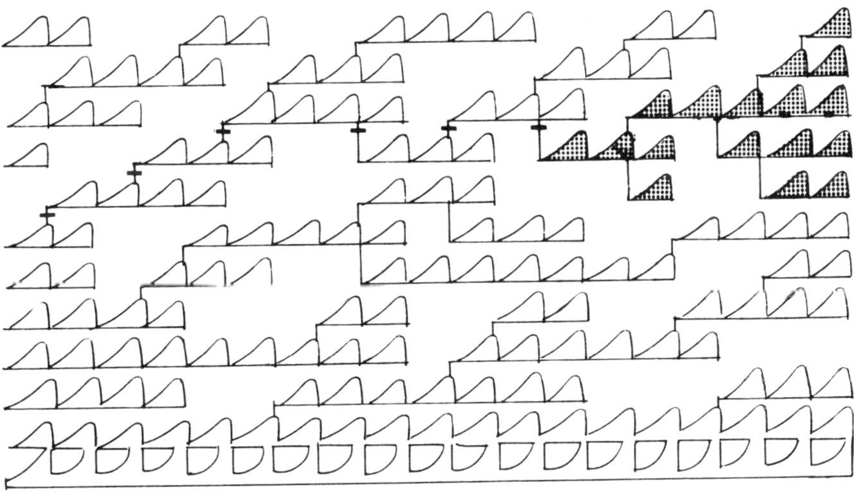

Figure 18.5. Wright's diagram of breeding structure in a species in which the subpopulations (demes) in certain localities are subject to frequent extinction with reestablishing by rare migrants. Different geographical territories are distinguished vertically. Generations proceed from left to right horizontally. The shaded group represents a large population whose entire ancestry has passed through small groups of migrants (bottlenecks) six times in the period shown. (Redrawn from Wright, 1940, Fig 2).

the selection coefficients are often sufficient for the postulated operation of Phase 3.

The conditions most favorable for one phase may not be the most favorable for other phases. The most favorable population structure for Phase 1 is so small in number that certain regions are liable to frequent extinction with reestablishment by rare migrants (Wright, 1940). A structure consisting of larger, more stable local populations is more favorable for Phases 2 and 3. During the last few thousand years of the period since the agricultural revolution, human evolution has proceeded by Phase 2, and especially Phase 3, enormously more than by Phase 1. In general, a mixture of population structures in time and space would seem more favorable than any consistent pattern (Wright 1977, p 473).

Chromosomal evolution in the primates offers support for the Shifting Balance theory. Polymorphism for chromosomal rearrangements is rare within species but common between species. There is a strong tendency for accumulation of a given type of chromosome rearrangement in particular primate families; inversions in Pongidae and Hominidae, fissions and shifts in Cercopithecidae, and Robertsonian translocations in several prosimian families (Dutrillaux, 1988). Humans differ from chimpanzees by 11 chromosomal rearrangements, from gorillas by 15 rearrangements (Dutrillaux 1980, 1988). These rearrangements are sufficiently disadvantageous that it is hard to explain their evolution by natural selection. The only plausible mechanism is that translocations in very small subdivided populations, where selection pressures are low, sometimes drift into high frequencies. It is precisely the set of balance of evolutionary processes postulated for the Shifting Balance theory that may account for observed chromosomal differences in higher primates.

The most favorable conditions for Phase 1 happen to be precisely those giving a relatively high probability of fixation in the local population of a translocation that is semisterile as a heterozygote (Wright, 1941b). The chance of fixation in animal populations with exclusive sexual reproduction is of the order 10^{-3} if the effective population number is 10, but of the order 2×10^{-6} in populations of 20. Because polymorphy for chromosomal translocations is rare in human and other primate populations but common between closely related primate species, it is probable that many species have their origin in the fixation of a translocation in a very small deme (Wright, 1977, p 473). Lande (1979) has verified and extended Wright's observations on the importance of small population size and genetic drift in the fixation of chromosomal rearrangements.

The essential differences between the genetical evolutionary theories of Fisher, Kimura, Nei, and Wright are biological and not mathematical, although Wright's theory differs in mathematical detail (Crow, 1986). Wright always tried to make mathematics reveal and expose rather than ignore and conceal underlying biological complexity. Crow (1955) helped to unify the general theory by extending Fisher's Fundamental Theorem of Natural Selection to include the creative role played by interpopulation selection, one of the most attractive features of Shifting Balance models:

$$dM/dt = V_g + V_m + 2(da/dt) \qquad [18.1]$$

"the rate of increase in fitness of a population is equal to the average genic (additive genetic) variance of the subpopulations plus the genotypic (total genetic) variance of the means of the subpopulations" (where M is the average fitness of the entire population, V_g is the additive genetic variance of the subpopulation, V_m the total variance of the means of the subpopulations, and the last term on the right measures the effects of changes in homozygosity, fitness coefficients, and in the environment).

Electrophoretic comparisons of the kinds and frequencies of alleles at 43 loci in chimpanzee and human proteins show that the two species differ more in the frequencies of homologous alleles at a locus (26/43 = 60.5%) than in substitution of different alleles (17/43 = 39.5%). Thus Nei's (1987, p 420) criticism of Shifting Balance theory in stressing change in frequency rather than kind (substitution) of alleles is not appropriate for hominoid evolution specifically.

Cavalli-Sforza, et al. (1988) computed the most extensive and intensive genetic tree (phenogram) now available. Their tree used data in 42 pooled populations studied for 120 alleles by average linkage analysis of Nei's genetic distance, with the bootstrap statistical technique used to test the reproducibility of the sequence of steps in the tree (Figure 18.6).

The first dichotomy in the tree separates Africans from non-Africans at a genetic distance of 0.0297 ± 6.8. The next split separates two superclusters—the North Eurasian and the Southeast Asian. The North Eurasian then divides into (i) Caucasoids and (ii) Northeast Asians and American Indians.

The Southeast Asian supercluster separates into (i) mainland and insular Southeast Asians, (ii) the Pacific Islanders, and (iii) Australians and New Guineans. All the early bifurcations in the tree are statistically significant. The tree has six terminal clusters that correspond to major geographical races (as defined by Garn, 1971). The period of differentiation between the six races is about twice as long (as measured by genetic distance) as the period of differentiation within the races. The average of the genetic distances before the race-forming dichotomies is 0.020 and that after is 0.010.

The average genetic distance depths in the North Eurasian and Southeast Asian supercluster is 1.69 times greater than the maximum depth between the six African populations. The divergence of the New Guinea–Australia cluster is 1.53 times deeper than that between the most divergent of the six African populations.

Nei and Livshits (1989) calculated the genetic distances between Africans, Asians, and Europeans (the most widely recognized trinity of major human races) using 84 protein, 33 blood group, 8 leukocyte antigen and immunoglobulin loci, and 61 DNA markers, a total of 186 loci/markers. Their results are the first study to establish the evolutionary relationships of the three major human races at a statistically significant level. The overall genetic distance between Asians and Europeans (0.049 ± 0.007) is significantly lower

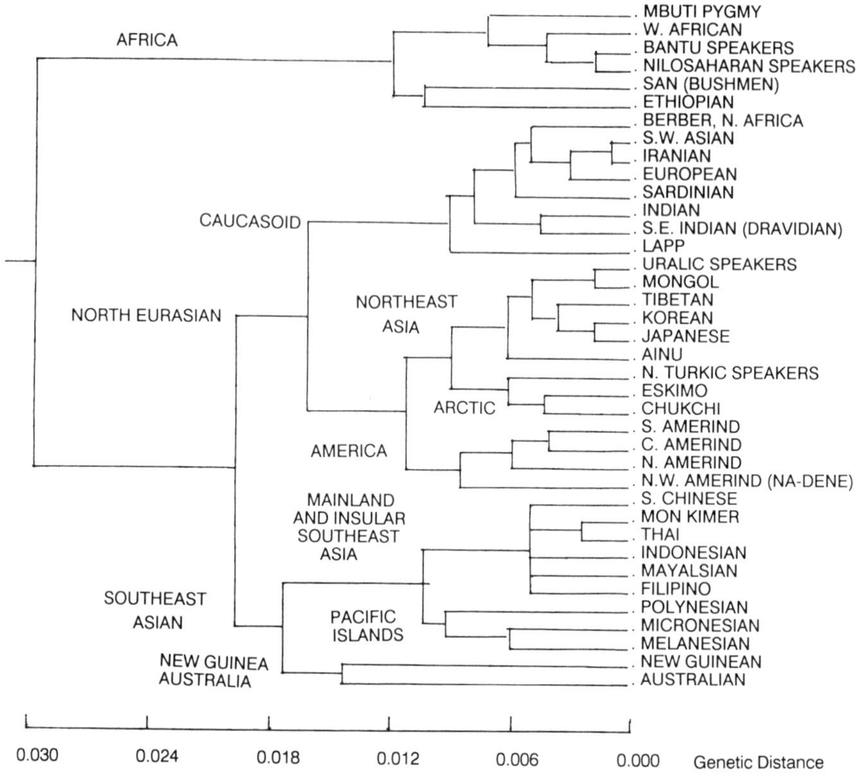

Figure 18.6. A genetic tree generated by average linkage analysis of 42 pooled, indigenous populations of the world based on 120 allele frequencies (redrawn from Cavalli-Sforza et al., 1988, Fig 1).

than that between Africans and Europeans (0.063 ± 0.011) or that between Africans and Asians (0.078 ± 0.013).

As Nei and Livshits point out (1989, p 279) the genetic relationships they and Cavalli-Sforza and associates obtain do not by themselves "reveal the place of origin of *H.s. sapiens.* This place must be determined from the fossil records."

Given uniform evolutionary rates, the Recent-African-replacement hypothesis predicts a very different topology of the genetic tree from that produced by Cavalli-Sforza and associates. The clusters most geographically remote from Africa (say Australia–New Guinea) should show strong determination by replacement migration and less depth in genetic distances because they have had less time for local genetic differentiation than the African cluster of populations. The observed topology is more compatible with the hypothesis of multiple regional transformation where local differentiation (subspeciation) may be expected to be as great in degree in southeast Asia–Australia-New Guinea as in Africa, while Africa as a whole shows the greatest genetic distances because some of the genes tested arose in populations belonging to *Homo erectus* in the period I million to 500,000 years ago whereas the local subspeciation identified by both morphological differences and marker gene frequencies arose differentially in local populations of archaic *Homo sapiens* in the period after 500,000 years ago.

Some evolutionary biologists assume that rates of gene flow in Pleistocene populations were too small to keep the human species intact in populations spread thinly over Africa and Eurasia for a million and more years. These same workers tend to assume that effective population size is large and rates of gene flow small in extant human populations (see Schull, 1972, for a well-reasoned discussion). Morton (1982) catalogued population number and gene flow in 28 populations and showed that the demographic parameters for effective population size, effective systematic pressure (including mutation, gene flow, and directional selection), and the standard deviation of short-range migration may be estimated from the Malécot (1973) parameters for isolation by distance, so that estimates of gene flow can be obtained from bioassay of kinship in populations lacking genealogies or direct counts of rates of gene flow. Morton and associates conclude that no known human population has an evolutionary size greater than 2,027 (Japan prefectures) or less than 49 (ethnic groups) and the effective systematic pressures lies between 0.032 and 0.174, so that gene flow dominates weak selection and mutation. For the 28 populations, mean gene flow is greater than 6% per generation. Bocquet-Appel (1985) argues that rates of gene flow must be of the order of 3% to 11% during the Paleolithic, in order for most adult individuals to find mates, a conclusion also supported by Yellen and Harpending (1972). For native Australia, Birdsell (1953, 1968) estimates gene flow as high as 15% per generation. If most gene flow is from close neighbors, there is a correlation, r, between the gametes of the out- and in-groups, and the measure of m must be replaced by $m(1-r)$ if m is to represent the actual amount of replacement by immigration (Wright, 1940).

The situation indicated by inbreeding and crossbreeding studies in modern Phase 3 human populations is represented diagrammatically in Figure 18.7 adapted from Morton et al. (1967, Fig 5). The small dotted circles symbolize local populations. The intermediate circles enclosing clusters of populations denote regional groups or minor races with inbreeding coefficient α_i, among which the equivalent inbreeding coefficient is α_s. The large circles represent major races with equivalent inbreeding coefficient α_t. The groups are arbitrarily delimited but represent to some unspecified degree a common phylogeny (Morton et al., 1967, p 138).

First generation hybrids between races are intermediate in size, mortality, and morbidity between the parental groups. Contemporary human populations do not represent coadapted genetic combinations that are disrupted by outcrossing. The equivalent inbreeding coefficient is 0.0005 for crosses of regional populations and 0.0009 for major races. However, Spielman et al. (1977), based on computer simulation and general historical considerations, infer that all estimates from genotype frequencies greatly underestimate the inbreeding coefficient for alleles in the founding population of American

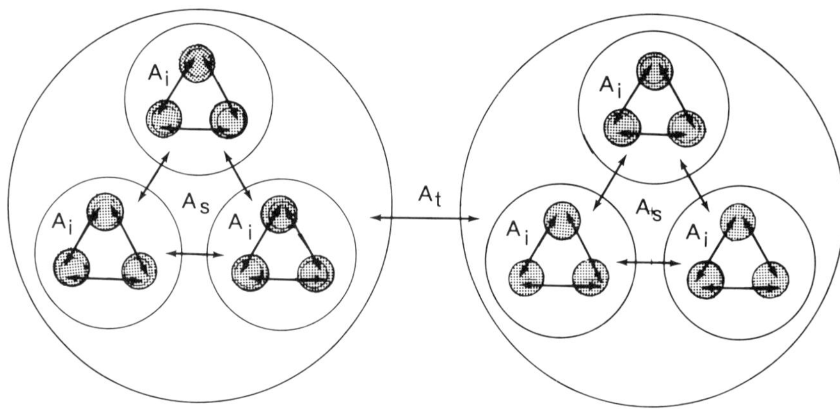

Figure 18.7. Genetic structure indicated by inbreeding and outcrossing studies in human populations. The small, dotted circles represent demes, the intermediate circles regional groups or minor "races," with inbreeding coefficient A_i within groups and A_s among groups, the larger circles symbolize subspecies or major "races," with equivalent inbreeding coefficient A_t. The equivalent inbreeding coefficient ($A_t = H/2B$) is the ratio of the outcross effect (H) to twice the inbred load (B). (Redrawn, with change of symbols, from Morton et al., 1967, Fig 5).

Indians. They surmise that in the highly subdivided tribal populations that prevailed before the origin of states, the probability of identity by descent for homologous alleles was roughly 0.5. They consider that the much lower estimates—0.005 to 0.01—represent only a very recent departure from the inbreeding intensity that predominated before 2,000 to 4,000 years ago.

During nearly all of the last two million years the ecological and demographic settings for human evolution are exactly those postulated by the three phases of the Shifting Balance theory (Weiss, 1986; Weiss and Maruyama, 1976). Before the Neolithic food-producing revolution starting about 10,000 years ago, all human populations had a tribal organization with total tribal size limited to an average of about 500 individuals. Nei and Imaizume estimate that human populations with overlapping generations have an effective population size (N_e) about 40% of the total population including nonreproductive individuals (Nei, 1987).

Chakraborty and Neel (1989) used data on electrophoretic variants at 27 protein loci to esimate the effective sizes (N_e values) of 12 relatively unacculturated American Indian tribes of Central and South America. Because of inbreeding, differential fertility, and lack of panmixia, N_e should be less than the current number of adults of reproductive age, but their estimated effective size was substantially greater (by a mean factor of 4.94) for 7 of the 12 tribes. The authors conclude that long-time interchange between tribes is probably more important than recent decrease in population numbers in accounting for the overestimate in the seven tribes. As expected, three of the tribes known to have increased their population size in the last few decades show estimated effective size substantially less than the census count of reproductive adults.

Short (1976), in a masterful review, showed that in known hunter-gatherer populations' overall fecundity is lower than in nearly all agricultural groups due to a combination of later age of menarche (16–18 years in contrast to 12–13), adolescent infertility (by about 3 years due to menstrual cycles without ovulation), and long birth intervals (3–4 years caused by suppression of ovulation during lactation), all this reducing the maximum number of children to about 5 of which 2 to 3 survived into the reproductive period. Maximum observed human fecundity, for example, as among the Hutterites, could result in a population doubling every 16 years. But if Short's estimates are accurate, during the first million years human populations increased at an annual rate of the order of 0.001% due mainly to time of the reproductive period and lactational suppression of ovulation rather than high rates of mortality from infectious diseases or infanticide. Based on Short's analyses, May (1978) points out that during her reproductive life span of about 23 years a Paleolithic hunter-gatherer female would experience 15 years of lactational amenorrhea, just under 4 years of pregnancy, and just under 4 years of menstrual cycle. In contrast modern females in developed countries, with menarche at 13 and menopause at 50, spend under 2 years in pregnancy, little time in lactation, and 35 years in menstrual cycles.

During much of the Paleolithic, Phase 1 was in operation with random genetic drift tending to increase differentiation between demes and decrease diversity within demes. Bocquet-Appel (1985) showed that the coefficient of biological heterogeneity between groups (H_i of Rao, 1980) increased during the 500,000 years of the Paleolithic in four measurements of the parietal bone in 27 fossil hominids from 5.67 to 8.52 and in four measurements of the mandible in 18 hominid fossils from 3.52 to 7.05. During the 10,000 years after the Neolithic revolution, Phases 2, with mass selection, and 3, with intergroup selection, were in effect. Henneberg et al. (1978), investigating 10 cranial measurements on 58 skeletal series in Europe from Neolithic to modern times, showed that the correlation between standardized measures of within-group variability ($m_{(s)}$) increased from -0.3 in the Neolithic to $+0.3$ in modern times. The same measurements on the same European skeletal series showed that between-group variability as expressed by standard deviations of mean values for separate series decreased in males from 1.00 in the Neolithic, Bronze, and Iron ages to 0.87 in moderns, and in females from 1.00 to 0.50.

Wright (1931) considered intergroup selection as a primary process controlling evotutionary change. Subdivided populations produce partially isolated and differentiating demes among which selection could operate. He was impressed by the marked changes in milk yield in cows and egg production in chickens resulting from intergroup selection of male relatives, a result that cannot be explained by the action of strictly individual selection. Wade (1977) verified Wright's concept of intergroup selection using experimental studies on populations of the flour beetle *Tribolium castaneum*.

CLADISTIC METHODS

The ideas and the terminology of the cladistic method of Hennig (1966) were warmly and widely accepted by many paleoanthropologists. Many believe that Hennigean methods of phylogenetic inference, that include multiple character analysis of both fossil and extant taxa, identification of shared and derived characters, joined with consideration of temporal and paleogeographical distribution of fossils offer the best approach for interpreting phylogeny (see Carroll, 1988, for a well-reasoned treatment of cladistics and fossils). But several workers, while still claiming to use "cladistics," depart fundamentally from the ideas and methods of Hennig.

There are now several varieties of the modified methodology, perhaps best identified by city-museum labels, as the New York/AMNH (G. Nelson, E. Gaffney, N. Platnick, N. Eldredge, E. Delson, I. Tattersal), London/BMNH (P. Andrews, C. Stringer, C. Patterson), and Pittsburgh/CMNH (J. Schwartz). To evolutionary geneticists, the most suspicious modified cladists are those that ignore the time position of fossils and make anti-historical or non-historical inferences from time-free cladigrams or "Hennigian combs" (arguments for by C. Patterson and against by C. Paul, P. Florey, and A. Charig are in Joysey and Friday, 1982, and against in Brace (1988), Jarvik (1980), Panchen (1982), and Tartarinov (1984).

Perhaps the most noteworthy output of a modified cladistic method is the conclusion of Schwartz (1984) that the Asian orangutan is more closely related to humans than are the African chimpanzee and gorilla, a conclusion falsified by massive data from molecular studies (Holmquist et al., 1988).

Cann et al. (1987, p 35) cite the modified cladistic claim of Andrews (1984) that four apomorphic characters remove Asian *Homo erectus* as a possible ancestor of modern humans. The characters are frontal (sagittal) keel present, thick cranial vault bones, angular torus (of the parietal) present, and inion $\neq$ endinion (of occipital). Andrew's argument is fallacious if *H. erectus* evolved to *H. sapiens* by anagenic species, as claimed by supporters of multiregional transformation rather than species splitting, because he then has no proper Hennigan sister- or out-group to establish polarity. His logic is then circular. It is like claiming that either ponies or Percherons can be ancestors of members of the next generation of horses, but not both.

Foley (1987) is a good example of a wrong modified cladistic method. His cladogram of hominid evolution (Figure 18.8) has 19 taxa differing in divergence time of some 10 million years all lined up at the recent terminals of the cladogram. His conclusion is that anatomically modern *Homo sapiens* evolved only in Africa. On the next back he warns that the validity of his cladogram depends on the correct identification of character polarity and the absence of parallelism. Yet on the very next page his cladogram accepts falsely derived characters of Andrews, Stringer, and Wood and exhibits no less than six cases of independent parallel evolution of archaic *Homo sapiens* in different regions! "Regional features in fact cannot possibly be regarded as 'archaic' or 'primitive' because they characterize a geographic area and not

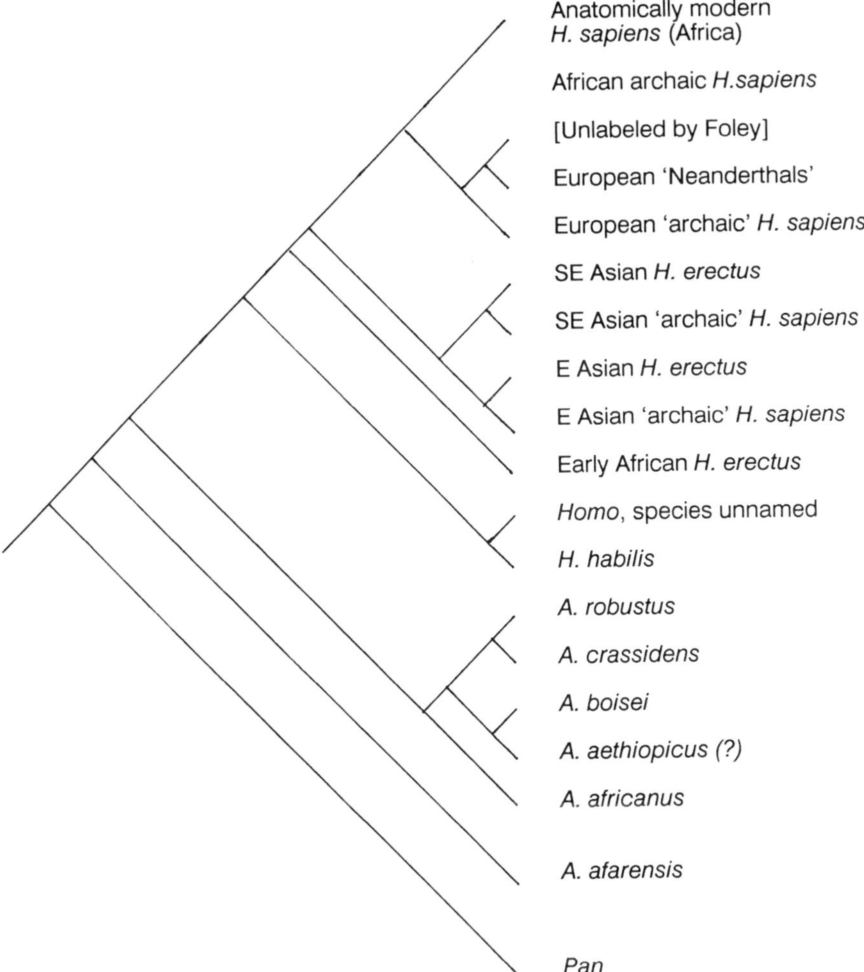

Figure 18.8. Foley's cladogram of hominid evolution (redrawn from Foley, 1987, Fig 2).

an evolutionary stage, and the morphological features that characterize one geographic race of a species cannot be validly regarded as more primitive than features of another contemporary race of the same species" (Wolpoff, 1990b).

The validity of arguments from modified cladistic methods that ignore time is clearly in question.

MULTIVARIATE STATISTICAL ANALYSES

Howells (1969) lucidly presents the advantages of multivariate analysis in anthropometric studies. Clearly the multivariate methods of Fisher,

Mahalanobis, Hotelling, and Rao are an advance beyond the univariate or bivariate methods of Pearson and the early *Biometrika* group, especially in treating fossil samples of one. Yet, as the papers in van Vark and Howells (1984) show, serious problems remain.

Kowalski (1972) argued that multivariate techniques often do not allow the effective description and communication of the informational content of a body of data. He agrees with John Tukey that "data analysis can gain much from formal statistics, *but only if the connection is kept adequately loose*." Rao (1966) called attention to an apparent paradox concerning multivariate tests of significance using some data sampled from two bivariate normal populations assumed to have common variances and covariance. Taken separately, both cases give a t value significant at the 5% level. Taken simultaneously, the value of Hotelling's T^2 corresponds to a probability as high as 12%. Healy (1969) gives a graphical demonstration of Rao's paradox and notes that its consequences are particularly alarming in problems of discrimination and classification.

In Howell's (1973) widely and properly acclaimed monograph, the results of the multivariate analysis of cranial variation in male and female samples from 16 populations are disturbing. Whether clustered by successive mergers (his Fig 4) or successive splits (his Fig 5), Howells found males and females from the same breeding population are classified differently in 1/2 (mergers) or 3/4 (splits) of the major clusters corresponding to African, European, Asian, and Southwest Pacific regions.

Because both methods used by Howells in the most careful study available produce contradictory trees for males and females, we should reserve judgement when, for instance, Rightmire (1979) applies a modified version of Howells' analysis to the Border Cave sample of one.

CRANIAL POLYLATERALS

In order to avoid, just for the purpose of this chapter, the difficulties mentioned above of multivariate statistical analysis in phylogenetic studies on fossil hominid skulls, I will use a simple, multivariate geometrical analysis, depending on some properties of general polygons (see Selby, 1971). Of initial interest is the general quadrilateral that outlines the brain case in the midsagittal plane. General quadrilaterals lack the symmetrical properties that define the eight special quadrilaterals such as squares, rectangles, parallelograms, kite-shapes, trapeziums, and rhombuses. By adding adjoining triangles the cranial quadrilateral is expanded to pentilaterals and hexilaterals. And, although we will not consider them here, by adding the third dimension, increasingly complex polygons may be made to approximate more and more closely the measurable surface topography of the skull, including the regionally diagnostic frontal, simotic, rhinial, and premaxillary triangles.

Let the following landmarks be defined in reference to the midsagittal plane of the cranium shown in Figure 18.9 denoting points and angles with

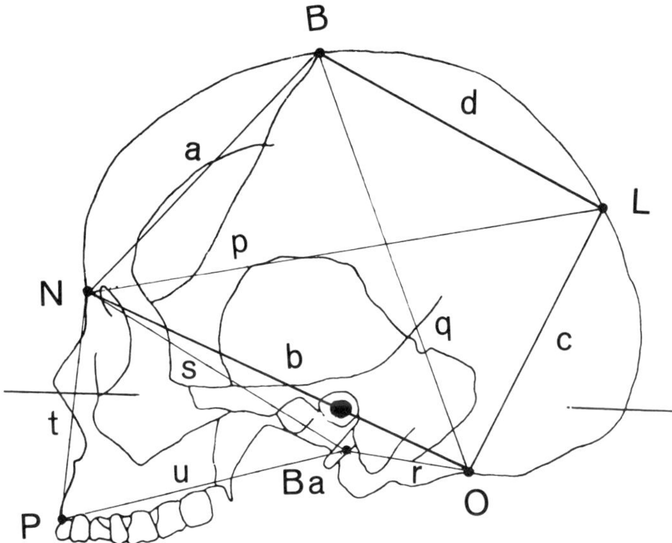

Figure 18.9. Landmarks for measurement of midsagittal cranial polylaterals. The landmarks are: N nasion, B bregma, L lambda, O opisthion, Ba basion, and P prosthion. The chords of the brain case are: a nasion-bregma, b nasion-opisthion, c lambda-opisthion, and d bregma-lambda. The diagonals are: p nasion-lambda and q bregma-opisthion. The chord of the foramen magnum is: r basion-opisthion. And the cords of the facial triangle are: s nasion-basion, t nasion-prosthion, and u prosthion-basion.

vertices at the points: N = nasion, O = opisthion, L = lambda, B = bregma, Ba = basion, and P = prosthion.

The distances between these points reflect the relative volume of the frontal, parietal, temporal, and occipital lobes of the brain (von Bonin, 1963; Kochetkova, 1978) and have long been recognized as differentiating fossil skulls of *Homo erectus* from *Homo sapiens* regardless of total brain volume (Weidenreich, 1941; Coon, 1962).

The general quadrilateral is determined by: 4 sides and 1 angle, or 4 sides and 1 diagonal, or 3 sides and 2 included angles, or 3 sides and 3 angles. Adjoining triangles may be constructed from sides and angles if 3 independent sides or angles are known. Thus some missing measurements can be obtained by calculation, and the accuracy of some doubtful measurements can be judged from other more reliable measurements. For example, the diagonals nasion-lambda (here labeled p) and bregma-opisthion (q) often are not given in the original craniometric reports of human fossil skulls. These are obtained from

$$p = \sqrt{[a^2 + b^2 - 2ab\cos N]}, \text{ and} \qquad [18.2]$$

$$q = \sqrt{[c^2 + d^2 - 2cd\cos 0]}. \qquad [18.3]$$

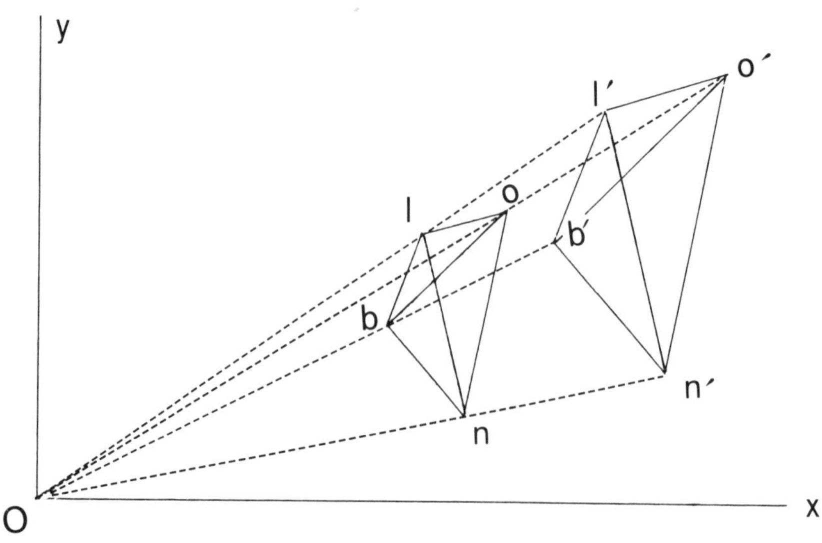

Figure 18.10. Generation of similar quadrilaterals by enlargement, and a transformation of similitude (see Ledermann and Vajda, 1985, Part A, pp 6–8). The degree of geometrical similarity of two cranial quadrilaterals may be measured by the ratios of their chords and diagonals, nb/n′b′, no/n′o′, ... nl/n′l′, bo/b′o′.

The cranial polygons of two skulls from the same population tend to be more alike in shape and size than those from two distantly related populations. Similarity is a geometrical idea like that of congruence but more general (Ledermann and Vajda, 1985). Consider a quadrilateral (Figure 18.10) with vertices labeled N, O, L, B and the corresponding points of a second quadrilateral labeled N′, O′, L′, B′. The transformation $x' = kx$, $y' = ky$, in rectangular coordinates, where k is the scale factor such that N′O′/NO = O′L′/OL = B′L′/BL = N′B′/NB = k, is called a transformation of similitude or a homeothetic transformation.

We will examine cranial polygons determined by 10 midsagittal measurements defined in Figure 18.9 (a, b, c, d, p, q, r, s, t, u). In this preliminary study we will consider the average similitude of these 10 measurements in 4 fossil skulls of *Homo erectus*, 4 of Archaic *Homo sapiens*, and 4 of anatomically Modern *Homo sapiens*, 1 each from Africa, Europe, China, and Australia. All are adult males. The names of the 12 skulls, their chronological age, and the references for the measurements used, are listed in Table 18.1. All measurements are in millimeters.

In addition, Table 18.1 gives the 10 measurements of the cranial hexilaterals, area of the quadrilaterals and hexilaterals (here and elsewhere areas are in square millimeters), area of the facial triangle (N-P-Ba), facial triangle index (the ratio of the area of the facial triangle to that of the cranial pentilateral), perimeter of hexilaterals, and perimeter index (that compares the perimeter of a circle [with area equal that of the cranial hexilateral] to the area of the

Table 18.1. Data on cranial hexilaterals based on 10 measurements of 12 fossil skulls from 4 old-world regions.

Fossil Skull and → Reference → / Measurements ↓ and Indices ↓	Koobi Fora, ER-3733 Rightmire, 1990	Arago Reconstruction Spitery, 1992	Zhoukoudian Reconst. Weidenreich, 1943	Sangarin 17 Thorne & Wolpoff, 1981	Eliye Springs 11693 Bräuer & Leakey, 1986	Petralona Stringer, et al., 1979	Maba Woo [Wu] & Peng, 1959	Ngandong 12 Reconst. Weidenreich, 1951	Fish Hoek Howells, 1989a,b	Mladeč 1 Howells, 1989a,b	Upper Cave 101 Howells, 1989a,b	Keilor Howells, 1989a,b
1 Evolutionary Grade	*Homo erectus*				Archaic *Homo sapiens*				Modern *Homo sapiens*			
2 Chronological age (Kiloyears BP)	1200	450	400	710	50?	200	50?	200	18	34	18	12.9
3 Nasion—bregma sagittal chord	104.0	150.0	115.0	105.6	116.5	111.0	116.0	112.0	122.0	115.0	117.2	114.0
4 Nasion—episthion sagittal chord	141.3	146.6	154.0	150.0	135.0	133.6	145.6	157.0	121.5	144.2	137.4	141.3
5 Lambda—opisthion sagittal chord	88.0	99.0	84.0	84.7	93.0	91.0	88.5	90.0	90.0	106.0	91.5	104.0
6 Bregma—Lambda saggital chord	82.0	101.2	96.2	113.6	116.5	105.0	107.0	97.0	124.0	117.0	119.2	119.0
7 Nasion—lambda diagonal	148.6	181.2	175.0	185.5	187.5	172.5	181.0	173.0	187.1	182.8	183.6	188.5
8 Bregma—opisthion diagonal	140.1	136.3	127.8	132.5	131.3	135.5	139.5	137.0	137.2	156.5	144.0	151.1
9 Basion—opisthion chord	37.0	40.9	39.4	35.1	40.4	41.5	38.0	45.0	37.0	42.0	36.9	40.0
10 Nasion—basion chord	107.0	110.0	108.2	111.0	96.5	110.5	109.2	113.9	94.0	104.2	103.5	109.0
11 Nasion—prosthion chord	86.0	89.3	70.8	74.5	65.0	93.2	76.9	72.9	60.0	68.0	65.9	69.9
12 Prosthion—basion chord	118.0	135.0	96.0	126.0	108.0	117.0	108.0	129.0	98.0	108.0	100.0	109.0
13 Area hexilateral (× 10,000)	1.549	1.808	1.499	1.620	1.606	1.840	1.689	1.636	1.666	1.823	1.716	1.908
14 Area Facial Triangle (× 1,000)	4.433	4.875	3.339	4.105	2.930	4.824	3.905	4.131	2.730	3.408	3.312	3.613
15 Facial triangle Index	40.08	28.81	28.66	33.93	22.32	35.53	24.09	33.78	19.60	22.98	23.90	23.36
16 Perimeter hexilateral	515.0	568.4	496.2	544.4	526.6	558.7	534.4	545.9	521.0	556.0	538.7	556.0
17 Perimeter Index	85.68	83.85	87.48	82.88	85.30	86.00	86.22	83.06	87.82	86.10	86.21	88.06
18 Diagonal Index	94.28	75.22	73.03	71.43	72.22	78.55	77.07	79.19	73.33	85.61	78.41	80.16

cranial hexilateral) giving a measure of compactness of cranial design, and the diagonal index of the cranial quadrilateral, that gives a measure of the shape of the brain case in midsagittal section.

Table 18.2 presents (above the diagonal) the 66 pairwise average similitudes (the mean absolute pairwise differences of the 10 measurements of the cranial hexilaterals) of the 12 groups. The 66 pairwise differences in chronological ages (in kiloyears) of the fossil skulls are given below the diagonal. The correlation between pairwise similarities and age differences is only moderate, $r = 0.249 \pm 0.121$, as is expected in data sets of this sort.

Skulls that are identical in all 10 measurements have a mean similitude of zero. The more dissimilar the two skulls in comparison the larger the similitude quantity. The mean similitude of the 66 comparisons is 9.997 ± 3.177. The maximum difference in hexilateral similitude (18.54) is between Arago, a European *Homo erectus* and Fish Hoek, a gracile African modern. The two fossils closest in similitude of this set (2.88) are Mladeč, an Upper Paleolithic European, and Keilor, an native Australian of 9 millennia ago. The similitude of the two skulls with maximum difference in chronological age is 14.40 between ER-3733 and Keilor that differ in dates by 1487.1 ky. Those with minimum difference in dates (300 years) are Fish Hoek and Upper Cave 101 that differ in similitude by 9.41.

Table 18.3 presents the similitudes in cranial hexilaterals of archaic *Homo sapiens* to anatomically Modern *Homo sapiens* and to *Homo erectus* in the four regions. This is the most important result of the present study. In both China and Australasia, archaics are closer to erecti of their own region than to erecti of Africa or Europe. And in China and Australasia, moderns are closer to archaics of their own region than to archaics of Africa or Europe. The Eliye Springs archaic in Africa is not closer to African erectus, but rather to Sangarin 17 of Java; but, the archaic in Africa is closer to the African modern Fish Hoek than to the moderns of other regions. (The same is true where Kabwe or Ndutu are used to represent the African archaic form, except that in the case of Kabwe the nearest erecti is Zhoukoudian.) This may reflect the great difference in time and morphological grade between the Eliye Springs African archaic and the African ER-3733 erectus. Petralona used here to represent the European archaic is closer to both erecti and moderns of non-European regions. (The same is true where the Le Ferrassie 1 Neanderthal is used to represent the European archaic form.)

Of course, these results include only four fossils of each grade, and I have deliberately selected those that show maximal similitude *both within and between* regions. But the results for regional continuity are quite robust, especially for China and Australasia, less so for Africa, and least so for Europe (this statement is based on an analysis of 18 additional archaic and modern fossil skulls from the four regions that I plan to present, using additional measurements, and more fossils, in a future publication). It is the strong case for regional continuity in China and Australasia that is decisive in support for the regional transformation hypothesis, and negation of the African replacement hypothesis.

Table 18.2. Pairwise similitudes (S), 12 groups, above diagonal; pairwise difference in chronological ages in kiloyears (below) diagonal.

	Specimen	A	B	C	D	E	F	G	H	I	J	K	L	
A	ER-3733	0	14.31	10.75	11.62	14.72	8.54	9.71	10.72	17.22	15.35	14.17	14.40	A
B	Arago	1050	0	13.50	10.63	13.73	9.43	9.58	9.89	18.59	13.54	12.12	13.15	B
C	Zhoukoudian	1100	50	0	7.93	9.65	8.67	6.00	7.83	11.07	10.72	10.70	10.31	C
D	Sangarin 17	790	260	310	0	8.89	8.73	6.18	6.97	11.94	10.01	8.57	8.70	D
E	Eliye Springs	1440	390	340	650	0	8.82	6.31	11.04	4.58	7.55	8.83	7.12	E
F	Petralona	1250	200	150	460	190	0	6.41	7.46	13.26	11.39	9.27	10.35	F
G	Maba	1450	400	350	660	10	200	0	6.41	10.01	6.66	5.54	6.30	G
H	Ngandong 12	1300	250	200	510	140	50	150	0	15.94	12.05	11.69	11.69	H
I	Fish Hoek	1482	432	382	696	42	232	32	182	0	10.95	9.91	10.11	I
J	Mladeč	1466	416	366	676	26	216	16	166	16	0	6.38	2.88	J
K	Upper Cave 101	1481.7	431.7	381.7	691.7	41.7	231.7	31.7	181.7	0.3	15.7	0	4.30	K
L	Keilor	1487.1	437.1	387.1	697.1	47.1	237.1	37.1	187.1	5.1	21.1	5.4	0	L

Table 18.3. Mean similitudes (S) in cranial hexilaterals of archaic *Homo sapiens* with *Homo erectus* and anatomically modern *Homo sapiens* in 4 old-world regions based on 10 measurements of 12 fossil skulls.

S	*Homo erectus*	Archaic *Homo sapiens*	Modern *Homo sapiens*	S
Africa				
14.72	ER-3733	Eliye Springs 11693	Fish Hoek	4.58
13.73	Arago Reconstruction		Mladeč	7.55
9.65	Zhoukoudian Reconstruction		Upper Cave 101	6.83
8.89	Sangarin 17 Reconstruction		Keilor	7.11
Europe				
8.54	ER-3733	Petralona	Fish Hoek	13.26
9.43	Arago		Mladeč	11.39
8.67	Zhoukoudian		Upper Cave 101	9.27
8.73	Sangarin 17		Keilor	10.35
China				
9.71	ER-3733	Maba	Fish Hoek	10.01
9.58	Arago		Mladeč	6.66
6.00	Zhoukoudian		Upper Cave 101	5.54
6.18	Sangarin 17		Keilor	6.30
Australasia				
10.72	ER-3733	Ngandong 12 Reconstruction	Fish Hoek	15.94
9.89	Agago		Mladeč	12.05
7.83	Zhoukoudian		Upper Cave 101	11.69
6.97	Sangarin 17		Keilor	11.69

Stringer and Andrews (1988, Table 1, p 1264) accurately indicate that the Recent African Replacement hypothesis requires (a) "continuity of pattern only from the late Pleistocene appearance of *H. sapiens* to present," with (b) "Transitional fossils restricted to Africa," and (c) "modern regional characters of low antiquity at peripheries (except Africa)." The existence of several tens of well preserved, accurately dated fossil skulls in China and Australasia, some greater than 200,000 years in date, all with total morphological patterns that are regionally specific and different from African morphologies, flatly contradict these three requirements and thus falsify the African Replacement hypothesis.

Everyone agrees that *Homo erectus* originated in Africa and, about 1 million years ago, colonized from there the European and Asian areas then uninhabited by hominids.

The major archaeological indications, like the morphological testimony, are in dispute. My opinion, as a sometime nonprofessional archaeological fan and rooter, is that the Middle and Upper Paleolithic data summarized by Clark and Lindly (1989), Lindly and Clark (1990), and Marshack (1989) favor with convincing strength the regional transformation hypothesis in reference to Eastern Europe, Western Asia, China, and Australasia. A major defect of the African Replacement hypothesis is the complete lack of substantial archaeological evidence for the postulated, relatively rapid, large-scale, long-distance migration from Africa to East Asia. As Weiss (1990) put it "... hominids have social structure, territoriality, and other reasons to resist being displaced by interlopers. If a [replacement] population did indeed expand from Africa < 500,000 years ago, an explanation of how this was done on the ground is needed...."

THE EVIDENCE OF MITOCHONDRIAL DNA

Stringer and Andrews (1988) accept as support for the Replacement hypothesis the claim by Cann et al. (1987) that the mtDNA of all living humans originates exclusively from one woman who lived about 200,000 years ago in Africa. I have argued (Spuhler 1988, 1989) that the significant regional variation and great time depth in human mitochondrial DNA differences contradicts a single, relatively recent origin for all modern humans. The observed data show the mtDNA types vary significantly not only between but also within regional African, Asian, Near Eastern, European, and Australian populations and that many types are not only unique to a region but also very old in that region. Using Nei's rate of 0.71 substitutions per million years, estimates of the latest mtDNA ancestor range from 162,000 years for Caucasians and 246,000 years for Asians to 402,000 years for Africans. These results show that mtDNA polymorphism existed prior to racial divergence and that the current geographic subspecies of anatomically modern humans cannot be correlated with a single recent African speciation event.

The single African Replacement model is not contradicted if the time depth of significant mtDNA difference in regions outside Africa is more recent than about 180,000 years ago. The Multiregional Transformation model is supported if regions outside Africa exhibit differences that are significantly older than 180,000 years (Cann et al., 1987; Spuhler, 1989). The details of the expectation of mtDNA divergence times under the two models are not well understood. The coalescence time is the time at which all sampled genes trace back to a single ancestral gene. Weiss (1990) emphasizes that the coalescent is not equal to the actual splitting time if the splitting population was polymorphic, as is probable for hominid mtDNAs in the middle Pleistocene. Thus, depending on effective population size, rates of gene flow, and other features of the mating system, with a true 100,000 year split, 350,000 years may be the expected coalescent time. Under a phyletic transformation model, although *Homo erectus* spread out of Africa 1,000,000 years ago, the coalescent may be more recent due to deme-to-deme gene flow. This is the gene tree analog of the argument by Weiss and Maruyama (1976) for the effect of migration in reducing divergence times based on gene frequencies. It is notably difficult to make accurate estimates of N_e and m in living human populations; much more so for the past. Thus applied models predicting the joint operation of effective population size and migration rates on Pleistocene coalescence times are highly speculative.

Wilson's Berkeley group persists in setting the divergence time between chimpanzees and humans at about 5 million years ago, while I agree with Gingerich (1985) that the date 9.2 ± 1.2 myBP is more reliable. Had Sarich and Wilson used a curve with better statistical fit to the same albumin data (Read and Lestrel, 1970) they would have estimated 10 plus myBP in agreement with Gingerich (1985) and the mid-range estimate of 10 myBP accepted by several molecular evolutionists (see Britten, 1986).

Recently I (Spuhler, 1988) fitted a linear regression line to the mtDNA distances for humans, chimpanzees, gorillas, orangutans, and lar gibbons published by Brown et al. (1982) to get a rate of 0.8% per million years, in close agreement with Nei's (1985) estimate based on orangutan divergence time estimated at 13 mya. It is important to note that, unlike the case for chimpanzees and gorillas, we do have reliable fossil evidence for the orangutan divergence time. Nei's arithmetic is that the nucleotide substitution rate, $k = 0.093/(13 \times 10^6)$ is 0.715 per nucleotide site per million years.

If we accept Nei's or my estimate of mutation rates, we get an estimated date for Cann et al.'s (1987) ancestral mtDNA African molecule (I dislike naming a molecule "Eve") of about 350,000 years ago that tells us our mtDNA common ancestress belonged to the species *Homo erectus* and not *Homo sapiens*.

CONCLUSION

The rich biological anthropological record generally observes (with few exceptions that are saltations) gradual change in morphological sizes and shapes within and between living regional populations of modern humans and past populations from the Lower Paleolithic to the Neolithic.

These historical modifications are better interpreted by the Shifting Balance theory of Sewall Wright than by Punctuated Equilibrium theory, because Wright's theory explains both anagenic species transformation and species bifurcation, as well as the observed diversification within and between hominid populations, past and present, at the levels of the deme, region, subspecies, and species.

The ample hominid fossil morphological evidence from Africa, China, and Australasia supports the Regional Transformation hypothesis on the origin of modern humans. Regional differences and continuity are observed in the fossil series from *Homo erectus*, through archaic *Homo sapiens*, to all of us.

The combined morphological, archaeological, and molecular information explains the origin of modern humans by regional evolutionary transformation with continual unification of the species by gene flow and local isolation by distance.

IN MEMORIAM

James Spuhler died in the late fall of 1992 after a protracted illness. His contributions to physical anthropology were many and seminal. He anticipated the relationship between human genetics and physical anthropology long before this became fashionable. Indeed, to him should go most of the credit for encouraging this merging of perspectives that has proved profitable to both areas of endeavor. But the ultimate measure of a person is the trust, goodwill, forbearance, self-restraint, and compassion shown toward others. By these measures no one could have found Jim wanting. To those of us who were privileged to know him well, we are richer through the life he lived and shared with us. His death is a loss to science; but the grief we feel is for the departure of a man who inspired admiration and affection in all who knew him.

REFERENCES

Andrews P (1984) An alternative interpretation of the characters used to define *Homo erectus*. *Courier Forsch Inst Senckenberg* 69:167–175.

Birdsell JB (1953) Some environmental and cultural factors influencing the structuring of Australian Aboriginal populations. *Amer Nat* 87:171–207.

Birdsell JB (1968) Some predictions for the Pleistocene based on equilibrium systems among recent hunter-gathers. In Lee RB, DeVore I (eds), *Man the Hunter*. Chicago, Aldine-Atherton, pp 229–240.

Bocquet-Appel J-P (1985) Small populations: Demographic and paleoanthropological inferences. *J Hum Evol* 14:683–691.

Brace CL (1988) Punctualism, cladistics and the legacy of medieval neoplantonism. *Hum Evol* 3:121–138.

Brace CL, Rosenberg KR, Hunt KD (1987) Gradual changes in human tooth size in the Late Pleistocene and Post-Pleistocene. *Evolution* 41:705–720.

Bräuer G, LeaKey RC (1986) The ES-11693 cranium from Eliye Springs, West Turkana, Kenya, *J Hum Evol* 15:289–312.

Britten RJ (1986) Rates of DNA sequence evolution differ between taxonomic groups. Science 231:1393–1398.

Brown WM, Prager EM, Wang A, Wilson AC (1982) Mitochondrial DNA sequences in Primates: Tempo and mode of evolution. *J Mol Evol* 18:225–239.

Campbell BG (1965) The nomenclature of the Hominidae. *Occas Pap Roy Anthropol Inst* 22:1–34.

Cann RC, Stoneking M, Wilson AC (1987) Mitichondrial DNA and human evolution. *Nature* 325:31–36.

Carroll RL (1988) *Vertebrate Paleontology and Evolution.* New York, W.H. Freeman and Company.

Cavalli-Sforza LL, Bodmer WF (1971) *The Genetics of Human Populations.* San Francisco, W.H. Freeman and Company.

Cavalli-Sforza LL, Piazza A, Menozzi P, Mountain J (1988) Reconstruction of human evolution: bring together genetic, archaeological, and linguistic data. *Proc Natl Acad Sci USA* 85:6002–6006.

Chakraborty R, Neel JV (1989) Description and validation of a method for simultaneous estimation of effective population size and mutation rate from human population data. *Proc Natl Acad Sci USA* 86:9407–9411.

Charlesworth B (1979) Evidence against Fisher's theory of dominance. *Nature* 78:848–849.

Charlesworth B, Lande R, Slatkin M (1982) A Neo-Darwinian commentary on macroevolution. *Evolution* 36:474–498.

Clark GA, Lindly JM (1989) The case for continuity: observations on the biocultural transition in Europe and Western Asia. In Mellers P, Stringer C, *The Human Revolution.* Princeton, NJ, Princeton University Press, pp 626–676.

Coon CS (1962) *The Origin of Races.* New York, Alfred A. Knopf.

Crow JF (1955) General theory of population genetics: synthesis. *Cold Spring Harbor Symp Quant Biol* 20:54–59.

Crow JF (1986) *Basic Concepts in Population, Quantitative, and Evolutionary Genetics.* New York, W.H. Freeman.

Crow JF, Engels WR, Denniston C (1990) Phase Three of Wright's Shifting Balance theory. *Evolution* 44:233–247.

Dobzhansky T (1937) *Genetics and the Origin of Species.* New York, Columbia University Press.

Dobzhansky T (1950) Discussion. *Cold Spring Harbor Symp Quant Biol* 15:106–107.

Dobzhansky T (1963) Review of *The Origin of Races*, By Carleton S. Coon. *Sci Amer* 208(2):169–172.

Dutrillaux B (1980) Chromosomal evolution of the great apes and man. *J Reprod Fert* (suppl 1) 28:105–111.

Dutrillaux B (1988) Chromosome evolution in primates. *Folia Primatol* 50:134–135.

Eldredge N Gould SJ (1972) Punctuated equilibria: an alternative to phyletic gradualism. In Schopf TJM (ed) *Models in Paleobiology.* San Francisco, W.H. Freeman and Company, pp 82–115.

Fisher RA (1928) The possible modification of responses of the wildtype to recurrent mutation. *Amer Natur* 65:115–126.

Fisher RA (1930) *The Genetical Theory of Natural Selection.* Oxford, Clarendon Press.
Foley R (1987) Hominid species and stone-tool assemblages: how are they related? *Antiquity* 61:380–392.
Garn SM (1971) *Human Races,* 3rd ed. Springfield IL, Charles C Thomas.
Gingerich PD (1978) Paleontology and phylogeny: patterns of evolution at the species level in early Tertiary mammals. *Amer J Sci* 276:1–28.
Gingerich PD (1985) Nonlinear molecular clocks and ape-human divergence times. In Tobias PV (ed) *hominid Evolution: Past, Present and Future.* New York, Alan R. Liss, pp 411–416.
Goldschmidt R (1940) *The Material Basis of Evolution.* New Haven, Yale University Press.
Gould SJ (1987) *The Flamingo's Smile: Reflections in Natural History.* New York, W.W. Norton & Company.
Haldane JBS (1924) A mathematical theory of natural and artificial selection, Part I. *Cambridge Phil Soc* 23:19–41.
Haldane JBS (1932) *The Causes of Evolution.* New York, Harper and Brothers Publishers.
Hammond M (1982) The expulsion of Neanderthals from human ancestry: Marcellin Boule and the social context of scientific research. *Soc Stud Sci* 12:1–36.
Healy MJR (1969) Rao's paradox concerning multivariate tests of significance. *Biometrics* 25:411–413.
Hecht MK (1974) Morphological transformation, the fossil record, and the mechanisms of evolution: A debate. Part I. *Evol Biol* 7:295–303.
Hennig W (1996) *Phylogenetic Systematics.* Urbana, University of Illinois Press.
Henneberg M, Piontek J, Stralko J (1978) Natural selection and morphological variability: the case of Europe from Neolithic to modern times. *Curr Anthrop* 19:67–72.
Holmquist R, Miyamoto MM, Goodman M (1988) Higher-primate phylogeny—Why can't we decide. *Mol Biol Evol* 5:201–216.
Howells WW (1969) The use of multivariate techniques in the study of skeletal populations. *Am J Phys Anthrop* 31:311–314.
Howells WW (1973) Cranial variation in man: a study of multivariate analysis of patterns of difference among recent human population. *Pap Peabody Mus Archaeol Ethnol Harvard Univ* 67:1–259.
Howells WW (1989a) Minidisk with data on 57 measurements for 28 male and female series, and 524 Test (non-series) skulls including 26 fossil specimens collected during 1968–1980 by Professor W.W. Howells, Peabody Museum, Harvard University.
Howells WW, (1989b) Skull shapes and the map. Craniometric analyses in the dispersion of modern Homo. *Pap Peabody Mus Archaeol Ethnol Harvard Univ* 79:1–189.
Jarvik E (1980) Lungfishes, tetrapods, paleontology, and pleisomorphy. *Syst Zool* 30:378–384.
Josey Ka, Friday AE (eds) *Problems of Phylogenetic Reconstruction. The Systematics Association Special Vol No. 21.* London, Academic Press.
Kimura M (1968) Evolutionary rate at the molecular level. *Nature* 217:624–626.
King WBR (1864) The reputed fossil man of the Neanderthal. *Quart J Sci* 1:88–97.
King M-C, Wilson AC (1975) Evolution at two levels in humans and chimpanzees. *Science* 188:107–116.

Kochetkova VI (1978) *Paleoneurology*. New York, John Wiley & Sons.
Kowalski CJ (1972) A commentary on the use of multivariate statistical methods in anthropometric research. *Am J Phys Anthrop* 36:119–132.
Lande R (1979) Effective deme size during long-term evolution estimated from rates of chromosome rearrangements. *Evolution* 33:234–251.
Ledermann W, Vajda S (1985) *Handbook of Applicable Mathematics*, Vol V: *Combinations and Geometry, Parts A and B*. New York, John Wiley & Sons.
Lindley J, Clark G (1990) On the emergence of modern humans. *Curr Anthrop* 31:59–63.
Malécot GM (1973) Isolation by distance. In Morton NE (ed) *Genetic Structure of Populations*. Honolulu, University of Hawaii Press, pp 72–75.
Marshack A (1989) Evolution of the human capacity: The symbolic evidence. *Yrbk Phys Anthrop* 32:1–34.
May RM (1978) Human reproduction reconsidered. *Nature* 272:491–495.
Mayr E (1942) *Systematics and the Origin of Species*. New York, Columbia University Press.
Mellers P, Stringer C (1989) *The Human Revolution*. Princeton, Princeton NJ, University Press.
Morton NE (1982) Estimation of demographic parameters from isolation by distance. *Hum Hered*. 32:37–41.
Morton NE, Chung CS, Mi M-P (1967) Genetics of interracial crosses in Hawaii. *Monog Hum Genet* 3:1–158.
Neel JV (1984) Human evolution: many small steps, but not punctuated equilibria. *Pers Biol Med* 28:75–103.
Nei M (1985) Human evolution at the molecular level. In Ohta T, Aoki K (eds) *Population Genetics and Molecular Evolution*. Tokyo, Japan, Science Press, pp 41–64.
Nei M (1987) *Molecular Evolutionary Genetics*. New York, Columbia University Press.
Nei M, Livshits G (1989) Genetic relationships of Europeans, Asians and Africans and the origin of Modern *Homo sapiens*. *Hum Hered* 39:276–281.
Oppenoorth WFF (1932) *Homo* (*Javanthropus*) *soloensis*. Ein Plistocene mensch van Java. *Wet Meded Dienst Mijnb Ned-Ind* 20:49–75.
Osgood WH (1909) *Revision of the Mice of the American Genus Peromyscus*. North American Fauna Bull. No. 28. Washington, DC, U.S. Department of Agriculture.
Panchen AL (1982) The use of parsimony in testing phylogenetic hypotheses. *Zool J Linn Soc* 74:305–328.
Piontek J (1979) Procesy mikroewolucyjne w europejskich populacjach ludzkich. Universytet im. Adama Mickiewicza w Poznaniu, Ser. Antropol. Nr. 6.
Provine WB (1986) *Sewall Wright and Evolutionary Biology*. Chicago, University of Chicago Press.
Rao CR (1966) Covariance adjustment and related problems in multivariate analysis. In Krishnaiah PR (ed) *Multivariate Analysis*. New York, Academic Press, pp 87–103.
Rao CR (1980) Diversity and dissimilarity coefficients: a unified approach. Institute for Statistics and Applications, University of Pittsburgh, Technical Report No. 80–10.
Rensch B (1929) *Das Prinzip Geographische Rassenkreise und das. Probleme der Artbildung*. Berlin, Borntraeger.
Read DW, Lestral PE (1970) Primate phylogeny and immunology: a critical appraisal. *Science* 168:578–580.

Rightmire, GP (1979) Implications of Border Cave skeletal remains for later Pleistocene human evolution. *Curr Anthrop* 20:23–26.

Rightmire GP (1990) The evolution of *Homo erectus. Comparative Anatomical Studies of an Extinct Human Speices.* Cambridge, Cambridge University Press.

Romer AS (1966) *Vertebrate Paleontology*, 3rd ed. Chicago, University of Chicago Press.

Santa Luca AP (1980) The Ngandong fossil hominids. *Yale University Publications in Anthropology* 78:1–175.

Schull WJ (1972) Primitive populations—some contributions to the understanding of human population genetics. In Grouchy J de, Ebling F, Henderson I (eds) *Human Genetics.* Amsterdam, Excerpta Medica, pp 112–123.

Schwartz JH (1984) The evolutionary relationships of man and the orangutan. *Nature* 308:501–505.

Schwidetsky I (1967) Vergleichend-statistische Untersuchungen zur Anthropologie des Neolithikums. *Homo* 18:174–198.

Schwidetsky I (1972) Vergelichend-statistiche Untersuchungen zur Anthropologie der Eisenzeit. *Homo* 23:245–272.

Selby SM (1971) *Standard Mathematical Tables*, 19th ed. Cleveland, Chemical Rubber Company.

Short RV (1976) The evolution of human reproduction. *Proc Roy Soc Lond B*, 195:3–24.

Simpson GG (1944) *Tempo and Mode of Evolution.* New York, Columbia University Press.

Sing CF, Boerwinkle E, Moll, PP, Templeton AR (1988) Characterization of genes affecting quantitative traits in humans. In Weir BS, Eisen EJ, Goodman MM, Namkoong G (eds) *Proceedings of the Second International Conference on Quantitative Genetics.* Sunderland, MA, Sinauer Associates, pp 250–269.

Smith FH, Spencer F (eds) (1984) *The Origins of Modern Humans: A World Survey of the Fossil Evidence.* New York, Alan R. Liss.

Spielman RS, Neel JV, Li FHF (1977) Inbreeding estimation from population data: Models, procedures and implications. *Genetics* 85:355–371.

Spitery E (1982) L'occipital de l'Homme de Tautavel: essai de reconstitution. In *L'Homo erectus et la place de l'Homme de Tautavel parmi les Hominidés Fossils* (Tome 1) Nice, Palais des Expositions, pp 89–109.

Spuhler JN (1984) The evolution of apes and humans: Genes, molecules, chromosomes, anatomy, and behaviour. Butler Memorial Lecture, The University of Queensland, 16 August 1984. St. Lucia, University of Queensland Press.

Spuhler JN (1988) Evolution of mitochondrial DNA in monkeys, apes, and humans. *Yrbk Phys Anthropol* 31:15–48.

Spuhler JN (1989) Raymond Pearl Memorial Lecture, 1988: Evolution of mitochondrial DNA in human and other organisms. *Amer J Hum Biol* 1:509–528.

Stebbins GL, Ayala FJ (1981) Is a new evolutionary synthesis necessary? *Science* 213:967–971.

Stringer CP (1974) Population relationships of later Pleistocene hominids: a multivariate study of available crania. *J Archaeol Sci* 1:317–342.

Stringer CS (1983) Some further notes on the morphology and dating of the Petralona hominid. *J Hum Evol* 12:731–742.

Stringer CP, Howell FC, Melentis JK (1979) The significance of the fossil hominid skull from Petralona, Greece. *J Archaeol Res* 6:235–253.

Stringer CP, Andrews P (1988) Genetic and fossil evidence for the origin of modern humans. *Science* 239:1263–1268.

Suzuki H (1969) Microevolutionary changes in the Japanese population from the Prehistoric Age to the present day. *J Fac Sci*, Univ Tokyo, Sec. V (Anthropol) 3:279–308.

Tartarinov LP (1984) Cladistic analysis, morphology and paleontology in reconstruction of vertebrate phylogeny. *Proc 27th Int Geol Cong* 2:171–182. Utrecht, VNU Science Press.

Thorne AG, Wolpoff MH (1981) Regional continuity in Australasian Pleistocene hominids. *Amer J Phys Anthrop* 55:337–349.

Tobias PV (1971) *The Brain in hominid Evolution.* New York, Columbia University Press.

Turner GG II (1987) Late Pleistocene and Holocene population history of East Asia based on dental variation. *Amer J Phys Anthrop* 73:305–321.

van Vark GN, Howells WW (eds) (1984) *Multivariate Statistical Methods in Physical Anthropology.* Dordrecht, D. Reidel Publishing Company.

von Bonin G (1963) *The Evolution of the Human Brain.* Chicago, University of Chicago Press.

Wade MJ (1977) An experimental study of group selection. *Evolution* 31:134–153.

Weidenreich F (1941) The brain and its role in the phylogenetic transformaton of the human skull. *Trans Amer Philos Soc* (new series) 31:321–442.

Weidenreich F (1943) The skull of *Sinanthropus pekinensis*: a comparative study of a primitive hominid skull. *Paleontol Sinica* (new series), D 10.

Weidenreich F (1951) Morphology of Solo man. *Anthrop Pap Amer Mus Nat Hist* 43:205–290.

Weiss KM (1986) In search of times past: gene flow and invasion in the generation of human diversity. In Mascie-Taylor CGN, Lasker GW (eds) *Biological Aspects of Human Migration.* Cambridge, Cambridge University Press, pp 130–166.

Weiss KM (1990) Gene separation times and effective separation times in the study of modern human origins. AAAS Annual Meeting, Abstracts of Papers, 15–20 Feb 1990, p 20. Washington, DC, Amer Assoc Adv Sci.

Weiss KM, Maruyama T (1976) Archeology, population genetics and studies of human racial ancestry. *Amer J Phys Anthrop* 44:31–50.

Wolpoff MH (1980) *Paleo-Anthropology.* New York, Alfred A. Knopf.

Wolpoff MH (1984) Evolution in *Homo erectus*: the question of stasis. *Paleobiology* 10:389–406.

Wolpoff MH (1990a) Personal communication.

Wolpoff MH (1990b) Mitochondria and the origin of the races. Paper read at the Annual Meeting of the American Association for the Advancement of Science, New Orleans, 18 Feb 1990.

Woo J-K, Peng R (1959) Fossil human Skull of early Paleolithic stage found at Mapa, Shaoquan, Kwangtung Province. *Vert PalAsiatica* 3:176–182.

Woodward AS (1921) A new cave man from Rhodesia, South Africa. *Nature* 108:371–372.

Wright S (1931) Evolution in Mendelian populations. *Genetics* 16:97–159.

Wright S (1932) The roles of mutation, inbreeding, crossbreeding and selection in evolution. *Proc 6th Internatl Cong Genet* 1:356–366.

Wright S (1940) Breeding structure of populations in relation to speciation. *Am Nat* 74:232–248.

Wright S (1941a) Review of *The Material Basis of Evolution*, by R. Goldschmidt. *Sci Monthly* 53:165–170.

Wright S (1941b) On the probability of fixation of reciprocal translocations. *Amer Nat* 75:513–522.

Wright S (1950) Population structure as a factor in evolution. In Gruneberg H (ed) *Moderne Biologie: Festschrift für Hans Nachtsheim* Berlin, F.W. Peter, pp 274–287.

Wright S (1977) *Evolution and the Genetics of Populations, Vol. 3: Experimental Results and Evolutionary Deductions.* Chicago, University of Chicago Press.

Wright S (1978) *Evolution and the Genetics of Populations, Vol. 4: Variability Within and Among Natural Populations.* Chicago, University of Chicago Press.

Wright S (1982a) Character change, speciation, and the higher taxa. *Evolution* 36:427–443.

Wright S (1982b) The shifting balance theory and macroevolution. *Ann. Rev Genet.* 16:1–19.

Wu Xinzhi (1981) A well-preserved cranium of an archaic type of early *Homo sapiens* from Dali, China. *Scienta Sinica* 24:530–541.

Yellen J, Harpending H (1972) Hunter-gatherer populations and archaeological inference. *World Archaeol* 4:244–253.

Index